RETHINKING GEOPOLITICS

Cold War geopolitics may be dead, but struggles over space and power are more important than ever in a world of globalizing economies and instantaneous information. Using insights from contemporary cultural theory, the contributors address questions of political identity and popular culture, state violence and genocide, speed machines and militarism, gender and resistance, cyberwar and mass media – connecting each question to a generalized rethinking of the spaces of politics at the global scale. *Rethinking Geopolitics* argues that the concept of geopolitics needs to be conceptualized anew as the twenty-first century approaches.

Critical geopolitics has emerged from the work of a number of scholars in the fields of geography and international relations who, over the last decade, have sought to investigate geopolitics as a cultural and political practice, rather than as a manifest reality of world politics. Challenging conventional geopolitical assumptions, the diverse chapters include analyses of: postmodern geopolitics, historical formulations of states and cold wars, the geopolitics of the Holocaust, the gendered dimension of Kurdish insurgency, political cartoons concerning Bosnia, representations of the Persian Gulf, the Zapatistas cyberpolitics, conflict simulations in the US military, and the emergence of a new geopolitics of global security.

Exploring how popular cultural assumptions about geography and politics constitute the discourses of contemporary violence and political economy, *Rethinking Geopolitics* brings the ideas of a new generation of scholars to a wide audience for the first time.

Gearóid Ó Tuathail is Associate Professor of Geography at Virginia Tech, USA and **Simon Dalby** is Associate Professor of Geography at Carleton University, Canada.

RETHINKING GEOPOLITICS

Edited by
Gearóid Ó Tuathail
and Simon Dalby

Routledge
Taylor & Francis Group

LONDON AND NEW YORK

First published in 1998
by Routledge
2 Park Square, Milton Park, Abingdon, Oxon, OX14 4RN

Simultaneously published in the USA and Canada
by Routledge
711 Third Avenue, New York, NY 10017

Transferred to Digital Printing 2006

Routledge is an imprint of the Taylor & Francis Group, an informa business

Typeset in Galliard by
J&L Composition Ltd, Filey, North Yorkshire

British Library Cataloguing in Publication Data
A catalogue record for this book is available from the British Library

Library of Congress Cataloging-in-Publication Data
Rethinking geopolitics/edited by Simon Dalby and Gearóid Ó Tuathail.
p. cm.
Includes bibliographical references and index.
1. Geopolitics. I. Ó Tuathail, Gearóid. II. Dalby, Simon.
JC319.R44   1998
320.1′2—dc21   97–50287

ISBN 978-0-415-17250-9 (hbk)
ISBN 978-0-415-17251-6 (pbk)

Publisher's Note
The publisher has gone to great lengths to ensure
the quality of this reprint but points out that some
imperfections in the original may be apparent

To Mordechai Vanunu
and all those imprisoned
for speaking truth
to geopolitical power.

CONTENTS

CONTENTS

FIGURES

TABLES

CONTRIBUTORS

Carlo J. Bonura Jr is a doctoral student of political theory at the Department of Political Science, University of Washington. He received his B.S. in political science from Arizona State University. His research focuses on ethnic minorities and questions of nationalism along the Malaysian–Thai border. He is currently pursuing research in these areas in Thailand.

David B. Clarke is Lecturer in Human Geography at the University of Leeds and an Economic and Social Research Council Fellow. He has published on a variety of geographical topics, has edited *The Cinematic City* (Routledge, 1997), and is currently working on a book on the geography of the consumer society, *Commodity, Sign and Space* (Blackwell, forthcoming).

Simon Dalby is Associate Professor of Geography at Carleton University in Ottawa. His research interests are in environmental security and critical geopolitics. He is author of *Creating the Second Cold War* (1990) and co-editor (with Gearóid Ó Tuathail and Paul Routledge) of *The Geopolitics Reader* (Routledge, 1998)

James Der Derian is Professor of Political Science at the University of Massachusetts at Amherst, and author of *On Diplomacy: A Genealogy of Western Estrangement* (1987) and *Antidiplomacy: Spies, Terror, Speed, and War* (1992); editor of *International Theory: Critical Investigations* (1995); and co-editor (with Michael Shapiro) of *International/Intertextual Relations: Postmodern Readings of World Politics* (1989). His next book, *Virtual War*, is virtually forthcoming.

Klaus Dodds was educated at the University of Bristol and is now a lecturer in Geography at Royal Holloway College, University of London. He was previously a lecturer at the University of Edinburgh. His main research interests are critical geopolitics, war and the media, and the international politics of Antarctica.

Marcus A. Doel is a lecturer in Geography at Loughborough University. His research interests centre on the relationship between space and social theory.

He is the author of *Poststructuralist Geography: The Harsh Law of Space* (forthcoming), and the co-editor of *Business, Trade and Economic Development in Pacific Asia* (forthcoming) and *Fragmented Asia: Regional Integration and National Disintegration in Pacific Asia* (1996).

Jouni Häkli is acting Professor of Human Geography at the University of Joensuu. He holds a doctoral degree from the University of Tampere and has been a visiting researcher at the University of Southern California. His research deals with territory, discourse and spatial identity within geographic frameworks. His recent work has been published in *Fennia, Capitalism, Nature, Socialism* and *Political Geography*.

Timothy W. Luke is Professor of Political Science at Virginia Tech and the author of numerous books and articles. His work engages the problematic of informationalization and how it has transformed social structures, political institutions, environmental politics, notions of art, practices of education, and the nature of geopolitics. His latest books are *Eco-Critique* (University of Minnesota, 1997) and *Departures from Marx* (University of Illinois, 1998).

Paul Routledge is a lecturer in Geography at Glasgow University. He received his Ph.D. from Syracuse University and has held visiting positions at Harvard University and at Bristol University. He is author of *Terrains of Resistance* (Praeger, 1993) and has undertaken research on social movements in the developed and developing world.

Kim Rygiel is a recent masters graduate in International Affairs from the Norman Paterson School of International Affairs at Carleton University. Her research interests are in Turkey, ethnic conflict, gender and nationalism.

Joanne P. Sharp has a Ph.D. from Syracuse University and is currently a lecturer in Geography at Glasgow University. Her research interests include popular geopolitics and national identity, particularly in the context of twentieth century US cultural politics. She has recently co-edited the feminist geography collection *Space/Gender/Knowledge*, with Linda McDowell.

James Derrick Sidaway has a bachelor's degree in Geography and Development Studies from Bulmershe College, Reading, England; a masters in International Studies from the University of Reading, England, and a Ph.D. in Geography from Royal Holloway College, University of London. His general interests are in 'East–West' and 'South–North' relations and his current research focuses on transformations in Southern Africa, Portuguese and Spanish geopolitical discourses and the sociology of geographical knowledge. He has been a lecturer at the School of Geography, University of Birmingham, since 1993.

Matthew Sparke is an Assistant Professor in Geography in the Jackson School of International Studies at the University of Washington. His intellectual interests concern the intersection of the social and cultural dynamics of identity formation with the economic and political geographies of contemporary capitalism. He is the author of *Negotiating Nation-States: North American Geographies of Culture and Capitalism* (University of Minnesota Press, forthcoming), and is currently working on a new project concerned with the changing geographies of trans-border regions in both Europe and North America.

Anders Stephanson was educated at Gothenburg, Oxford and Columbia University, where he now teaches history. His latest book is *Manifest Destiny: American Expansionism and the Empire of Right* (Hill & Wang, 1995). At present, he is writing a book on the history of US diplomatic history as a discipline.

Gerard Toal (Gearóid Ó Tuathail) is Associate Professor of Geography at Virginia Tech. His research interests range from the history of geopolitics to international political economy, the US foreign policy/mass media relationship to information technology and education. He is the author of *Critical Geopolitics* (University of Minnesota, 1996) and a co-editor of *An Unruly World?: Globalization, Governance and Geography* (Routledge, 1998) and *The Geopolitics Reader* (Routledge, 1998).

INTRODUCTION: RETHINKING GEOPOLITICS

Towards a critical geopolitics

Gearóid Ó Tuathail and Simon Dalby

Is geopolitics dead? At first glance the end of the Cold War, the deepening impacts of 'globalization' and the de-territorializing consequences of new informational technologies seem to have driven a stake into the heart of geopolitics. As the Berlin Wall fell in 1989, so also crumbled a pervasive and persuasive order of geopolitical understanding about meaning and identity across global political space. Particularistic and parochial yet nevertheless hegemonic, Cold War geopolitics was always too simplistic a cartography to capture the heterogeneity and irreducible complexity of world politics in the second half of the twentieth century. Yet the very ideological directness of Cold War reasoning was its strength. It drained international affairs of its indeterminancies and lived off its ability to reduce the organic movements of history to a perpetual darkness of 'us' versus 'them.' It provided strategic elites with a discourse that they could instrumentalize to further their bureaucratic careers within the military–industrial–academic complex created by the Cold War. It provided political leaders with scenes for demonstrating hardheaded statesmanship, comforting and easy applause lines, and a workable model of 'gamesmanship' in international affairs. Last, but not least, it provided the public with a recognizable and gratifying fantasy story of heroes and villains fighting for the fate of the world in obscure and exotic locales across the globe. Cold War geopolitics, in short, was a powerful and pervasive political ideology that lasted for over forty years. It was also premised upon an extraordinary double irony. It simultaneously denied both geographical difference and its own self-constituting politics (Ó Tuathail 1996).

While regional variations of the Cold War script live on in certain locations – in US–Cuban relations, for example, and on the Korean peninsula – the days of Cold War geopolitics as the spellbinding 'big picture' of world politics, the global drama that eclipsed all others, have ended. Strategic analysts have been searching ever since for a new global drama to replace it, launching 'the end of history,' 'the clash of civilizations' and 'the coming anarchy' among others as new blockbuster

1

visions of global space, only to see them fade before the heterogeneity of international affairs and proliferating signs of geographical difference. Political leaders have struggled to articulate visions of the new world (dis)order amidst the overwhelming flux of contemporary international affairs, while those in the culture industries have invented a plethora of flexible new enemies and more implacable dangers to bedazzle, entertain and gratify the public. In a world of perpetual speed and motion, convulsed by globalization, saturated by information, and entranced by ephemeral media spectacles and hyperbole, geopolitics seems decidedly old-fashioned and out of place. Indeed, in the search for a new paradigm of world politics a number of strategists and politicians have proclaimed the end of geopolitics altogether, its eclipse and supersession by geo-economics, speed or eco-politics (Ó Tuathail 1997a). In many analyses, geopolitics has been left for dead.

This volume is not dedicated to resurrecting traditional themes of geopolitics. Rather we are concerned to radicalize its components, 'geo' and 'politics,' so that the self-evident character of the sign 'geopolitics' can be problematized and pluralized. Conceptualized in a critical way as a problematic of geo-politics or geographical politics, this volume seeks to radicalize conventional notions of geopolitics through a series of studies of its proliferating, yet often unacknowledged and under-theorized, operation in world politics past, present and future. The 'geopolitics' we seek to analyse is not the mummified remains of Cold War understandings of the concept but the plural traces of geopolitics that have long been with us in the practices of world politics. Geopolitics, for us, engages the geographical representations and practices that produce the spaces of world politics (Agnew 1998). Rather than accepting geopolitics as a neutral and objective practice of surveying global space – the conventional Cold War understanding of the concept – we begin from the premise that geopolitics is itself a form of geography and politics, that it has a con-textuality, and that it is implicated in the ongoing social reproduction of power and political economy. In short, our perspective is a critical one, our practice a *critical geopolitics* (Dalby 1991, Ó Tuathail 1996).

Critical geopolitics has emerged out of the work of a number of scholars in the fields of geography and international relations who, over the last decade, have sought to investigate geopolitics as a social, cultural and political practice, rather than as a manifest and legible reality of world politics. Critical geopolitics is informed by postmodern critiques that have placed the epistemological limits of the ethnocentric practices underpinning Cold War geopolitics in question. Dissonant and dissident voices have articulated feminist, post-colonial and post-structuralist perspectives on the power strategies of Cold War discourse itself, on its privileging and marginalizing, its inclusions and exclusions, on, in sum, the *geo-politics of geopolitics* itself. Informed by this variety of postmodernisms, which all point beyond orthodox representations, critical geopolitics has advanced five arguments that, in various ways, inform the chapters of this book.

First, geopolitics is a much broader cultural phenomenon than is normally described and understood by the geopolitical tradition of 'wise men' of statecraft (Parker 1985). As the geographical politics that enframes all foreign policy practices, geopolitics is not a specific school of statecraft but rather can be better understood as the spatial practices, both material and representational, of statecraft itself. Consequently, the critical study of geopolitics must be grounded in the particular cultural mythologies of the state. Critical geopolitics confronts and analyses the *geopolitical imagi-nation* of the state, its foundational myths and national exceptionalist lore (Agnew 1983) (see Figure 0.1).

The founding and specification of the state as a national community is a geopolitical act. This involves making one national identity out of many, establishing a boundary with an outside and converting diverse places into a unitary internal space. It also involves forging scattered and heterogeneous histories into a transcendent and providential duration (Dijkink 1996). These practices of nationhood involve ensembles of acts to create nation-space and nation-time, the projection of imaginary community, the homogenization of nation-space and pedagogization of history. The geopolitical imagi-nation is an ongoing and precarious project involving all three. It is certainly at work in the projecting of a visual order of space, usually in the form of cartographic surveys and national atlases, across an uneven and broken landscape that is being territorialized with lines delimiting administrative provinces and an official inside and outside. But it is also at work in the founding constitution of community and the renegotiation of boundaries of citizenship and belonging.

Furthermore, it is at play and under contestation in the multicultural struggles over the (re)consolidation of tradition, and the representation and remembrance of history. Counter-narratives of the nation are forms of critical geopolitics:

> Counter-narratives of the nation that continually evoke and erase its totalizing boundaries – both actual and conceptual – disturb those ideological manoeuvres through which 'imagined communities' are given essentialist identities. For the political unity of the nation consists in a continual displacement of the anxiety of its irredeemably plural modern space – representing the nation's modern territoriality is turned into the archaic, atavistic temporality of Traditionalism. The difference of space returns as the Sameness of time, turning Territory into Tradition, turning the People into One.
>
> (Bhabha 1994: 149)

Critical geopolitics bears witness to the irredeemable plurality of space and the multiplicity of possible political constructions of space. Thus, and this is the second argument characterizing critical geopolitics, it pays particular attention to the boundary-drawing practices and performances that characterize the everyday life of states. In contrast to conventional geography and geopolitics, both the material borders at the edge of the state and the conceptual borders designating this as a boundary between a secure inside and an anarchic outside are objects of investigation. Critical geopolitics is not about 'the outside' of the state but about the very construction of boundaries of 'inside' and 'outside,' 'here' and 'there,' the 'domestic' and the 'foreign' (Walker 1993). As Campbell (1992) has argued, the study of foreign policy involves more than the study of conventional inter-state relations. States are not prior to the inter-state system but are perpetually constituted by their performances in relation to an outside against which they define themselves. Foreign policy involves the making of the 'foreign' as an identity and space against which a domestic self is evoked and realized. 'The construction of the "foreign" is made possible by practices that also constitute the "domestic." In other words, foreign policy is a "specific sort of *boundary-producing political performance*"' (Ashley 1987: 51). In describing the struggle between the Soviet Union and the United States as 'not simply geopolitical' Campbell (1992: 26) suggests that territorial geopolitics is contextualized and sustained by a more pervasive cultural geo-politics. In other words (following Campbell's (1992: 76) capitalized distinction between 'foreign policy' and 'Foreign Policy'), a primary and pervasive *foreign policy geo-politics* makes the secondary, specialist and conventionally understood *Foreign Policy Geopolitics* of elites possible.

The essays in this volume demonstrate that there is no geopolitics that is ever 'simply Geopolitical.' Geopolitics is already about more boundaries than those on a map, for those boundaries are themselves implicated in conceptual boundary-drawing practices of various kinds. Critical geopolitics is concerned as much with maps of meaning as it is with maps of states. The boundary-drawing practices we seek to investigate in this volume are both conceptual and cartographic, imaginary and actual, social and aesthetic. Critical geopolitics is particularly interested in analyzing the interdigitation of all these practices, in examining how certain conceptual spatializations of identity, nationhood and danger manifest themselves across the landscapes of states and how certain political, social and physical geographies in turn enframe and incite certain conceptual, moral and/or aesthetic understandings of self and other, security and danger, proximity and distance, indifference and responsibility.

Third, critical geopolitics argues that geopolitics is not a singularity but a plurality. It refers to a plural ensemble of representational practices that are diffused throughout societies. While not denying the conventional notion of geopolitics as the practice of statecraft by leaders and their advisors, critical geopolitics complements this with an understanding of geopolitics as a broad social and cultural phenomenon. Geopolitics is thus not a centered but a decentered set of practices with elitist and popular forms and expressions. A three-fold typology of geopolitical reasoning is useful in loosely distinguishing the *practical geopolitics* of state leaders and the foreign policy bureaucracy from the *formal geopolitics* of the strategic community, within a state or across a group of states, and the *popular geopolitics* that is found within the artifacts of transnational popular culture, whether they be mass-market magazines, novels or movies.

Each of these different forms of geopolitics has different sites of production, distribution and consumption. Linked together, as seen in Figure 0.1, they comprise the geopolitical culture of a particular region, state or inter-state alliance. In understanding 'the geopolitical' as a broad socio-cultural phenomena it is important to appreciate both that geopolitics is much more than a specialized knowledge used by practitioners of statecraft and that the different facets of its practices are interconnected in various ways to quotidian constructions of identity, security and danger. Geopolitics saturates the everyday life of states and nations. Its sites of production are multiple and pervasive, both 'high' (like a national security memorandum) and 'low' (like the headline of a tabloid newspaper), visual (like the images that move states to act) and discursive (like the speeches that justify military actions), traditional (like religious motifs in foreign policy discourse) and postmodern (like information management and cyberwar). While its conventionally recognized 'moment' is in the dramatic practices of state leaders (going to war, launching an invasion, demonstrating military force, etc.), these practices and the much more mundane practices that make up the conduct of international politics are constituted, sustained and given meaning by multifarious representational practices throughout cultures.

Fourth, critical geopolitics argues that the practice of studying geopolitics can never be politically neutral. Critical geopolitics is a form of geopolitics but one that seeks to disturb the objectivist perspectivism found in the history of geopolitics and in the practices of foreign policy more generally. It is a 'situated knowledge' that intervenes to disturb the 'god trick' of traditional geopolitics, which

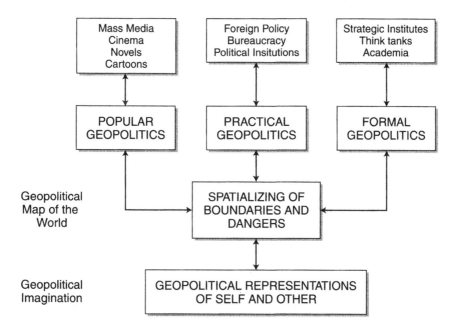

Figure 0.1 A critical theory of geopolitics as a set of representational practices.

claimed to re-present effortlessly the drama of international politics as an intelligible spectacle without interpretation. This conceit, while certainly not particular to the geopolitical tradition, is a consistent feature of geopolitical texts from Mackinder to Kissinger and from Bowman to Brzezinski. Yet it is a conceit that is persistently being undone in the course of exposition and analysis, for writings that deny their interpretative status open themselves up to deconstruction. Classical geopolitics is a form of geopolitical discourse that seeks to repress its own politics and geography, imagining itself as beyond politics and above situated geographies in a transcendent Olympian realm of surveillance and judgement. The response of critical geopolitics is to insist on the situated, contextual and embodied nature of all forms of geopolitical reasoning.

One means of doing this is to insist on the gendered nature of geopolitical writings and interpretative acts, demonstrating how practices of statecraft are also practices of man-craft-ing (e.g. the political leader using military action to demonstrate his toughness, as Israeli Labor leader Shimon Peres did in unleashing Israeli warplanes against guerilla and civilian targets in southern Lebanon during his election battle with Likud leader Benjamin Netanyahu in April and May of 1996, an election he nevertheless lost) and how acts of geo-graphing are also acts of bio-graphing (the intellectual whose geopolitical representations are self-fashionings evoking 'the hardheaded geopolitician', or the tabloid newspaper whose jingoism is part of a strategy of defining itself as 'patriotic'). Geopolitics, whether high or low, is invariably complicitous with certain hegemonic forms of masculinity (Dalby 1994). In Mackinder's case, that masculine subjectivity is a privileged English imperial manhood, while in Kissinger's case it is an elitist *émigré* cosmopolitanism (Kearns 1997; Isaacson 1992). In the cases of Oliver North and Timothy McVeigh, that masculine subjectivity is an insecure and ultra-patriotic warrior masculinity (Gibson 1994).

Fifth, and finally, in conceptualizing geopolitics as 'situated reasoning' a critical perspective also seeks to theorize its broader socio-spatial and techno-territorial circumstances of development and use. Historically, the question of geopolitics has always been the question of states and their societies, technological networks and their relationship to territoriality (Matellart 1996). As a practical rationality devoted to thinking about space and strategy in international politics, geopolitics has historically been deeply implicated in what Foucault (1991) terms the 'governmentalization of the state.' Questions such as 'What is the path to national greatness for the state?' (a key question for Alfred Mahan), 'What is the best relationship of a state to its territory and how can the state grow?' (a fundamental question for Friedrich Ratzel), and 'How can the state be reformed so that its empire can be strengthened' (Mackinder's question) were the practical governmental questions motivating the founders of what we know as 'classical geopolitics.' The history of this practical problem-solving statist knowledge is bound up with the formation of states and empires and the techniques of power that made it possible for them to develop discrete objectifiable territories and societies for management and control.

Geopolitics itself is part of the drive to create 'the right disposition of things' within states and societies through the adoption of certain visualization tech- nologies (like cartography and social sciences such as geography), the establish- ment of certain techno-territorial networks (railways, telegraph cables, automotive highways, national media and now digital information superhigh- ways), the implementation of certain governmental reforms (customs unions, tariff reforms, military spending programs) and the pursuit of certain military strategies and technologies (naval buildups, strategic lines of communication, defensive perimeters and strategic bases). Critical geopolitics, thus, situates its engagement with geopolitics within the context of literatures on the historical expansion of states (Giddens 1987; Mann 1993), techniques of governmentality (Barry, Osborne and Rose 1996) and histories of technology and territoriality (Mumford 1967; DeLanda 1991; Virilio 1997).

Inevitably, given that these five arguments radically problematize the mean- ing, location and stability of that which is considered 'geopolitics,' there is a tremendous diversity of influences and approaches, topics and themes within what we loosely call 'critical geopolitics.' The different essays gathered together in this volume are reflective of that diversity but they are all united in a common commitment to rethink geopolitics in creative and critical ways. They further extend the critical analysis of geopolitics begun in special issues of the geography journals *Society and Space* (1994) and *Political Geography* (1996), while supple- menting the themes presented in our co-edited (with Paul Routledge) intro- ductory volume *The Geopolitics Reader* (Ó Tuathail, Dalby and Routledge 1998). This volume is not meant to be a survey of the new conditions of geopolitics in the late 1990s; it does not discuss in detail such phenomena as the expansion of NATO, the problem of failed states, the geopolitics of finance, or the regional impacts of globalization. Rather, its focus is on *the conditions of possibility of geopolitical truth, knowledge and power.* From the more formal analytical styles of Kim Rygiel and Jouni Häkli through Timothy Luke's innovative prose to James Der Derian's journalistic immersion in the vertiginous simulations of cybercor- porations, we have attempted to include a variety of stylistic modes of thinking critically about geopolitics. How one might analyse, engage and critique geopo- litical practices is not an intellectual and political given. Neither is substance completely divorced from style. To rethink geopolitics necessarily requires a multiplicity of perspectives to unpack the many practices that involve questions of geopolitical power/knowledge.

Three themes thread their way through the chapters in this volume. The first is the theme of modern geopolitics and the state. As Agnew (1998) and others have suggested, the modern geopolitical imagination came into its own at the time of the consolidation of the modern inter-state system after the Treaty of Westphalia. Geopolitics was a form of state geo-power, its gaze a governmental one interested in 'the right disposition of things so as to lead to a convenient end' (Foucault 1991: 93). Geopolitics, in other words, was not essentially a practice concerned with international space but a practice concerned with both

7

domestic and international space, and the maintenance of the boundaries, material and otherwise, between them.

In the first chapter, Ó Tuathail reviews Agnew's conception of the 'modern geopolitical imagination' as a means of posing the question as to whether the boundary-challenging condition of postmodernity is inducing a 'postmodern geopolitics' beyond the modern geopolitical map. A rethinking of geopolitics, he suggests, is long overdue, for the existing spatial ontology that informs and enframes geopolitical thinking is under erasure by postmodernity. Organizing speculative theorizing on the postmodern into a schematic table of contrasts between idealized forms of modern and postmodern geopolitics, Ó Tuathail underscores how the dimensionality and practices of geopolitics are being transformed by globalization and informationalization. Yet, rather than endorsing any simple transition from the modern to the postmodern, Ó Tuathail complicates matters by raising the question of what Latour (1993) terms the 'non-modern,' the actually existing hybridity and impurity of our organizing ontological understandings. Through the lens of actor-network theory, geopolitics looks quite different from conventional understandings of it, the latter understandings being implicitly reliant upon assumptions about 'Man' in control and subordinate 'machines' as mere 'tools.' Ó Tuathail argues that contemporary geopolitical practices are in actuality quite messy (con)fusions that are neither essentially modern nor postmodern.

Perhaps the most notorious and infamous episode in the history of geopolitics and the state is the story of German geopolitics. In their chapter on the Holocaust, Doel and Clarke ignore what is conventionally taken to be German geopolitics – the nationalistic and militaristic school of geopolitical reasoning lead by Karl Haushofer during the interwar period – in order to focus on a more pervasive and deadly German geopolitics, the Nazi geopolitics, which sought to purify the German nation as it sought *Lebensraum* in the East. While the story of *Lebensraum*, the term coined by Friedrich Ratzel and appropriated by Hitler to articulate an expansionist and imperialist territorial project, is well known to political geographers, it has invariably been separated from the genocidal practices of the Nazis against the Jews and others, the others that made the Nazi vision of the German nation distinct. Exploring the varying ways in which the Holocaust is made meaningful through the threefold figure of singularity – it as an exception, an extremity and, in a Derridean-inspired alternative, an experience of serial erasure or 'seriasure' – Doel and Clarke insist on an openness to its unknowable density. In considering the spacing of the Holocaust, they reiterate critical geopolitics' refusal of the distinction between conceptual and physical spaces. The story of the Holocaust is marginal to conventional histories of German geopolitics (Parker 1985) because the interdependence of territorial and racial spacing in Nazi 'foreign policy' is ignored. Killing centers like Auschwitz as sites of spatial purification for the Nazis were literally producing the conceptual and aesthetic Aryan nation the Nazis imagined. *Lebensraum* and *Entfernung* (the removal of Jews and others from the German lifeworld) were two sides of the same murderous geopolitics. Any consideration of the Holocaust that fails to take account of its spacing, Doel and Clarke conclude, is seriously impoverished.

Anders Stephanson's chapter of notes on the Cold War ostensibly takes up the question of this unusual term's origin but quickly becomes a wide-ranging reflection on the genealogy of war and the territorialization of its conceptualization within broader systems of belief. Noting the embeddedness of notions of cold war in spy novels and popular culture generally, Stephanson finds the concept signifying an absolutist non-recognition of one's antagonist as a worthy opponent and a consequent refusal of dialogue and diplomacy with this opponent. Precisely defined in this manner, Stephanson claims that the Cold War actually ended in the early 1960s, with Reagan's revitalization of its themes in the 1980s a shallow form of posturing. Stephanson also argues that its origins can be found in Roosevelt's policy of 'unconditional surrender' towards Nazi Germany, a policy with Civil War precedents and roots in US notions of national exceptionalism more generally. Practising the important refusal of distinction between physical and conceptual space, Stephanson traces how certain states, like the USSR and the United States, territorialized ecumenical philosophies. Arguing that the Cold War was a US project, Stephanson finds its logic in the American tradition of refusing negotiation in times of war, for wars were considered absolute moral struggles between good and evil. The global struggle between the USA and the USSR after World War II was quickly spatialized in these terms (and in universalized American Civil War terms as a struggle between the enslaved world and the free world), but this geo-ideological struggle never became actual war. The Cold War, Stephanson concludes, was a contradictory unity of non-war and non-recognition, a continuation of war by all means other than war – in third spaces – between two antagonists that refused to recognize themselves recognizing one another. When that non-recognition slipped somewhat and feeble diplomacy began, the Cold War, Stephanson suggests, ended.

Carlo Bonura extends the concern with the spatialities in political thinking into the contemporary academic debates about political culture and the methods whereby it might be described and measured. Drawing on work in contemporary international relations theory and critical geopolitics, he shows that the taken-for-granted nature of the cartographic practices of American political science rest on what John Agnew (1998) calls the 'territorial trap.' The social construction of sovereignty is, he argues, crucial because the incorporation of particular cultures, their articulation as nations, is the converse practice of the construction of the possibility of international relations. In the process, Bonura once again emphasizes the importance of 'remembering' the spatial practices of politics precisely where they are so frequently 'forgotten' because they are simply so obvious. Political identity is conjoined with geographical location and specified in terms of states, even in many cases where researchers claim to be studying global phenomena under such rubrics as geocultural areas. The technical apparatus of 'political' culture is also tied into the assumptions about American-led global political arrangements, which perpetuate the differences between national identities and subsume complex political transitions within a foreign policy objective of fostering democracy in supposedly homogeneous areas.

Kim Rygiel's account of the contemporary Turkish state's attempts to construct a unitary 'modern' state in the face of ethnic and religious diversity emphasizes the practical impact of attempts to impose a national culture in locales where such efforts have generated opposition. The understandings of nation and space as unitary in the Turkish pursuit of secular modernity has led to bloodshed in the eastern part of the state and politicized populations caught in the crossfire between rebels and the state's security forces. The gendered implications of this process are noteworthy too, with the politicization of dress and in some cases women even joining the PKK guerilla movement to avoid arranged marriages and conventional lifestyles. The processes of cultural homogenization and state security are spatial strategies of inclusion and sovereignty assertion, but as Rygiel's analysis shows, these are neither socially simple processes nor are they undertaken without in this case ongoing violence on a large scale. But the point that is most important in much of this is that there is no simple designation of Kurdish identity that can be reduced to some set of fixed cultural attributes. In part, the definition of Kurdishness is one that is constructed in opposition to the violence of attempts at homogenization.

Jouni Häkli's chapter concerns itself with the deployment of a modern geopolitical gaze upon the surface of a 'Finland' that is being made the home of a Finnish nation in the nineteenth century. In a concrete case study of the visualization techniques stressed by Agnew, Häkli provides an account of how geography became an empirical knowledge with optical consistency for young Finnish nationalist and later government officials, enabling them to territorialize a country of historical and administrative provinces. Central to Häkli's argument is an ongoing tension between the deep space and popular geo-graphs (space writings) of provincial life and the administrative geo-graphs produced by governmental institutions and expert discourse. The former, he claims, always exceed the latter. Popular geo-graphs are part of an exorbitantly lived social spatiality that can never be fully captured by governmental practices and discourses; they are 'silent and scattered occasions for resistance to the official projections of territory.'

The second theme that threads its way through many of the essays in this volume concerns contemporary crises of identity and popular geopolitics. The chapters by Sharp and Dodds explore the power of popular visual images in creating a geopolitical unconscious that helps to enframe and inform foreign policy debate. Films provide a ready vocabulary for representing geopolitical scenes, scenarios and subjectivities. Use of a good script line can remake the subjectivity of a politician. Dogged by a 'wimp' or 'feminized' image, George Bush remade his image into a hard manly one by appropriating Clint Eastwood's 'make my day' line. Sharp discusses post-Cold War American movies and traces a generalized phenomenon of 'remasculinization' in popular American film entertainment in the themes of 'good men overcoming chaos and disorder in the international realm' – typified by Tom Clancy's Jack Ryan – and 'heroic men struggling against the tyranny of a feminized state.' The disastrous consequences of such identities is a theme picked up later in Matt Sparke's chapter on Timothy McVeigh.

Klaus Dodds analyses popular geopolitics in a series of images that would appear at first glance to be a highly unlikely site for such analysis. He looks at a number of cartoons drawn by Steve Bell on the theme of the violence in Bosnia in the 1990s and the ambiguities of the 'Western' response to the suffering of populations undergoing ethnic cleansing and living through warfare and siege. Dodds demonstrates how the themes of exclusionary identities and cultural homogenization are implicated in the construction of geopolitical frameworks. Bell's cartoons helped to expose the inadequacy of the Western geopolitical framework towards Bosnia and the moral distancing that this framework involved. As visual critique, Bell's cartoons helped to un-enframe the Bosnia constructed by policy makers, making it a place of stark moral responsibility, a part of our universe of obligation once again. Bell's cartoons reinforce the point that geopolitical images are in all facets of popular culture, not just in the planning seminars of national security bureaucracies and foreign ministries.

Matt Sparke extends these themes in considering the (con)fusions of many different forms of geopolitics evident in the case of the Oklahoma City bombing. Problematizing the inclusionary/exclusionary dynamic of a geopolitical system he terms 'Heartland Geopolitics' – itself a (con)fusion of physical space and idealized space – Sparke traces its double displacement by the Oklahoma City bombing as domestic not foreign terrorism, and by the history of one of the convicted bombers, Gulf War veteran Timothy McVeigh. Projected initially upon a foreign Orientalist otherness, the Oklahoma City bombing turned out to be the work of domestic terrorists who, in a mark of the many-layered dimensions of Heartland Geopolitics and its attendant patriot system, represented themselves as authentic patriotic insiders striking a blow against the imposing 'foreign outsiderness' of the federal government – represented as ZOG or the Zionist Occupied Government in some of the racist, anti-Semitic far right literature – particularly enforcement agencies like the Federal Alcohol, Firearms and Tobacco Bureau, which had a branch division in the Murrah Federal Building in Oklahoma City.

Mapped as a 'crazed outcast' after the bombing, lead suspect Timothy McVeigh was actually an inside product of the US government's own patriot system, the US military, in which he served as a gunner during the Gulf War. Sparke traces the subsequent displacement of the legislative clamour to 'do something about terrorism' in Congress, mostly on to minority death row inmates and illegal immigrants, as a shoring up of the 'heart of whiteness' of Heartland Geopolitics. Ironically, McVeigh is a self-styled defender of this implicitly racist Heartland imagi-nation, a Ramboesque figure who chose to do the dirty work that no one wanted to do in order that 'white America' would be awakened to the threat posed to it by its corrupted and 'feminized' state. Geopolitics, as Sparke's essay makes clear, is everywhere.

The Persian Gulf is another region where popular, practical and formal geopolitics have long intertwined. Beginning with the Gulf War of 1991, Sidaway's chapter traces the emergence of the 'Persian Gulf' as a region of US strategic anxiety in the 1970s, particularly during the administration of Jimmy Carter. Popular representations of Zbigniew Brzezinski's Cold War 'arc of crisis' vision in magazines such as *Time* help to establish the region's geo-strategic significance in the Western mind. Although nominally a East versus West vision, Brzezinski's geo-strategic representation also had an important North–South dimension that persisted after the end of the Cold War in 1990. Sidaway's chapter is a useful reminder that post-World War II geopolitics involved much more than the Cold War and that geopolitics is never far removed from geo-economics. It is also a reminder of how decades of media representation of a region as strategically vital in due time makes war in that region eminently more 'natural' and 'inevitable.'

The third and final theme in this volume extends concerns with popular culture and geopolitical identity further by focusing explicitly on informationalization and cyber-geopolitics. Paul Routledge examines the case of the Zapatista 'insurgency' in southern Mexico. Beginning symbolically on 1 January 1994, the Zapatistas were a guerrilla movement with a difference that sought to use global media vectors to advance their cause through info-war more than through real warfare. Symbolically challenging the image of Mexico expensively imagineered for the Salinas administration by American public relations firms during the North American Free Trade Agreement in 1993, the Zapatistas deftly captured the world media's attention and used this initial attention to disarm the Mexican state symbolically and construct an effective global communications infostructure that disintermediated the Mexican state and its offical media (Ó Tuathail 1997b). In so doing, the case of the Zapatistas suggests new forms of political practice in the informationalized spaces of the global media. It also raises the important question of how practitioners of critical geopolitics can engage in a constructive dialogue with forces of opposition.

James Der Derian investigates the worlds of simulation and the construction of virtual realities in which wars can be planned, played and analysed by militaries facing numerous possible contingencies in the complex spaces of the contemporary world. He notes that real potential conflicts are in danger of being overwhelmed by the technologically mediated hyper-realities and hyper-identities made possible by virtual spaces. An important theme in his work is the juxtaposition of popular culture with the scenarios of warfare. Commercial video games and training exercises for the marines are often one and the same product. Popular geopolitics and practical geopolitics reproduce each other, and images of danger from military scenarios become part of the discursive economy of popular imaginations invoked in political discussions of foreign policy. But, in an inversion of the conventional assumptions of military secrecy and corporate openness, his efforts to talk to the most high-profile imagineers of popular culture in Disneyland met a blank wall of silence, while the corporations supplying the military are anxious to demonstrate their wares. In the process of inventing scenarios and stories of future conflicts, the simulations tell only some stories of the history of, in this case, American warfare. The simulations that reinvent American identities do so by remembering only some of the violent past. The iconography of success is enshrined in the narratives that structure the identities that play the simulations. In the process, Der Derian concludes that all sorts of new dangers may be created.

Timothy Luke's chapter explores some of the potential political implications of the rapid growth of electronic and especially 'digital' communications of cyberspace. Using numerous neologisms, he considers the change in power in a world in which material flows across boundaries are replaced by electronic flows, atoms replaced by bits in the evaluation of political boundaries: virtual life replacing real political life. Thinking about possible new political identities in the language of the atom state fails to grasp the contemporary accelerations and interconnections in the virtual life of cyberspace, where geography is now a matter of laser flows and digital images. Luke offers a cautionary word on the assumptions of universal access to cyberspace, pointing out that only some people in some parts of the world have access to computers and the money to gain access to Internet servers, on a planet where 70 per cent of humans do not even have a simple telephone service. The digital nation is one that may transgress state boundaries, but it remains the virtual home of a small elite fraction of the world's overall population. Nonetheless, with the rapidly growing interconnections on the Net and the expansion of computer-using populations, the ability of states to maintain control over information and communication is becoming increasingly limited as cultural identities and technological capabilities collapse some of the traditional notions of space and political identity.

Simon Dalby's concluding chapter works at the largest scale, the globe itself, arguing that the contemporary languages of geopolitics are involved with the specification of the planet itself as apparently threatened and in need of securing and management. The ecosphere frequently enters into geopolitical discussions in ways that perpetuate many of the earlier geopolitical practices of modernity. By specifying the planet as threatened by environmental degradation, the precise cause of the degradation is often obscured and the managerial ethos of governmentality. This is reinforced by the use of powerful information technologies to monitor the physical properties of the planet, invoked to ensure that the political order premised on modern modes of consumption continues uninterrupted. The culture of consumption is taken for granted as the starting point for geopolitical specifications of danger to the culture of modernity. But inverting the logic of security by looking at the specific localities of the 'South' that are supposedly the cause of environmental insecurities subverts the normal direction of geopolitical gaze and turns it back on the culture of expertise that can know a planet in such a manner. By turning the analysis back on the producers of geopolitical texts he argues, in parallel with many of the critiques in earlier chapters, that the attribution of blame for insecurity caused by, in this case, environmental degradation, to external Others, obscures the role of the global political economy in causing insecurity in numerous places. Although not a formal conclusion to the volume in the conventional sense, this analysis of the 'Pogo Syndrome', which obscures political responsibilities, in part through the cartographic representations of contemporary geopolitics, reprises many of the themes from earlier chapters.

This introduction is subtitled 'towards a critical geopolitics' because we do not understand this book as a statement of a fixed and finished project. Critical geopolitics is very much work in progress, a proliferation of research paths rather than a fully demarcated research field. Its continued development is dependent, we believe, on an intellectual openness to new forms of critical social theory from across the social sciences and humanities, and to a relinquishing of conventional disciplinary attitudes and delimiting borders. We look forward to new variants of critical geopolitics that will address the connections between political economy and geopolitical practices, cultural studies and popular geopolitics, gendered identities and geopolitical discourse, psychoanalysis and geopolitical imaginations, actor-networks and geopolitical cyborganizations, cyber-war and virtual geopolitics, globalization and the restructuring of geopolitical regions. We present the essays in this book in the belief that they provide some preliminary steps towards these and other future variants of critical geopolitics. We hope that these new variants and voices can extend the problematization of geopolitical practices to challenge the assumptions that practitioners have for so long taken for granted.

References

Agnew, J. (1983) An excess of 'national exceptionalism': Towards a new political geography of American foreign policy, *Political Geography Quarterly*, 2: 151–166.

Agnew, J. (1998) *Geopolitics: Re-Visioning World Politics*. London: Routledge.

Agnew, J. and Corbridge, S. (1995) *Mastering Space*. London: Routledge.

Ashley, R. K. (1987) Foreign policy as political performance, *International Studies Notes* 13: 51–54.

Barry, A., Osborne, T. and Rose, N. (1996) *Foucault and Political Reason*, Chicago: University of Chicago Press.

Bhabha, H. (1994) *The Location of Culture*, London: Routledge.

Campbell, D. (1992) *Writing Security: United States Foreign Policy and the Politics of Identity*, Minneapolis: University of Minnesota Press.

Dalby, S. (1991) Critical geopolitics: discourse, difference and dissent, *Environment and Planning D: Society and Space*, 9: 261–283.

Dalby, S. (1994) Gender and critical geopolitics: Reading security discourse in the new world disorder, *Environment and Planning D: Society and Space*, 12: 595–612.

DeLanda, M. (1991) *War in the Age of Intelligent Machines*, New York: Zone Books.

Dijkink, G. (1996) *National Identity and Geopolitical Visions*, London: Routledge.

Foucault, M. (1991) Governmentality, in G. Burchell, C. Gordon and P. Miller (eds) *The Foucault Effect: Studies in Governmentality*, Chicago: University of Chicago Press, 87–104.

Gibson, J. W. (1994) *Warrior Dreams*, New York: Hill & Wang.

Giddens, A. (1987) *The Nation-State and Violence*, Berkeley: University of California Press.

Isaacson, W. (1992) *Kissinger: A Biography*, New York: Simon & Schuster.

Kearns, G. (1997) The imperial subject: Halford Mackinder and Mary Kingsley, *Transactions, Institute of British Geographers*, 22: 450–472.

Latour, B. (1993) *We Have Never Been Modern*, Cambridge: Harvard University Press.

Mann, M. (1993) *The Sources of Social Power. Volume II: The Rise of Classes and Nation-States, 1760–1914*, Cambridge: Cambridge University Press.

Matellart, A. (1996) *The Invention of Communication*, Minneapolis: University of Minnesota Press.

Ó Tuathail, G. (1996) *Critical Geopolitics*, Minneapolis: University of Minnesota Press and London: Routledge.

Ó Tuathail, G. (1997a) At the end of geopolitics? Reflections on a plural problematic at the century's end, *Alternatives*, 22: 35–56.

Ó Tuathail, G. (1997b) Emerging markets and other simulations: Mexico, the Chiapas revolt, and the geofinancial panopticon, *Ecumune* 4: 300–317.

Ó Tuathail, G., Dalby, S. and Routledge, P. (1998) *The Geopolitics Reader*, London: Routledge.

Parker, G. (1985) *Western Geopolitical Thought in the Twentieth Century*, New York: St Martin's Press.

Virilio, P. (1997) *Open Sky*, London: Verso.

Walker, R. B. J. (1993) *Inside/Outside International Relations as Political Theory*, Cambridge: Cambridge University Press.

1

POSTMODERN GEOPOLITICS?

The modern geopolitical imagination and beyond

Gearóid Ó Tuathail

Glocalization, it would appear, can implode geopolitics.

Luke (1994: 626)

The challenge for a mode of representation adequate to our post-modern times is . . . to articulate an understanding of world politics attuned to the need to move beyond the sovereignty problematic, with its focus on geopolitical segmentarity, settled subjects, and economistic power, that appreciates the significance of flows, networks, webs, and the identity formations located therein but does not resort simply to the addition of another level of analysis or of more agents to the picture.

Campbell (1996: 19)

A certain amount of mess is perhaps the most general characteristic of human society, past and present.

Mann (1996: 1964)

We live in complicated and confusing times, in spaces traversed by global flows and warped by the intensity and speed of information technologies. Whether we term it late modernity or postmodernity, it is a condition that is unevenly yet unmistakably eroding our inherited ontologies and fixed imaginations of 'how the world works.' Our conveniently conventional geopolitical imagination, which envisions and maps the world in terms of spatial blocs, territorial presence and fixed identities, is no longer adequate in a world where space appears to be left behind by pace, where territoriality is under eclipse by telemetricality, and where simple settled identities are blurring into networks of complex unsettled hybridity. The postmodern condition seems to problematize and unsettle the modern geopolitical map; its disturbs its time-worn conditions of possibility, its conventional geographical rhetoric, its traditional territorial objects, and onto-logical purities. Does, therefore, postmodernity give us a new geopolitics?

The need profoundly to rethink constellations of knowledge like 'geopolitics' on the eve of the new millennium is a consequence of everyday global practices and networks, which are regularly calling geopolitics as we have known it into question: economic globalization, global media flows, the Internet, transnational webs of crime, the hyper-real universe of information perpetually conditioning the practices of statecraft in the late twentieth century. This chapter seeks to rethink geopolitics by engaging in a critical dialogue with the theoretical schemata of John Agnew, Timothy Luke and others on the historical past, confusing present, and speculative future of geopolitics as a sign of the representations of space and the spatial practices underpinning world politics. Agnew (1998: 5) suggests that 'the history of modern world politics has been structured by practices based on a set of understandings about "the way the world works" that together constitute the elements of the modern geopolitical imagination.' This geopolitical imagination, which has its beginnings in sixteenth-century Europe, has structured and conditioned world politics ever since. Though the balance of power between the dominant world powers has changed down the centuries, as has the nature of the international economy, Agnew claims that the modern geopolitical imagination 'still remains prevalent in framing the conduct of world politics' (1998: 6). Yet Agnew himself (1998), Luke and others describe a contemporary state of affairs that puts this observation into question.

Outlining first Agnew's reading of the modern geopolitical imagination, this chapter seeks to complement Agnew's categorization of modern geopolitics using Luke's (1994, 1995, 1996) speculative writings and those of others to suggest the outlines of a *postmodern geopolitics* that disturbs yet, I wish to suggest in the conclusion, has not fully transcended the modern geopolitical imagination. Like the works it engages, this chapter is inevitably historically sweeping and theoretically speculative. It reviews and clarifies the historical schemata and ideal constructs that have been used to explain and understand our contemporary geopolitical conjuncture. As heuristic abstractions and ideal types, these schemata and classifications are far from perfect. They tend to smooth out the messy historicity and complex spatiality of geopolitical discourses and practices, attributing a deep logic and underlining coherence to these that they may not necessarily have. Nevertheless, these schemata have an undeniable heuristic and pedagogic value, provocatively clarifying yet also doubtlessly simplifying the dense history, confused present, and possible future forms of geopolitics. While the contrast between a modern and a postmodern geopolitics can lead to an unnecessary and misleading logic of dichotomization, it is nevertheless incumbent upon critical geopoliticians to theorize how the modes of representation and conditions of practice of geopolitics are changing on the eve of the twenty-first century. Geopolitics, as I have suggested elsewhere, is best studied in its messy contextual specificity (Ó Tuathail 1996). Engaging that contextual specificity today requires speculative theorization of the condition of postmodernity and the multiple transformations it is inducing in the contemporary forms and practices of geopolitics.

Modern geopolitics

The term 'geopolitics' dates from the late nineteenth century but has become in the late twentieth century a widely used signifier for the spatiality of world politics. John Agnew, on his own and together with Stuart Corbridge, has sought to give the concept some rigor and specificity, offering what is perhaps the most comprehensive historical and materialist theory of modern geopolitics in recent years (Agnew and Corbridge 1995; Agnew 1998). Blending the Marxian political economy of the Italian Communist Antonio Gramsci and the idiosyncratic writings on space of the French philosopher Henri Lefebvre with a qualified anti-textualist critical geopolitics, Agnew provides a general theory of geopolitics that treats it both as practices and ideas, as a materialist world order and as a discursive set of understandings and enframing rules. The result is what both Agnew and Corbridge once termed 'geopolitical economy,' a hybrid of geopolitics and political economy (Agnew and Corbridge 1989).[1]

From Lefebvre, Agnew and Corbridge take the distinction between spatial practices and representations of space (Lefebvre 1991: 38–39).[2] Spatial practices for them 'refer to the material and physical flows, interactions, and movements that occur in and across space as fundamental features of economic production and social reproduction' (Agnew and Corbridge 1995: 7). Spatial practices are the everyday material practices across space that help to consolidate worldwide orders of political economy. Representations of space 'involve all of the concepts, naming practices, and geographical codes used to talk about and understand spatial practices.' Implicitly, spatial practices are a pre-discursive materiality, while representations of space are ideology and discourse. Haunting this schema, of course, is the longstanding and unsustainable Marxist distinction between practices and discourse. Aware of this yet nevertheless dependent upon it, Agnew and Corbridge stress the 'dialectical relations' between the categories as a means of handling this recurrent divide.

Building on these distinctions, Agnew and Corbridge make a crucial distinction between geopolitical order and geopolitical discourse, the first a worldwide political economy of spatial practices, while the second is a congealed hegemonic organization of representations of space. Their notion of hegemony, derived from Gramsci and supplemented by the work of Robert Cox (1987), places great emphasis on the ensemble of rules and regulations enmeshing and conditioning actors in world politics. They specify a geopolitical order thus:

> In our usage 'order' refers to the routinized rules, institutions, activities and strategies through which the international political economy operates in different historical periods. The qualifying term 'geopolitical' draws attention to the geographical elements of a world order. . . . Orders necessarily have geographical characteristics. These include the relative degree of centrality of state territoriality to social and economic activities, the nature of the hierarchy of states (dominated by one or a

number of states, the degree of state equality), the spatial scope of the activities of different states and other actors such as international organizations and businesses, the spatial connectedness or disconnectedness between various actors, the conditioning effects of informational and military technologies upon spatial interaction, and the ranking of world regions and particular states by the dominant states in terms of 'threats' to their military and economic 'security.'

(Agnew and Corbridge 1995: 15)

Emphasizing the historical emergence of a 'society of territorial states' and modern rules about 'power politics' after the Napoleonic Wars, Agnew and Corbridge specify three different geopolitical orders: the British geopolitical order (1815–1875), the geopolitical order of inter-imperialist rivalry (1875–1945), and the Cold War geopolitical order (1945–1990) (Table 1.1). There is a certain slippage in Agnew and Corbridge's schema between historical periods, geopolitical orders, and the condition of hegemony, a function, I have argued elsewhere, of the imprecision of the Gramscian notion of hegemony, when a condition of hegemony does or does not exist (Ó Tuathail, 1995). They note that 'a geopolitical order is always partial and precarious' (p. 19) but nevertheless specify their geopolitical orders as permanent, discrete, identifiable periods of time. While they allow a geopolitical order without a hegemon (a dominant state), they do not conceive of a geopolitical order without a condition of hegemony. Geopolitical order is hegemony. A non-hegemonic geopolitical order is not admitted as a possibility. The current post-Cold War epoch is described as a hegemony without a dominant state hegemon, a geopolitical order dominated by powerful countries like Germany, Japan and the United States, integrated by worldwide markets and regulated by transnational institutions and organizations like the European Union, the World Trade Organisation, the International Monetary Fund, and the World Bank (Agnew and Corbridge 1995: 193). The hegemonic ideology of this epoch is transnational liberalism, the belief that universal progress lies in the expansion and extension of capitalist markets across the globe.

Geopolitical discourse, for Agnew and Corbridge, is the discourse by which intellectuals of statecraft, both formal theorists and practitioners, spatialize world

Table 1.1 Modern geopolitics (after Agnew 1998; Agnew and Corbridge 1995).

Spatial practices	*Representations of space*
Geopolitical Order	Geopolitical Discourse
British Geopolitical Order, 1815–75	Civilizational Geopolitics
Inter-Imperial Rivalry, 1875–1945	Naturalized Geopolitics
Cold War Geopolitical Order, 1945–90	Ideological Geopolitics
Transnational Liberalism, 1991–?	Enlargement Geopolitics

politics. It refers to the reading and writing of a geography around the international political economy. It involves the 'deployment of representations' of space which guide the spatial practices central to a geopolitical order' (*ibid.*: 47). Rejecting what they see as the idealism of the textualist approach and the determinism and functionalism of geopolitics-as-ideology, they stress the contingent relationship between thought and practice: 'modes of representation are implicit in practice but are subject to revision as practice changes. Spatial practices and representations of space are dialectically interwoven. In other words, the spatial conditions of material life are shaped through their representations as certainly as representations are shaped by the spatial contours of material life' (*ibid.*: 47).

Just as certain organizations of spatial practices become hegemonic geopolitical orders, so also do certain dominant modes of geopolitical representation become hegemonic geopolitical discourses, epistemological enforcers that suggest to people how they should live, think, and imagine how the world works. While acknowledging that hegemonic orders of geopolitical discourse are, like all conditions of hegemony, fluid, contingent, and perpetually shifting in response to challenge, they nevertheless identify three relatively stable and sweeping historical modes of geopolitical representation, which correspond to the three geopolitical orders already identified: civilizational geopolitics (1815–1875), naturalized geopolitics (1875–1945) and ideological geopolitics (1945–1990) (see Table 1.1). Although Agnew and Corbridge are not explicit about it, the dominant representations of space in the contemporary period could be termed, after the Clinton administration's strategy of enlarging the community of so-called 'market democracies' (a questionable construct that is deeply riven by contradictions and tensions of many kinds), *enlargement geopolitics* (Ó Tuathail and Luke 1994). In all cases, 'the practical geopolitical reasoning of political elites is the link between the dominant representations of space and the geopolitical order of dominant spatial practices' (Agnew and Corbridge 1995: 48).

Enframing and conditioning all these historical and hegemonic modes of geopolitical representation are even more abstract and sweeping macro-historical principles that define 'modern geopolitical discourse.' In *Mastering Space*, the beginnings of modern geopolitical discourse are traced back to the encounters between Europeans and non-European others during the 'Age of Discovery.' While previous empires and social orders long had notions of 'otherness,' Agnew and Corbridge claim that the 'singular trait of modern geopolitical discourse' is its representation of 'others as "backward" or permanently disadvantaged if they remained as they are' (*ibid.*: 49). Europe's others were fixed for all time in a state of inferiority to Europe. They were represented as Europe's past, as the external barbarity and savagery that defined the civilization of Europe. Geographical difference was translated into a temporal schema of backward and modern. To travel beyond Europe, therefore, was to travel back in time, to earlier backward stages in the evolution of human civilization (Doty 1996; Gregory 1994; Grovogui 1996).

These ideas about an overarching modern geopolitical mentality are considerably expanded in Agnew's *Geopolitics: Re-Visioning World Politics* (1998). In this work, Agnew identifies a series of meta-theoretical characteristics of 'the modern geopolitical imagination' that has its beginning during the Renaissance. Though he continues to emphasize the 'singular trait' noted above, the new 'primary feature' or 'most distinguishing feature' of the modern geopolitical imagination, he argues, is a 'global visualization' without which world politics would not be possible. The development of the philosophy and cartographic techniques of global visualization in Europe from the sixteenth century onwards made modern geopolitics possible, for it enabled the seeing of the world as a unitary, albeit still incomplete, whole. The technical invention of perspective made possible the consideration of the world-as-a-picture from a single-eye vantage point. Cartesian philosophy rendered this monocular eye a point of objectivity upon the world. This objective seeing of the world as a unified homogeneous whole led to its differentiation by Europeans into a horizontal hierarchy of places. Ancient binary geographies and hierarchies were recycled to differentiate the globe into vast swathes of fixed and essentialist space.[3] Local difference was cartographically purified and translated into pervasive global danger.

A second characteristic of the modern geopolitical imagination for Agnew reiterates his earlier argument: time is turned into space. Blocks of space are isolated and 'labeled with essential attributes of different time-periods relative to the idealized historical experience of one of the blocks' (1998: ch. 2). This spatial mode of representing the 'modern' generates those binary geographies that have been persistently part of geopolitics since the Renaissance: developed and backward, modern and traditional, West and East, the Occident and the Orient. Each hegemonic geopolitical order gives its own particular meaning and value to these terms. Hegemonic states, according to Taylor (1996), are laboratories of modernity producing hegemonic visions of modern politics, economy, culture, and ordinary everyday comfort. They project an idealized vision of their present – seventeenth-century Amsterdam, nineteenth-century Manchester or mid-twentieth-century Los Angeles – to the rest of the world as its future. This practice tends to organize the geography of the world ethnocentrically into a hierarchy of spaces defined in terms of their degree of modernity, progress and development *vis-à-vis* the ordinary modernity of the hegemon. The Cold War division of the globe into three worlds and Walt Rostow's modernization-as-development theories on the stages of economic growth were merely the latest manifestations of this longstanding feature of the modern geopolitical imagination.

A third characteristic of the modern geopolitical imagination is its state-centered representation of global space, what Agnew (1994) terms the 'territorial trap.' This state-centered approach to world politics has evolved in both practical and formal geopolitical reasoning over the centuries. According to Agnew, it is underpinned by three geographical assumptions: '(1) that states have an exclusive power within their territories as represented by the concept of sovereignty; (2) that "domestic" and "foreign" affairs are essentially separate realms in which different rules obtain; and (3) the boundaries of the state define the boundaries of society such that the latter is "contained" by the former' (1998: ch. 3 1). All of these assumptions are historically questionable and tenuous, yet they nevertheless function in the practices of everyday statecraft to give world politics a geopolitical segmentarity and territorially defined sets of boundaries and identities (Murphy 1996). While these identities or imaginations of in-stated space are often precarious and contested they are nevertheless enforced by a complex of state institutions, international organizations, and everyday social practices. Geopolitics is made not given.

A fourth component of the modern geopolitical imagination isolated by Agnew is the pursuit of primacy by dominant states in the inter-state system. Although nominally equal sovereign entities, states in the modern inter-state system are in reality radically different from each other in geographic location, territorial extent, natural resource endowment, social organization, political leadership, and power potential. These differences have long been classified and conceptualized by geopoliticians within the context of relative struggles for power between states. The pursuit of primacy, at the local, regional and global scales, by dominant states has generated discourses that have sought to explain and justify state militarism, territorial expansionism, overseas imperialism and warfare as inevitable consequences of the uneven distribution of power potential across the globe and timeless 'laws' of competition between states under conditions of anarchy for finite resources. In the late nineteenth century and throughout the twentieth century, the 'realist' language of power politics blended with the so-called 'scientific' language of emergent modern disciplines like geography and biology to create geopolitical discourses and practices that were strongly social Darwinist in tone, locating and explaining various orders of civilizational, racial and statist hierarchy in a primordial 'state of nature.' The geopolitical assumptions that, first, 'power flows from advantages of geographical location, size of population and natural resources' and, second, 'that power is entirely an attribute of territorial states that attempt to monopolize it in competition with other states' are, Agnew (ch. 4) correctly notes, no longer plausible, and their redundancy is evidence of the limits and historical contingency of the modern geopolitical imagination.

So characterized, geopolitics can be described as a particular mode of representing global space. The practices of global visualization that produce the world-as-a-picture are dependent upon an unproblematized Cartesian perspectivism, a supposed view from nowhere that in practice historically represented particularistic views from Europe and the West generally as objective renditions of global space. The hierarchical organizations of global space into essentialist blocs are dependent upon the deep logocentrism of the Western tradition, which has sought to discipline contingency by appeal to the underlying truths of science, history, and nature. The containment of the dynamic currents of global space by territorial geometries and spatial dichotomies mobilizes a metaphysics of presence that makes borders, divisions, and frontiers possible.

Revealing this dependence of modern geopolitics on an order of philosophical commitments and conceits is not to absolutize geopolitics as discourse (*contra* Agnew 1998). Geopolitics is state philosophy, a technology of govern-mentality. It was conceived and nurtured in the imperial capitals of the Great Powers, in their learned academies, in the map and war rooms of ambitious expansionist states. A parochial imperialist gaze that represented lands beyond the horizon as spaces of destiny, it helped to colonize the globe with networks of communication, logistics of war, and ethnocentric models of territorial organization (Matellart 1996). The modern geopolitical imagination is a legacy of the imposition of European territorial forms across the globe from the sixteenth century, an order of power over the Earth that sought to discipline its infinite spaces – internal and external, mountain and valley, land and sea – around sovereign presence and immanent logos. Global space was stamped by essential presence (and absence), organized into natural regions and hierarchies, graded for its inherent value and worth, and marked as the destined property of providential authorities.

Yet, this order of geo-power and its epistemological imperialism has not gone without challenge from alternative subjugated forms of organizing space and graphing the geo (Gregory 1994). In recent decades, the modern geopolitical project has appeared more precarious than before as globalization has rearranged the interconnectivity and functional boundaries of the world political map (Luke 1996b). Today, the fraying of the modern geopolitical project is becoming more and more evident as the daily practices of 'global life' slip territorial bounds and accelerate beyond the modern map, prompting declarations of the 'end of geopolitics' (Ó Tuathail 1997a). It is to the fraying lines and edges of the modern map, to the irruptions of the postmodern within our still nominally modern world politics, that we now turn.

Postmodern geopolitics

A series of distinct yet nevertheless related tendencies have served in recent years to generate considerable speculation about the 'end of the modern' in contemporary world politics. The first is the long relative decline of American hegemony in world politics, an inevitable process that has had many symbolic turning points: the end of the Bretton Woods system of pegged exchange rates, the oil crises of the 1970s, the US withdrawal and *de facto* defeat in Vietnam (Cox 1987). The second is the concurrent and also long-term increasing relative intensity of economic globalization, a phenomenon that is hardly new but that has *appeared* in the last decade to be a profound structural change away from a predominantly statist international political economy towards a deterritorialized global economy (Kofman and Young 1996, Mittelman 1996).[4] Again, many processes and events are read as symbolic of an inevitable and unstoppable 'globalization': the emergence of global financial space, the widespread adoption of flexible specialization production methods, the explosion of transnational

investment in the United States, the implementation of the NAFTA, the burgeoning US trade deficits with Japan and now China (Harvey 1996; Greider
1997; Leyshon 1996). The third tendency involves the oft-described 'revolutionary changes' wrought by the establishment, adoption, and ever-increasing
diffusion of new information technologies throughout the interstices of societies, economies, and polities: facsimile machines, satellite technologies, personal
computers, cable television, and, in recent years, networked computers, wireless
communications, and the Internet (Tapscott 1996). In keeping with McLuhan's
famous declaration that the 'medium is the message,' many theorists have, with
considerable justification, argued that these technologies have radically remade
the bonds, boundaries and subjectivities of actors, societies, economies, and
polities as they have unfolded across global space, itself transformed by the
process (Poster 1995; Morley and Robins 1995). All three tendencies in
combination with others – increasing ease of transnational transportation and
mass travel, the consolidation of transnational media empires, continued
transnational migration – have generated a widespread fourth tendency, the disembedding of societies from their nominal territorial roots, the shrinkage and
collapse of traditional conceptions of scale, and the emergence of a fluid experience of 'global life' (Appadurai 1996). In a world where traditional centers no
longer hold, technologies of time–space compression are colliding modern scales
into each other and generating postmodern local/global fusions that many have
termed 'glocalization' (Agnew and Corbridge 1995: 188–207; Robertson
1995).

Does glocalization, as Luke (1994: 626) suggests, implode geopolitics? One
means of exploring this question is to trace the emergence of new forms of
imagining global space in the condition of postmodernity, new modes of representation that Campbell (1996), like many others, identifies with *flows, networks,
and webs* (Appadurai 1996; Castells 1989; Shapiro and Alker 1996). Describing
the eroding of once discrete national economies by flows of transnational commerce, Robert Reich (1991) identifies 'global webs' as the emergent economic
geometry of the contemporary epoch. Corporate nationality is becoming
increasingly irrelevant as formerly centralized corporations restructure themselves into web-like organizations with global reach. Power and wealth flows to
those groups with the most valuable skills in problem solving, problem identifying and strategic brokering. 'As the world shrinks through efficiencies in
telecommunications and transportation, such groups in one nation are able to
combine their skills with those of people located in other nations in order to
provide the greatest value to customers located almost anywhere' (*ibid.*: 111).
Contemporary information technologies are fundamental to this new geometry
of power. 'The threads of the global web are computers, facsimile machines,
satellites, high-resolution monitors, and modems – all of them linking designers,
engineers, contractors, licensees, and dealers worldwide' (*ibid.*: 111).

Manuel Castells (1996) pushes this further, suggesting that the dominant
functions and processes of the information age are inducing a new *network soci-*

ety. While networks have long existed, 'the new information technology paradigm provides the material basis for its pervasive expansion throughout the entire social structure.' Networks, he argues, 'constitute the new social morphology of our societies, and the diffusion of networking logic substantially modifies the operation and outcome of processes of production, experience, power and culture' (1996: 469). They are making new types of spatial practices possible. Being part of a network, a set of interconnected nodes, is crucial to the exercise of power in the information age. Switches connecting networks are the privileged instruments of power. 'The switchers are the power holders' (*ibid.*: 471). Yet the switchers are powerful only by virtue of the network that 'induces a social determination of a higher level' than that of any social interest expressed through or located at any node or point along the network. Echoing earlier arguments (Castells 1988), he declares that 'the power of flows takes precedence over the flows of power' (1996: 469).

Castells' technologically driven analysis subsumes all the 'new' geo-graphing tropes of postmodernity – flows, webs, connectivity, and networks – within a schema that is ultimately eclectic and *ad hoc*. Bruno Latour's (1993, 1997) notion of the network is more ontologically radical than Castells' grab-bag conception. Challenging the operation of what he terms the 'Modern Constitution,' which legislates an ontology that holds that (1) 'even though we construct Nature, Nature is as if we did not construct it' and (2) 'even though we do not construct Society, Society is as if we did construct it,' Latour (1993, 32) claims that we have never strictly been modern, for we do not abide by the terms of the Modern Constitution. A vast middle kingdom of hybrids, of quasi-objects and quasi-subjects, of cyborgs and monsters, is the proliferating product of the socio-technical networks that make up the unacknowledged nonmodern world. So numerous and multiple have these nature–society–object–discourse amalgamations become that they have strained the acts of purification and translation needed to keep the Modern Constitution intact. Our Enlightenment ontologies struggle to make sense of a world where humans and nature are so intimately interdigitated with scientific and technological systems of all kinds.

The subjects, objects, and actors our postmodernity has thrown up are all impure, hybrid, boundary creatures. Our world, he suggests, is a made up of collectives of humans and nonhumans. It is best described as composed of 'actor-networks,' which are more than the technical or social networks isolated and described by Castells. Actor-network theory, Latour (1997) writes, 'claims that modern societies cannot be described without recognizing them as having a fibrous, thread-like, wiry, stringy, ropy, capillary character that is never captured by the notions of levels, layers, territories, spheres, categories, structure, systems. . . . Literally there is nothing but networks.' Thinking in terms of networks, according to Latour, problematizes proximity/distance and local/global distinctions, in short geography as we have conventionally known it. The science of geography, of mapping, measuring and triangulating physical space, is useless, according to Latour, for actor-network theory, for it seeks to define universal measures of proximity, distance, and scale based on physical measurements. Proximity, distance, and scale, however, are defined by the connectivity of a network. 'The notion of a network helps us to lift the tyranny of geographers in defining space and offers us a notion which is neither social nor "real" space, but associations' (*ibid.*). If geography is reconceptualized as connectivity not space, traditional 'real space' geography is merely one network among multitudes.

Using fragments from these and many other theorists – Marx, Mumford, Lukacs, Baudrillard, and Virilio – Luke (1994, 1995, 1996c, this volume) outlines a suggestive McLuhan-like three-stage narrative for conceptualizing the shifting relationship between humans and nature, and the transformative environments and orders of time–space these generate. Luke begins with first nature, an order of time–space where the relationship between humans and nature is largely unmediated by complex technological systems. In this ideal schema, the principle of spatial ordering is organic and corporeal. 'The wetware of the human body measured space, marked distance, metered time, and defined order with infinite variation in the contemporary manifestations of each traditional society' (Luke 1996c: 123). The enveloping environment and lifeworld is the natural biosphere. If first nature has a geopolitics, it is one organized by terrestrial visions and practices (see Table 1.2).[5]

Luke's schema is not strictly successionist; older orders of space are certainly succeeded and displaced by newer ones, but the older orders do not necessarily disappear. The social order of primordial communities in organic space prevailed before the invention and implementation of city and state building but also beyond it (*ibid.*: 124). Echoing Lukacs and Mumford, Luke describes second nature as the artificial technosphere manufactured and built by modern industrial capitalism from the eighteenth century onwards. Its spatial orderings are engineered, its lifeworld the artificial technosphere created by humans and mechanical machines, its landscapes those of cities and states, its identities those of nations, peoples, and ethnicities. In contrast to the localistic corporeal technologies of first nature, second nature is spatialized by evolving hardware complexes of railways, electrical grids, steamships, hard-surface roads, canals, and telegraph/telephone systems (*ibid.*: 125; Matellart 1996). Space is mastered

Table 1.2 Three geopolitical natures (after Luke 1994, 1995, 1996c).

First nature	Second nature	Third nature
Agrarian antiquity	Modern industrial capitalism	Postmodern informational capitalism
Natural biosphere	Artificial technosphere	Informational cybersphere
Earth and gods	Map and clock	Television and computer
Organic spatiality	Engineered spatiality	Cybernetic spatiality
Terrestriality	Territoriality	Telemetricality
Bioscape/ecoscape/ geoscape	Ethnoscape/metroscape/ plutoscape	Cyberscape/infoscape/ mediascape

by states and these hardware complexes. This, in sum, is the classic era of modern territorial geopolitics, of competition between distinct, bounded spatial entities for the domination of lands, oceans, and the resources of the Earth.

The most provocative aspect of Luke's schema is his elucidation of a distinct realm of third nature, where spatial orderings are generated by cybernetic systems. This is the domain of the informational cybersphere, its electronic landscapes the cyberscapes, infoscapes and mediascapes of postmodern informational capitalism. The forms and structures of second nature begin to buckle and disintegrate under the impact of fast capitalism and its globalizing infostructures. 'Systems of software, as cybernetic codes, televisual images, and informational multimedia, sublate the central importance of hardware. . . . A third nature of telemetricality emerges where informationalization rapidly pluralizes the spatialized operational potentialities of existing cultures and societies' (Luke 1996c: 127). Modern geo-graphing becomes postmodern info-graphing (Luke, this volume). Groups of people begin to join global webs, while the quickening space of flows erodes traditional divisions between the local, national, and global, creating a scalar dynamic of 'neo-world orders' composed of rearranged glocal space (Luke 1995). New networked social actors, quasi-subject cyborgs, and quasi-object 'humachines' within megamachinic collectives populate third nature and give it its functional ontologies, though not, *contra* Haraway, yet its politics, dominated still by mythic liberal categories, identities, and narratives (Haraway 1991; Luke 1996c, 1997).

All of these schematic theorizations have their problems. Reich has justly been criticized for exaggerating the erosion of national economies, the irrelevancy of corporate nationality, and globalization (Hirst and Thompson 1996: 96). Castells can be justly critiqued for his technological determinism, hasty eclecticism, and overly extended reductionist claims. Latour's schema threatens to dissolve all our inherited ontological notions into networks, inflating the concept, dehistoricizing it, and as a consequence generating only modest insight. Luke's schema can be accused of being too sweeping, abstract and intellectually isomorphic, an academic exercise with questionable relevance to the 'real' not 'hyper-real' dilemmas and dramas of world politics today (for counter-evidence see Luke 1991, 1993).

Yet, such schematic theorizations can be useful in clarifying immanent tendencies in contemporary affairs. Combining Agnew's arguments with the suggestive claims of Luke and others, I have constructed a table distinguishing a purified modern from an immanent postmodern geopolitics (see Table 1.3). The table is organized around five key questions central to the problematic of geopolitics *as practiced by dominant states in world politics*, with two sets of distinctions devoted to each. The questions are as follows:

1   How is global space imagined and represented?
2   How is global space divided into essential blocs or zones of identity and difference?

Table 1.3 Modern versus postmodern geopolitics.

Modern geopolitics	Postmodern geopolitics
Cartographic visualizations: maps	Telemetrical visualizations: GIS
Perspectivist theatre	Post-perspectivist simulations
Inside/outside, Domestic/international	Global webs, glocalization
East/West	Jihad/McWorld
Territorial power	Telemetrical power
Hardware ascendant: GPR	Software ascendant: C4I2
Territorial enemies	Deterritorialized dangers
Fixed, rigid posture	Flexible, rapid response
Geopolitical man	Cyborg collectives
States/Leaders	Networks/cyborgs

3 How is global power conceptualized?
4 How are global threats spatialized and strategies of response conceptualized?
5 How are the major actors shaping geopolitics identified and conceptualized?

While such an exercise has its limits, grappling with these five questions reveals some general trends and tendencies about the conditions of possibility of geopolitics at the end of the twentieth century that are worthy of critical attention. What the tabular distinctions highlight and elucidate are tendencies already finding expression in the practices of the US strategic complex of institutions, intellectuals, and actor-networks.

The first question points to the growing significance of telemetrical visualizations in contemporary world politics. It was no accident that the Bosnian War peace talks in 1995 were held at the Wright–Patterson Air Force Base in Dayton, Ohio, the place where the term 'bionics' was first coined and the site of some of the most advanced geographical information systems (GIS) and visualization technologies in the world (Gray 1997: 19). There the negotiating parties could visit the 'Nintendo' room, where they could see up-to-date three-dimensional maps of the disputed territories and settle precisely on lines of separation and demarcation. The technology, according to Secretary of State Warren Christopher, enabled the parties to 'fly' over the area and 'actually see what they were talking about' (quoted in Gray 1997: 19). But what the parties were 'actually seeing' was, of course, a simulation, a model of the real that became in Dayton more real than the real terrain itself.

The displacement in Dayton of maps by GIS, of modern cartographic representations of global space by postmodern telemetrical simulations, is symptomatic of a much broader technocultural transformation in how world politics is imagined and visually represented in the late twentieth century.[6] With globally positioned 24-hour news machines in perpetual operation, the drama of world politics has been turned into an information spectacle, a spectacle that takes its form from its virtual life in flow-mations. Perpetually projected and screened as televisual images and easily recognizable scripts – chaos in the streets, democracy in action, *coup d'état* in motion – world politics has long ceased to be the theatrical drama it was to geopoliticians in the first half of the twentieth century. It is now a hyper-reality of television spectacles and military simulations, a universe of information that encompasses and overwhelms all. CNN's spinning globe is a globe in informational spin. Residual yet redundant, the tropes of political realism can no longer cope with the dizzying world scene. Visions are eclipsed by vertigo (Ó Tuathail 1996). The speed, quantity, and intensity of information problematizes the very possibility of foreign policy as deliberative reflection and decision making (Luke and Ó Tuathail, forthcoming).

The second question foregrounds the disintegration of the Euclidian world of discrete nation-states imagined by so many political realists. Maintaining a distinct border between the inside and the outside, the domestic and the international was and still is always a matter of political performance (Walker 1993; Campbell 1992; Weber 1995), but it is today a performance that is becoming more complex and involved amidst the deterritorializing scale-scrambling consequences of globalization. In our postmodern condition of deterritorialization, Appadurai (1996) has argued, 'configurations of people, place, and heritage lose all semblance of isomorphism.' Contemporary cultural forms are 'fundamentally fractal, that is, as possessing no Euclidian boundaries, structures, or regularities' (*ibid.*: 46) The questions we need to ask in 'a world of disjunctive global flows,' he suggests, should rely on 'images of flow and uncertainty, hence *chaos*, rather than on older images of order, stability, and systematicness' (*ibid.*: 47). This is not to suggest that world politics has necessarily transcended the imaginary of the territorial state but it is to admit the disintegration of its traditional mythic Euclidian forms and to acknowledge strange new (con)fusions of delocalized trans-nations (*ibid.*), simulated sovereignty (Weber 1995), postmodern war (Gray 1997), deterritorialized currency (Kobrin 1997), and a glocalized networked economy of production and consumption (Burton 1997).

It can be argued that the questions many commentators and foreign policy analysts are asking today are no longer dependent upon such traditional binary conceptions of space as modern/backward, East/West – and the 'three worlds' of the Cold War that emerged out of them – as they are on new nominally postspatial binaries like Jihad versus McWorld (Barber 1996). McWorld represents the deterritorializing pace of globalization, MTV, Macintosh and McDonalds, the Utopia of free markets and fast food diffusing across the globe. Jihad represents the primordialist's reaction, the rally to fundamentalist myths, moral absolutes, and rocklike certainties in a boundary-collapsing world. Perhaps this narrative, and variations upon it, is the postmodern equivalent of the hierarchization of space that Agnew identifies with modern geopolitics. Certainly part of the appeal of Barber's dichotomy, irrespective of his own intentions, is its implicit recycling of longstanding Orientalist imaginary geographies. For transnational liberals, McWorld is the world's manifest destiny and Jihad, like Communism in an earlier age, is a dangerous 'disease of the transition' that has the potential to cause considerable unpleasantness. Jihad warriors are considered primitivists, deluded fanatics, and religious ideologues who want to turn back the clock and reverse the accelerating market destiny of history. While geographically concentrated in certain states like Iraq and Sudan, they are a pervasive danger throughout the world, even within the United States. Barber's argument, generated from the ontology of Western liberalism, is a critique of the disturbing implications of both McWorld and Jihad for democracy, but its terms could well be used as a discourse by McWorlders mobilizing against Jihaders everywhere (like President Clinton, who has cited Barber's thesis; Barber 1996: 299). Invocations of the threat posed by 'sons of globalization' like Patrick Buchanan, Vladimir Zhirinovsky and Jean-Marie Le Pen are emerging as an influential discourse of danger in contemporary world politics (Luke and Ó Tuathail, 1998; Rodrik 1997b).

The territorial versus telemetric contrast generated by the third question is easily overstated but it does, nevertheless, echo the discourse of some strategic analysts of global power today (e.g. Rosencrance 1996). In assessing power in the contemporary age, Nye and Owens (1996, 22) write that the significance of technology, education, and institutional flexibility has risen, whereas that of geography, population, and resources/raw materials (GPR), the traditional concerns of early twentieth-century geopolitics, has fallen. They suggest that the country that can best lead the information revolution will be more powerful than all others. For them that country is the United States. Its 'subtle comparative advantage' over its rivals is 'its ability to collect, process, act upon, and disseminate information, an edge that will almost certainly grow over the next decade' (*ibid.*: 20). This 'information edge' can 'help deter and defeat traditional military threats at relatively low cost.' It supposedly can improve the intellectual link between US foreign policy and military power as well as offer new ways to maintain leadership and cement alliances. Overall, America's information edge is a 'force multiplier,' adding greater potency to its hard military power and its soft economic and ideological power. *Software power* converts existing hard and soft power into power plus.

Blind to the unanticipated consequences of informationization (Levidow and Robins 1989; Rochlin 1997; Shenk 1997), Nye and Owens celebrate the role of ISR – intelligence collection, surveillance, and reconnaissance – and C4I – command, control, communications, computer processing, and intelligence; Gray (1997: 7) adds inter-operability, rendering it C4I2 – in providing the US military with 'dominant battlespace knowledge' in conflict situations. The ability of the US state to undertake real-time continuous surveillance of potential hotspots provides it with 'pre-crisis transparency' and an 'informational umbrella' that US decision makers can use, after the manner of its Cold War nuclear umbrella, as a weapon to be shared, if conditions warrant, with allies. 'Like extended deterrence,' America's informational capabilities 'could form the foundation for a mutually beneficial relationship.' Using information as a diplomatic instrument, the United States could provide 'accurate, real-time, situational awareness' to certain states, thus inducing and inclining them to work closely with the United States. Immanent to this reasoning is a condition where geo-graphing has already become info-graphing and where geopolitics becomes info-politics.

Nye and Owens conclude with an updated informational version of Henry Luce's mid-century articulation of American exceptionalism. Information, they declare, is 'the new coin of the international realm, and the United States is better positioned than any other country to multiply the potency of its hard and soft power resources through information' (1996: 35). The twenty-first century, not the twentieth, will turn out to be the century of America's greatest pre-eminence. With informationization, the old themes of American national exceptionalism can be replayed once more. Cyberspace is the latest frontier proliferating freedom and forging the American character (Dyson et al. 1994). The paradox, however, is that informationization deconstructs solid state presences and old-style frontiers. Instead of being a solid state presence, the 'United States' in a fully informationized world would become a node of global networks and webs, a switching point in functionality as it becomes a simulation of old codes in (hyper)reality. Furthermore, the 'American character' would be a fully cyborganized one. C4I2 is not simply a set of tools but a deterritorialized telemetrical civilization; it is part of 'the socio-economic-technical construction kit from which future societies will be assembled' (Rochlin 1997: 211).

New threats to America's software civilization, however, are on the horizon or, perhaps more accurately since dimensionality is being rearranged (Virilio 1997), in the flows and wires of the information age. The fourth question on the spatialization of threats and the conceptualization of response accents certain fashionable themes about flexibility and speed in contemporary strategic doctrine (Virilio and Lotringer 1983). After the Cold War, the meaning of security is essentially contested (Dalby 1997) and threats are increasingly represented as emanating not simply from territorial enemies, where containment imperatives remain in force, but from a plethora of deterritorialized dangers: stateless terrorism, cybernetic sabotage (Gray 1997), narco-terrorism, global corruption (Leiken 1996), infectious diseases (Garrett 1996), humanitarian crises (Luke and Ó Tuathail, 1997), environmental degradation (Dalby 1996), and the proliferation of weapons of mass destruction (Sopko 1996). Shadowy stateless terrorism increasingly targets complex interdependent systems, the spaces of flows – subways, world trade centers, skyscrapers, airports, network computers, switching nodes, databases, communications headquarters – of a McWorld that is engulfing and eroding (while also electronicizing) their most cherished myths.

In contrast to the transcendent containment imperative and fixed posture of Cold War strategy, what is required in response to persistent territorial concerns and proliferating deterritorialized threats is a geo-strategic doctrine premised on flexibility and speed. The 1995 *National Military Strategy of the United States* is subtitled 'a strategy of flexible and selective engagement' (Joint Chiefs of Staff 1995). Threats are described as widespread and uncertain, the possibility of conflicts probable, but their geographic sites are too often unpredictable. This document describes the current strategic landscape as characterized by four principal dangers that the US military must address: 'regional instability; the proliferation of weapons of mass destruction; transnational dangers such as drug trafficking and terrorism; and the dangers to democracy and reform in the former Soviet Union, Eastern Europe, and elsewhere' (*ibid.*: 1). What is remarkable about these threats is that none has a fixed spatial location; regional instability refers not only to the Middle East but also to Europe and Africa; proliferation and transnational dangers are global; even dangers to reform, which is the only danger explicitly linked to certain places, are potentially ubiquitous, as the 'elsewhere' indicates. Military strategy still has to negotiate territory and place but it has also, in interesting ways, become untethered from place and territory. Anywhere on the globe is now a potential battlefield. The document concedes as much, noting that 'global interdependence and transparency coupled with our worldwide security interests, make it difficult to ignore troubling developments almost anywhere on earth' (*ibid.*: 2).

Responding to threats which are potentially everywhere, the US military is now organized around two central strategic concepts: overseas presence and power projection. Overseas presence is the stationing of US military forces throughout the globe as well as the development of alliances with local and regional forces, the pre-positioning of equipment in certain sites, and the maintenance of a routine program of air, ground, and naval deployments across the surface of the planet. Power projection is the ability of the US military to organize the various elements of its overseas presence into a coherent, C4I2ized, multi-option, fighting force. It involves strategic mobilization and mobility with information coordination, speed, and flexibility fundamental to its operation. Swift, flexible power projection as a *fast geopolitics* buys time for liberal politics (Luke and Ó Tuathail 1998): 'the ability to project tailored forces through rapid, strategic mobility gives national leaders additional time for consultation and increased options in response to potential crises and conflicts' (Joint Chiefs of Staff 1995: 7). The logic of this strategy is the annihilation of space by military speed-machines in order to create flexible decision time in dromological crisis situations. Its institutional consequence is the restructuring of the US military as a globe-spanning collective of networks manned by cyborgs dedicated to space-destroying speed. This is described in the US military doctrine as 'strategic mobility enhancement,' its four components and imperatives being 'increased airlift capability, additional pre-positioning of heavy equipment afloat and ashore, increased surge capacity of our sealift, and improved readiness and responsiveness of the Ready Reserve Force' (*ibid.*: 7). Liberalism gives us our cyborgian way of life for the 'our' here is thoroughly cyborgian.

With collectives and cyborgs so obviously a part of the theorization and practice of geopolitics at the end of the twentieth century, the contrast foregrounded by question 5 is an unacknowledged and under-theorized one (see DeLanda 1991). Critical geopoliticians need to begin to recognize the pervasive yet unproblematized presence and anonymous functioning of collectives of humans and nonhumans in world politics (Luke 1997). Contemporary geopolitics obviously gives life and sustenance to military collectives and their networks, but do the networks of everyday collectives have secret geopolitical lives? Follow, Latour-style, our automobile network for just a short connection and we quickly encounter the very military nets we have just described and many other geopolitical quasi-objects and quasi-subjects: oil tankers, the House of Saud, autocyborgs, Fordism, petrol pump politics, George Bush, the Nigerian military, dromomechanics, Exxon, aircraft carriers, polluted beaches, and dying forests. What strange forms of life are revealed by Japanese transplants, strategic chokepoints, and 'what's good for GM or Exxon is good for America'? Proceed further into the network and one encounters the transcendent cyborg creature, Hydrocarbon Man and his megamachinic dromocracy, the Occidental Petroleum Worshipping Collective, a developed, voracious, and accelerating form of life that reacts primitively and violently to any threat, real or imagined, to its speedscapes and lifelines (Virilio 1995). Track the automobile actor-network and the Gulf War and many other wars soon reveal themselves (Yergin 1991). The network collective lives and expands as it kills and depletes. Some conscientious cyborgs within the collective can protest about its poisonous effects on what they still imagine is 'nature' and 'the human habitat,' but none will ever be powerful enough to control or dismantle the collective (Luke 1996a). It has us rather than us having it. It gives us its geopolitics.

Actor-networks do have geopolitical lives, and it is time to acknowledge and theorize these rather than chronicle the stories of Geopolitical Man, the Mackinder-like figure that eyes the globe and divines the secrets necessary for mastering it. Agency in geopolitics is now with the thoroughly cyborganized networks and not with the geopoliticians. New cyberorganized forms of geopolitical life are perpetually being conceived by our proliferating networks, expressing the fears and fantasies of competing and cooperative collectives. Perhaps Latour is right and that it is misleading to even talk about modern and postmodern geopolitics, for the world we inhabit – its 'we' acknowledged as an enhanced cyborgian identity encased within and enveloped by technological life support systems – is resolutely nonmodern. Maybe it is time to critically problematize *nonmodern geopolitics*.

Continuity and change in (post)modern geopolitics

Agnew and Corbridge have persistently emphasized both continuity and change in their studies of geopolitics. In *Mastering Space*, they declare that there is 'an obvious continuity running through modern geopolitical discourse in the continuing use of a language of difference expressed in terms of a temporal metaphor (modern/backward). However, the idioms and contexts of usage have changed dramatically over time' (1995: 51). Agnew returns to this theme in *Geopolitics*, noting that as a result of the dialectical interplay of spatial practices and representations of space the modern geopolitical imagination, 'while having an essential continuity, also shows dramatic shifts in content and form. . . . Within a general continuity . . . one can identify distinctive epochs in which the geo-graphical representations and practices implicitly in world politics have undergone important shifts' (1998: 6–7).

A conventional trope that is often a fudge, this theme of continuity and change nevertheless expresses a certain wisdom that sometimes eludes schematic theorizing about the modern and/or the postmodern. In playing the then/now game of designating the modern and its transformation into the postmodern, there is often an irresistible urge at work rounding up, branding and ordering the messy complexities of human history into clean and precise categories. Sometimes there is an appealing theoretical aesthetic at work, an admiration for theoretical contrasts, transcendent symmetries and elegant isomorphism rescued from the occluded density of history. Also implicated is the normalization of hyperbole, in this case manifested in academic writing, that is characteristic of postmodern culture generally (Shenk 1997).

Mann (1996: 1964) suggests that 'a certain amount of mess is perhaps the most general characteristic of human society, past and present.' Societies, he argues throughout his work, consist of multiple, entwined networks of interac-tion operating at a variety of scales. They are remarkably complex and should not be considered 'systems' with singular identities, clear boundaries, and an overarching essence. Though he does not question as Latour does, Mann's emphasis is on the networked nature of social relations, noting, in opposition to the nation-statism/globalism duality, that 'we do not today live in a society con-stituted "essentially" by the transnational or the global' (1996: 1960). This is also true of the categories we have been using in this chapter. We do not live in a world constituted essentially by modern or postmodern geopolitics but by conjunctural congealments of geopolitical theories and practices that are points of entry into the visual technics, transportational technologies, communicational capabilities, war logistics, political economy, state forms, global crises, spatial ontologies, and pervasive anxieties of our time. In fact, this very notion of 'our time' or 'the contemporary', with associated linear notions of past, present, and future, is inadequate amidst a condition where plural technologically mediated temporalities composed of spectral pasts, deferred and unsettled present tenses, and imperative and diabolical future tenses struggle to constitute and stabilize the 'now' (Derrida 1994; Virilio 1997). While the categories of modern and postmodern geopolitics have pedagogic merit, we should always be cognizant of how the density, hybridity, and impurity of contemporary socio-spatial and socio-temporal practices often escape the grasp of our theories. A modest note of caution, it is a point worth remembering as we struggle to untangle and describe the (con)fused, fragmented and fractal post/non/modern geopolitics of the twenty-first century.

Acknowledgements

Thanks to John Agnew, Timothy W. Luke and Nigel Thrift for their reactions to an earlier draft of this chapter.

Notes

1 This term seems to have disappeared. The closest to it is 'a geopolitical and economic order' in Agnew and Corbridge (1995: 9). It does not appear in Agnew (1998).

2 Agnew and Corbridge also transpose Lefebvre's confusing 'representational spaces' category into their schema somewhat awkwardly. In Lefebvre, the category corresponds to 'lived space,' idiosyncratically defined as the imaginative spaces conjured up by artists, writers, and philosophers. In Agnew and Corbridge (1995: 7), it becomes 'scenarios for future spatial practices or "imagined geographies" that inspire changes in the representation of space with an eye to the transformation of spatial practices.' In this rendering, the distinction between representations of space and representational spaces is never very clear.

3 Crosby (1997: 106) argues that the lines drawn by the Treaty of Tordesillas (1493 and 1494) and later by the Treaty of Zaragoza (1529) are evidence of Renaissance Europeans' confidence in the homogeneity of the Earth's surface, for they divided lands and seas not yet seen to either Spain and Portugal. Even though the Pope was instrumental in drawing these lines, the conception of Earthly space as homogeneous and potentially infinite is a departure from the traditional medieval conception of space as part of a sacred vertically hierarchical order.

4 Globalization has persistently been represented as ushering in a deterritorialized global economy but, as many commentators have pointed out, we still live in a triad (Europe–USA–East Asia)-dominated international political economy (Castells 1996; Rodrik 1997a; Hirst and Thompson 1996). The hyperbole associated with economic globalization can, in part, be explained by the functioning of globalization as an ideology closely associated with transnational liberalism (neoliberalism) (Cox 1996; Herod, Ó Tuathail and Roberts 1998).

5 Strictly speaking, geopolitics is a second nature phenomenon only in Luke's schema associated with the in-stating of space and the imposition of the modern territorial map across global space. In this chapter, however, I wish to retain a broader conception of geopolitics that identifies the geographical representations and practices associated with statecraft – ancient, modern, and postmodern – as its problematic.

6 Consider the fate of visualization in warfare today. Rather than poring over maps, today 'an aide would more likely find a field marshal pacing back and forth in an electronic command post, fiddling with television displays, talking to pilots or tank commanders on the front lines by radio, and perhaps even peeking over their shoulders through remote cameras' (Cohen 1996: 49–50).

References

Agnew, J. (1994) The territorial trap: The geographical assumptions of international relations theory. *Review of International Political Economy*, 1: 53–80.

Agnew, J. (1998) *Geopolitics: Re-Visioning World Politics*. London: Routledge.

Agnew, J. and Corbridge, S. (1989) The new geopolitics: The dynamics of geopolitical disorder, in R. J. Johnston and P. J. Taylor (eds) *A World in Crisis?: Geographical Perspectives*, second edition. Oxford: Blackwell, 266–288.

Agnew, J. and Corbridge, S. (1995) *Mastering Space.* London: Routledge.

Appadurai, A. (1996) *Modernity at Large: Cultural Dimensions of Globalization.* Minneapolis: University of Minnesota.

Barber, B. (1996) *Jihad vs McWorld.* New York: Ballantine.

Burton, D. (1997) The brave new wired world. *Foreign Policy,* 106: 23–38.

Campbell, D. (1992) *Writing Security.* Minneapolis: University of Minnesota Press.

Campbell, D. (1996) Political prosaics, transversal politics, and the anarchical world, in M. Shapiro and H. Alker (eds) *Challenging Boundaries: Global Flows, Territorial Identities.* Minneapolis: University of Minnesota Press.

Castells, M. (1989) *The Informational City.* Oxford: Blackwell.

Castells, M. (1996) *The Rise of the Network Society.* Oxford: Blackwell.

Cohen, E. (1996) A revolution in warfare. *Foreign Affairs,* 75: 37–54.

Cox, R. (1987) *Production, Power and World Order.* New York: Columbia University Press.

Cox, R. (1996) A perspective on globalization, in J. Mittelman (ed.) *Globalization: Critical Perspectives.* Boulder, Colo.: Lynne Rienner.

Crosby, A. (1997) *The Measure of Reality.* New York: Cambridge University Press.

Dalby, S. (1996) The environment as geopolitical threat: Reading Robert Kaplan's 'Coming Anarchy.' *Ecumune* 3: 472–496.

Dalby, S. (1997) Contesting an essential concept: reading the dilemmas in contemporary security discourse, in K. Krause and M. Williams (eds) *Critical Security Studies.* Minneapolis: University of Minnesota Press, 3–31.

DeLanda, M. (1991) *War in the Age of Intelligent Machines.* New York: Swerve.

Derrida, J. (1994) *Spectres of Marx.* New York: Routledge.

Doty, R. (1996) *Imperial Encounters.* Minneapolis: University of Minnesota Press.

Dyson, E., Gilder, G., Keyworth, J. and Toffler, A. (1994) A magna carta for the knowledge age. *New Perspectives Quarterly,* 11, 4: 26–37.

Garrett, L. (1996) The return of infectious disease. *Foreign Affairs,* 75: 66–79.

Gray, C. H. (1997) *Postmodern War: The New Politics of Conflict.* New York: Guilford.

Gregory, D. (1994) *Geographical Imaginations.* Oxford: Blackwell.

Greider, W. (1996) *One World, Ready or Not: The Manic Logic of Global Capitalism.* New York: Simon & Schuster.

Grovogui, S. N. (1996) *Sovereigns, Quasi Sovereigns, and Africans.* Minneapolis: University of Minnesota Press.

Haraway, D. (1991) *Simians, Cyborgs and Women.* New York: Routledge.

Harvey, D. (1996) *Justice, Nature and the Geography of Difference.* Oxford: Blackwell.

Herod, A., Ó Tuathail, G. and Roberts, S. (eds) (1998) *An Unruly World? Globalization, Governance and Geography.* London: Routledge.

Hirst, P. and Thompson, G. (1996) *Globalization in Question.* Cambridge: Polity.

Joint Chiefs of Staff (1995) *National Military Strategy of the United States of America.* Washington, DC.: US Government Printing Office.

Kobrin, S. (1997) Electronic cash and the end of national markets. *Foreign Policy,* 107: 65–77.

Kofman, E. and Young, G. (eds) (1996) *Globalization: Theory and Practice.* London: Pinter.

Latour, B. (1993) *We Have Never Been Modern,* translated by Catherine Porter. Cambridge, Mass.: Harvard University Press.

Latour, B. (1997) On actor-network theory: A few clarifications. Available from http://www.keele.cstt.cstt.latour.html.

Lefebvre, H. [1974] (1991) *The Production of Space*, translated by Donald Nicholson-Smith. Oxford: Blackwell.

Leiken, R. (1996) Controlling the global corruption epidemic. *Foreign Policy* 105: 55–73.

Levidow, L. and Robins, K. (1989) *Cyborg Worlds: The Military Information Society*. London: Free Association Books.

Leyshon, A. (1996) Dissolving difference? Money, disembedding and the creation of 'global financial space,' in P. W. Daniels and W. F. Lever (eds) *The Global Economy in Transition*. Harlow: Longman, 62–80.

Luke, T. (1991) The discipline of security studies and the codes of containment: Learning from Kuwait. *Alternatives*, 16: 315–344.

Luke, T. (1993) Discourses of disintegration, texts of transformation: Re-reading realism in the new world order. *Alternatives*, 18: 229–258.

Luke, T. (1994) Placing power/siting space: The politics of global and local in the New World Order. *Society and Space*, 12: 613–628.

Luke, T. (1995) New World Order or neo-world orders: Power, politics and ideology in informationalizing glocalities, in M. Featherstone, S. Lash and R. Robertson (eds) *Global Modernities*. London: Sage.

Luke, T. (1996a) Liberal society and cyborg subjectivity: The politics of environments, bodies, and nature. *Alternatives*, 21, 1–30.

Luke, T. (1996b) Governmentality and countragovernmentality: rethinking sovereignty and territoriality after the Cold War. *Political Geography*, 15: 491–508.

Luke, T. (1996c) Identity, meaning and globalization: Detraditionalization in postmodern time–space compression, in P. Heelas, S. Lash and P. Morris (eds) *Detraditionalization*. Oxford: Blackwell.

Luke, T. (1997) At the end of nature: Cyborgs, humachines and environments in postmodernity. *Environment and Planning A*, 29: 1367–1380.

Luke, T. and Ó Tuathail, G. (1997) On videocameralistics: The geopolitics of failed states, the CNN international and (UN) governmentality. *Review of International Political Economy*, 4: 709–733.

Luke, T. and Ó Tuathail, G. (1998) Flowmations, fundamentalism and fast geopolitics: 'America' in an accelerating world, in A. Herod, G. Ó Tuathail and S. Roberts (eds) *An Unruly World? Globalization, Governance and Geography*. London: Routledge.

Luke, T. and Ó Tuathail, G., forthcoming. Thinking geopolitical space: War, speed and vision in the writings of Paul Virilio, in P. Crang and N. Thrift (eds) *Thinking Space*. London: Routledge.

Mann, M. (1996) Neither nation-state nor globalism. *Environment and Planning A*, 28: 1960–1964.

Mattelart, A. (1996) *The Invention of Communication*. Minneapolis: University of Minnesota Press.

Mittelman, J. (ed.) (1996) *Globalization: Critical Reflections*. Boulder, Colo.: Lynne Rienner.

Morley, D. and Robins, K. (1995) *Space of Identity: Global Media, Electronic Landscapes and Cultural Boundaries*. London: Routledge.

Murphy, A. (1996) The sovereign state system as political-territorial ideal: historical and contemporary considerations, in T. Biersteker and C. Weber (eds) *State Sovereignty as Social Construct*. Cambridge: Cambridge University Press.

Nye, J. and Owens, W. (1996) The information edge. *Foreign Affairs*, 75: 20–36.

Ó Tuathail, G. (1995) Theorizing history, gender and world order amidst crises of global governance. *Progress in Human Geography*, 19: 260–272.

Ó Tuathail, G. (1996) *Critical Geopolitics.* Minneapolis: University of Minnesota Press and London: Routledge.

Ó Tuathail, G. (1997a) At the end of geopolitics? Reflection on a plural problematic at the century's end. *Alternatives*, 22: 35–56.

Ó Tuathail, G. (1997b) Emerging markets and other simulations: Mexico, Chiapas and the geo-financial panopticon. *Ecumune*, 4: 300–317.

Ó Tuathail, G. and Luke, T. (1994) Present at the (dis)integration: Deterritorialization and reterritorialization in the new wor(l)d order. *Annals of the Association of American Geographers*, 84: 381–398.

Poster, M. (1995) *The Second Media Age.* Cambridge: Polity.

Reich, R. (1991) *The Work of Nations.* New York: Knopf.

Robertson, R. (1995) Glocalization: Time–space and homogeneity–heterogeneity, in M. Featherstone, S. Lash and R. Robertson (eds) *Global Modernities.* London: Sage, 25–44.

Rochlin, G. (1997) *Trapped in the Net: The Unanticipated Consequences of Computerization.* Princeton, NJ: Princeton University Press.

Rodrik, D. (1997a) *Has Globalization Gone Too Far?* Washington, DC: Institute for International Economics.

Rodrik, D. (1997b) Sense and nonsense in the globalization debate. *Foreign Policy*, 107: 19–37.

Rosencrance, R. (1996) The rise of the virtual state. *Foreign Affairs*, 75: 45–61.

Shapiro, M. and Alker, H. (eds) (1996) *Challenging Boundaries: Global Flows, Territorial Identities.* Minneapolis: University of Minnesota Press.

Shenk, D. (1997) *Data Smog: Surviving the Information Glut.* New York: HarperCollins.

Sopko, J. (1996) The changing proliferation threat. *Foreign Policy*, 105: 3–20.

Tapscott, D. (1996) *The Digital Economy: Promise and Peril in the Age of Networked Intelligence.* New York: McGraw Hill.

Taylor, P. (1996) *The Way the Modern World Works.* New York: Wiley.

Virilio, P. and Lotringer, S. (1983) *Pure War.* New York: Semiotext(e).

Virilio, P. (1995) *The Art of the Motor.* Minneapolis: University of Minnesota Press.

Virilio, P. (1997) *Open Sky.* London: Verso.

Walker, R. B. J. (1993) *Inside/Outside.* Cambridge: Cambridge University Press.

Weber, C. (1995) *Simulating Sovereignty.* Cambridge: Cambridge University Press.

Yergin, D. (1991) *The Prize.* New York: Simon & Schuster.

2

FIGURING THE HOLOCAUST

Singularity and the purification of space

Marcus A. Doel and David B. Clarke

Introduction

The enormity of the death and destruction conveyed by the names Auschwitz and the Holocaust is an indelible stain on the fabric of the twentieth century. If one were forced to single out one sequence of events that necessitated an insistence on the 'dark side' of modernity, on the real and potential horrors that accompany rationalization, bureaucratization, adiaphorization, the will to power and order, and the will to purity and propriety, then the Holocaust would no doubt be it.

In order to contextualize what we wish to say in this chapter about the singularity of the Holocaust and its heterotopic spacing and tropology, it is perhaps worth beginning with the closing words of Weinberg and Sherwin's (1979: 22) historical overview of the Holocaust: 'THE TOTAL NUMBER OF DEAD FROM MAJOR GENOCIDAL OPERATIONS BY THE NAZIS IS ESTIMATED AS BEING BETWEEN TWELVE AND THIRTEEN MILLION.' This figure includes six million European Jews; four million Soviet POWs; two million non-Jewish Poles; 100,000–400,000 Gypsies; 2,000 Jehovah's Witnesses; an unknown number of homosexuals and Freemasons; and 50,000–250,000 of Germany's 'insane, mongoloids and retarded children . . . political dissidents and random victims' (Burleigh 1991). In addition, there were around a third of a million forced sterilizations between 1934 and 1939, ordered by Germany's Hereditary Health Courts for the maintenance of so-called 'racial hygiene' (Gellately 1990; Mason 1993). Yet of all the Nazis' many victims and atrocities (Wytwycky 1980), the Holocaust is associated almost exclusively with the murder of two-thirds of European Jewry, the vast majority of whom were systematically killed by the Nazi régime. And when one thinks of the Holocaust, one tends to call upon images of the 'Auschwitz universe' (Steiner 1987: 55), with its industrialized, bureaucratized, and rationalized machinery of (in)human slaughter; although it is important to recall that many were killed through starvation, mob violence, disease, shooting, mobile gassing, and death marches.

Both the magnitude and the ruthlessness of the Holocaust have aroused horror, incredulity, disbelief, disavowal, and even denial. Accordingly, in order to avoid becoming inoculated against the affects of the Holocaust—such as through commonplace and desensitized images of thought, clichéd phraseologies, and its translation into an all too familiar tropology—much of the Holocaust literature has adopted the concept of *singularity* when speaking of the Nazis' genocidal universe. Perhaps surprisingly, however, this singularity turns out to be threefold: it can mean a unique event, a limit case, and a serial erasure. And one could (s)play out this trimerous singularity in other terms, such as point, line, and trace; cutting, folding, and unfolding; exception, extremity, and *seriasure*. In this chapter we will have most to say about this last triplet of terms: the Holocaust figured as an *exceptional* event; the Holocaust figured as an *extreme* event; and the Holocaust (dis)figured as an experience of *seriasure*. And if this were not enough to strike the reader with anxiety and incredulity, what is perhaps even more surprising is the fact that the three modalities of singularity *do not add up* (Clarke *et al.* 1996; Doel 1996, forthcoming). From the off, then, we want to insist on the incommensurability, heterotopia, disjuncture, and diastema of both the Holocaust and its singularity. Unfolding and conveying the structuration and spatialization of this tortuous configuration forms the substance of the present chapter. In particular, we focus on the uneasy parallels between the 'racial' purification of space that underpinned the conceptual universe of the Third Reich, and the way in which the conception of singularity has (s)played itself out in various engagements with the 'Auschwitz universe.' We begin, therefore, by simply sketching these alternative conceptions of singularity, so that the reader will be able to get a sense of what is involved in the figuration of the Holocaust.

The understanding of singularity in terms of *exception* is exemplified in debates over the supposed uniqueness of the Holocaust, especially the extent to which it can be explained, understood, or even situated within a properly *historical* movement and *human(e)* language. Specifically, the question of uniqueness is at stake in the so-called 'Historians' Debate' (*Historikerstreit*) between those who dwell on the absolute singularity of Nazism and the Holocaust, and those who wish to embed, relativize, and, in some cases, normalize them within the flow of historical events (Baldwin 1990; Bosworth 1993; Braun 1994; Kansteiner 1994). With respect to this often subtle and complex debate, singularity is invariably presented in terms of *exception*. Nevertheless, whether one opts for 'separating,' 'isolating,' or 'containing' the Holocaust as an event that interrupts, eternally, the flow of so-called 'normal' history, or whether one opts for 'integrating,' 'accommodating,' and 'assimilating' it within such a flow, one will have risked repeating the gestures of expulsion, extermination, and cancellation. If the Holocaust is presented as being unique, then it risks being erased *from* history (it would no longer 'belong' to the historical); if it is presented as not being unique, then it risks being erased *in* history ('it' would disseminate into the labyrinthine minutiae of history's everyday life).

40

The framing of the Holocaust's singularity in the form of *extremity* stands in stark contrast to the conceptualization of the Holocaust as an exceptional, unique event. For not only do extreme events belong to the distribution of which they are a part—and are thereby relativized in relation to that distribution, yet without losing their singular status—extreme events also disclose what remains only latent or possible in other (less extreme) instances. Such is the complicity of skew, kurtosis, and extremity. Thus, whilst the (for the most part historical) discourses of exception have sought to *divide* the Holocaust from the structuration and flow of so-called 'normal(ized)' and 'standard(ized)' events, the (primarily sociological) discourses of extremity have sought to emphasize their (almost) absolute *proximity*. Extremity belongs to normality: it is not so much cut off as folded in. The norm(al) is interlaced and saturated with extremity; extremity is part and parcel of the calibration of the norm(al).

Finally, the articulation of singularity in terms of serial erasure is intended to convey the disjuncture of space–time. In a rather different context, Derrida (1991a: 424) dubs such a serial erasure 'the *seriasure*' [*sériature*], with all of the associated connotations of seismology, spectrography, and transpar(t)ition. In what follows we interlace this *seriasure* with the mathematical and physical meanings of singularity: the break-point at which a function takes on an infinite value, and the break-point in space–time where matter achieves an infinite density (such as the finality of a 'Final Solution' or the proper name 'Auschwitz'). As a transformer, singularity dissimilates the givens (Deleuze 1994; Lyotard 1990c). Strictly speaking, such break-points are wholly unrepresentable, except by allusion to all manner of black holes, vanishing points, superficial abysses, silences, and amassing effects (Baudrillard 1990a). In this way, rather than expelling singularity *outside of* the norm(al), or accommodating/assimilating it *within* the norm(al), the singularity of *seriasure* traces a certain disjuncture, dissimilitude, and differentiation *in* the event itself, and not merely *around* it. It relays an unpresentable break-point that will never be part of any memory, experience, or concept, maintaining within itself the disjuncture of an irreducible silence (Clarke and Doel 1994). Or again: whereas exception and extremity operate by way of a 'relation oriented from the unknown to the known or the knowable, to the always already known or to anticipated knowledge,' the ghostly drift of *seriasure* reverses this movement, in order that 'the known is related to the unknown, meaning to nonmeaning' (Derrida 1978: 270–271). On this basis, science can be made to submit 'to a radical alteration: without losing any of its proper norms, it is made to tremble, simply by being placed in relation to an absolute unknowledge' (*ibid.*: 268), which can be written 'science plus,' 'science + L,' or 'scilence.' Hereinafter: remember (not) to forget. For we will never have finished opening *ourselves* to the singularity of the Holocaust. Its disjuncture disjoins the presumed self-presence and self-identity of our thoughts and actions.

Through these three figures of singularity, then, what Steiner (1987: 55) has called 'the abyss of 1939–1945' still shakes and solicits us today. For however much we are obliged to try and document, engage with, and comprehend the events that took place on the continent of Europe throughout those years, we are equally obliged to become sensitive to their spacing. In the next section, we will develop further the three figures of singularity outlined above, not least because each evokes a very different ethico-political response to the Holocaust. The second section will then consider the significance of these conceptions for unfolding the spaces of the Holocaust, a task that has been neglected for too long (Charlesworth 1992, 1994; Clarke *et al.* 1996; Ohlin 1997; Lanzmann 1985).

Figuring the Holocaust

The historical consideration of the Holocaust has, until recently, been dominated by a polarization between 'functionalists' and 'intentionalists.' The former have viewed the Holocaust structurally, whilst the latter have stressed the unfolding of a 'Master Plan' to annihilate European Jewry (Kershaw 1993). The nub of this controversy has rested on whether the Nazis' 'Final Solution' to *their* 'problem' with 'the Jews' was driven by a systematic implementation of Hitlerite ideology (as the intentionalists argue), or emerged contingently in response to contextual factors such as the polycentric and competitive disarray of the Nazi bureaucracy, the shifting fortunes of war, and the variable nature and apparatuses of governance in Nazi-occupied territories (the functionalist position). There has also been a valuable attempt to negotiate an intermediate course between the functionalists and intentionalists (Browning 1991, 1992). Nevertheless, such approaches to the Holocaust are invariably played out in terms of uniqueness and exception, and often dwell on the extent to which Nazism and the Holocaust can and should be historicized. Many such attempts to normalize and relativize them have relied on four main tactical manoeuvres: comparison; continuity; amoral equivalence; and rhetoric. Attempts by revisionist authors such as Nolte and Hillgruber to reinscribe the Nazi era within a comparative framework have sought to situate it within so-called 'normal' history (Nolte 1985). However, and unlike extremity, this has not been in order to nudge the 'normality' of 'normal' history into the asymptotic curvature of a labyrinthine deviation (Baudrillard 1990b; Doel 1994; Doel and Clarke, forthcoming). On the contrary, it has been utilized as a means of asserting that the 'war against the Jews' was commensurate with other wars and genocides, especially those perpetrated against the Armenians and Kulaks by the forces of 'Judaeo-Bolshevism.' For example, Hillgruber (1986) has argued that the fate of the German army on the Eastern Front was no less tragic than the fate of the Jews caught up in the Holocaust, and that the saturation bombing of German civilians in Dresden by the Allies was commensurate with the killing of the Jews by the Nazis. Moreover, placing Nazism within a comparative framework clearly opens up the possibility of establishing a continuity between the actions of the Nazi régime and the actions of other, 'normal' states. In short, Nazism is in a sense re-legitimated, particularly through the ruse of attempting to present the Auschwitz universe as a reactive, defensive, and rational response to the threat of 'Judaeo-Bolshevism,' and the establishment of amoral equivalencies between heterogeneous events—to the point where the significance of the Holocaust is effectively normalized out of existence, even though nothing has been denied (Baldwin 1990; Joffee 1987; Maier 1986). Hence the appellation 'revisionist' or 'apologist' to such reinscriptions of the Holocaust (Eley 1988; Habermas 1989).

Little wonder, then, that rhetoric has played such a crucial role in the revisionist endeavour. On the one hand, there has been the presentation of the 'heroic struggle' of the Germany Army on the Eastern Front in the style 'of cheap war paperbacks,' as Habermas (1989: 218) puts it, whilst on the other hand, the Holocaust has been routinely presented through 'unrevised clichés' and 'the frozen language of the bureaucrat' (*ibid.*: 219). Habermas is particularly scathing about Hillgruber's (1986) *Two Kinds of Doom: The Smashing of the German Reich and the End of European Jewry*. '"Smashing" requires an aggressive opponent; an "end" takes place on its own' (Habermas 1989: 219). For Habermas, what is particularly at stake in the Holocaust is, significantly, what it discloses for the 'unfinished project of modernity';-especially the colonization of the intersubjective lifeworld by the criteria, principles, and practices of instrumental reason (Habermas 1987; *cf.* Lefebvre 1991). However, whilst for Habermas the project of modernity still holds its emancipatory promise, for others the impact of the Holocaust has been so great as to finally dislodge the shaky edifice of modernity as a whole (e.g. Lyotard 1988, 1990a). Nevertheless, the notions of disclosure and solicitation necessitate a shift in focus from exception to extremity, and from cutting and suturing to folding and unfolding.

One of the most effective statements on the Holocaust as an extreme event, which discloses the enormous range of actual and latent horrors of twentieth-century modernity, has been made by Bauman (1989). He has been at pains to place the Holocaust within a modernity that would disavow it; to insist that the Holocaust belongs to modernity—and not just as an attribute, affliction, or shadow that it could disown, rectify, or shake off, but in its very essence. At root, then, there is nothing contingent about the relation between modernity and the Holocaust. *À la* philosophical realism, contingency pertains only to the actualization of the Holocaust, and not to its virtuality (which should not be confused with its 'mere' possibility). For example, Bauman will have no truck with the notion that the Holocaust can be discounted as the application of modern 'solutions' to pre-modern 'problems.' As if the Holocaust could have been prevented through a higher and more penetrative dose of rationalization and/or modernization: 'every "ingredient" of the Holocaust . . . was normal, "normal" not in the sense of the familiar . . . but in the sense of being fully in keeping with everything we know about our civilization, its guiding spirit, its priorities, its immanent vision of the world' (Bauman 1989: 7–8). Thus, at the very least, the Holocaust is an aberration of the modern system, rather than something lying outside of it. By recasting Arendt's (1977) 'banality of evil' in terms of the 'rationality of evil,' Bauman insists that the Holocaust was internally related to modernity, and in particular to modern forms of racism and bureaucracy (cf. Lyotard 1990b). The refusal to believe that horror, evil, and crazed power bear absolutely no relation to the normal, rational, and progressive nature of modernity—an underlying current of much of the historical debate—is regarded as an in-built feature of modernity itself. So, while the Holocaust may not have been familiar with respect to the modernity that had gone before, it was no less familial for that. Consequently, Bauman refuses to defamiliarize or 'defamilialize' the Holocaust. In fact, he goes in the opposite direction. It belongs to all of those comfortable, 'ordinary modernities' that would disown it, to borrow a felicitous phrase from Taylor (1996).

Nazism was based on a kind of 'reactionary modernism' (Herf 1984; Bergen 1994)—an unstable mix of traditional 'folk' values and an appreciation of modern technology. An ambivalent *volkisch* nationalism was crucial to both the form taken by the Nazi régime itself and to the ambitions that paved the way for the Final Solution. It is in this light that Bauman (1989: 73) insists that whilst anti-Semitism *per se* has had a long and inglorious history, 'the exterminatory version of anti-Semitism ought to be seen as a thoroughly modern phenomenon.' For instance, the mob violence of *Kristallnacht* was not fundamentally dissimilar to the many pogroms scattered throughout the diasporic history of the Jews. Yet such violence was largely ineffectual, since it often reinforced popular opinion against the anti-Semitism espoused by the Nazis (Sabini and Silver 1980). And this was despite the fact that moves to isolate the legitimate means of violence effectively to the apparatuses of the state were largely unsuccessful in Weimar Germany, especially when compared with the contemporary experience of both Britain and France. However, even with such a caveat, the Final Solution needed to be more impersonal and mechanistic. It required 'a typically modern ambition of social design and engineering, mixed with the typically modern concentration of power, resources and managerial skills' (Bauman 1989: 77). The Final Solution needed modernity. It required the functioning of all of those modern bureaucracies, with their hierarchical and functional divisions of labour, and their instrumental rationalities, so as to enable the reduction of 'moral' responsibility to 'technical' responsibility—a process of adiaphorization: the production of indifference (Bauman 1989, 1993). Effective performance becomes one's only responsibility: consequential and intentional responsibility can be left to filter up the chain of command, which may or may not culminate in a central authority (Peukert 1994). One can sum up this systemic occlusion of moral responsibility by recalling that the Nazis 'did not see themselves as amoral barbarians but as moral agents acting within the moral code. . . . What is so disturbing is not their misperception of their legitimate obligations, but their commitment to the notion that the performance of duty is the supreme moral imperative' (Halberstam 1988: 41–42).

Thus, as Bauman (1989: 105) stresses, 'bureaucracy made the Holocaust. And it made it in its own image.' This is nowhere more apparent than with the death camps, which were run on the basis of an explicitly economic calculus. The cooperation of a host of agencies, perhaps most notably the railways, was crucial for the successful implementation of mass killing on the scale envisaged. The Holocaust was managed by the Economic Administration Section of the *Reichsicherheithauptamt*. 'It went about it the way all bureaucracies do: counting costs and measuring them against available resources, and then trying to determine the optimal combination' (Bauman 1989: 77). Similarly, industrial companies tendered to build and maintain the camps, competing on price, efficiency, and turnover time. However, whilst Berenbaum (1993: 108) notes that 'The Final Solution was a managerial triumph,' since although 'There was no budget for the program . . . the entire killing operation was run in the black,' he repeatedly emphasizes the enormity of the opportunity costs involved: 'For twelve years, the persecution and then the destruction of the Jewish people was a national priority, even at the cost of rational policy. Jewish workers were killed in spite of an acute labor shortage, and railway trains were made available to carry Jews to death camps even when every piece of rolling stock was needed to supply German troops on the eastern front' (Berenbaum 1993: 106; see also: Herbert 1993). Tens of thousands of workers and administrators kept the Holocaust in motion, most of whom appear to have carried out their tasks with a bewildering level of detachment and devotion to duty. The compilation of charts, statistics, timetables, specifications, costings, maps, itineraries, and so on provides a chilling illustration of the 'desk killer' (Arendt 1977; Milchman and Rosenberg 1992). Such is the banality and rationality of modern incarnations of evil. No longer exceptional, the elements out of which the Holocaust was composed fade into the normality of everyday life under the sign of modernity. It becomes, in fact, all too familiar: perhaps so familiar that for many the Holocaust has become a sort of de-realized, floating, and generic name for virtually any outbreak of mass killing that is not confined to the opposing war machines of 'nation-states.' One thinks, for example, of the names 'Bosnia' and 'Rwanda;' of the Cambodian 'Killing fields' and China's 'Cultural Revolution;' but also of the murderous terror wrought upon indigenous peoples by the agents of European colonialim and imperialism, as well as of environmental genocide, etcetera. Such a pale repetition invariably discloses a refusal and inability to think the specificity of each event. Drained of their singularity, intensity, and affect, these events are made to act as the dummy for the ventriloquy of a general discourse. Be that as it may, something resists the sociological reduction of the Holocaust to an exterminatory version of bureaucratic rationalization. It is to such an 'excess' that we now turn, by way of the third figure of singularity: *seriasure*.

Originally coined by Derrida as a disseminating quasi-concept aimed at conveying the serial erasure of faulty words, experiences, and concepts—everything that cannot be pinned down or fixed into place—we employ *seriasure* here with a sideways glance towards the mathematical and physical meanings of singularity: the break-points at which a function asymptotically approaches an infinite value and where matter takes on an infinite density or accelerates into the void. As we have already suggested, since such a break-point adsorbs all reference, sense, and meaning onto itself, not destroying or cancelling them, but (s)playing them out across the surface of a disadjusted space–time; it is wholly unrepresentable, except by way of allusion: black holes, vanishing points, superficial abysses, and silence (*cf.* Baudrillard 1990a).

To give a concrete illustration of these rather abstract concerns, consider the heaps of everyday objects that remain at Auschwitz and Majdanek: shoes, hair, glasses, toothbrushes, suitcases. These are Stains of the Real, which will never stop haunting the various attempts to close our imaginary and symbolic registers to the Holocaust's disjuncture. Here, as elsewhere, the unbearable silence of infinitely dense matter speaks louder than words. One member of the 'International Jewish Youth March of the Living' conveys the experience of *seriasure* upon encountering the piles of shoes at the site of the Majdanek death camp in the following terms:

> I glance at my own shoe, expecting it to be far different than those in this ocean of death, and my breath catches in my throat as I see my shoe, though lighter in color, is almost the same style as one, no two, three of the shoes I see: it seems as though every shoe here is my shoe. I wish I could throw my shoes into this pile, to grasp and feel each shoe, to jump into this sea, to become part of it, to take it with me.
>
> (quoted in Kugelmass 1994: 176)

Such is the dissimilation, disadjustment, and differentiation of the given, of the actual, of the present. *Seriasure*, then, relays a certain fissility *in* the event itself, and not merely *around* it. It conveys an irreducible disjuncture in space–time, traced out by a break-point of silence that will never be part of any memory, experience, or concept (*cf.* Carroll 1990). In a thoughtful consideration of Lanzmann's Holocaust film, *Shoah* (1985)—'*un film à ras de terre, un film de topographe, de géographe*'—Ohlin (1997: 1) sees in '*Shoah*'s refusal to visualize the past;' using, for example, archival footage of the death camps; not only a recognition of the limits of representation, but also a sensitivity to 'the places of the past, not *lieux de mémoire*, or repositories of the past, but charged pieces of the earth where the events happened.' 'Archival footage is only an iconic, that is, metaphoric picture of reality, not a part of it. But the index is metonymic: it is a piece of what happened' (*ibid.*: 17). The earth itself, in its banal presence, becomes a capacitor for untimely intensities and affects. Such is the singularity of a break-point 'that can transform a clean Israeli barber shop into Treblinka on the strength of a mere pair of scissors' (*ibid.*: 4).

In contradistinction both to those discourses that would hurl the singular outside of the norm(al), and those that would disperse it within the norm(al), *seriasure* serves to bear witness to the unrepresentable fact of the forgotten. With this latter, the Holocaust is no longer presented as something that falls outside the ebb and flow of so-called 'normal' events, but nor is it presented as the disclosure of the extremity of the norm(al) or that the extremity belongs to the norm(al). It suggests that the event remains open and untimely, augmenting and unfolding within itself a certain silence or hollowness—abyssal or otherwise—that has nothing to do with denial or forgetting to remember, or even with remembering to forget.

Even if there were a complete dearth of evidence, victims, and witnesses, and a certain blindness of the bystanders and judges—which there clearly is not—then these 'absences' would not at all prove that the Holocaust never took place. For they could just as easily be taken to be the signs of a perfect crime (*cf.* Baudrillard 1996; Lyotard 1988). Strictly speaking, then, the finalization of the Final Solution would have necessarily required not only the absence of all traces of the Holocaust, but also the absence of the absent itself. Neither negative nor positive, but cancelled, cancellated, and zeroed. Thus, the disappearance of the Jews and the Auschwitz universe would not have been sufficient. Only the disappearance of their disappearance (the negation of the negation) would have been. For example, consider the Nazi's classification of rail schedules that detailed the routing of trains to the death camps as 'Only for internal use.' 'The fact that . . . one cannot see . . . the word *geheim*, 'secret,' is astonishing to me,' exclaims Hilberg. 'On second thoughts . . . I believe that had they labeled it secret . . . they would have focused attention on the thing' (in Lanzmann 1985: 138–139). Such is the faulty fold of secrecy: absence presents, presence absents.

> We would do well, therefore, to ask precisely what is remembered by the ruins of crematoria at Auschwitz: the killing process or the attempted destruction of evidence by the Germans, an acknowledgment of their guilt? To what extent do the bins of artefacts at Auschwitz and Majdanek—the eyeglasses, hair, toothbrushes, suitcases—represent the absence of those who once animated them? Or to what extent do they remember the victims as the Nazis would have remembered them to us?
>
> (Young 1994: 24)

By way of the folding and unfolding of extreme normality and *seriasure*, the meaning, the place, and the identity of the Holocaust are all out of joint. 'This fissure is not one among others,' notes Derrida (1976: 200) in a very different context. 'It is *the* fissure: the necessity of interval, the harsh law of spacing.' However, one should not attempt to maintain together the disparate like a shattered windscreen or crazy paving, 'but to put oneself there where the disparate itself *holds together*, without wounding the dis-jointure, the dispersion, or the difference, without effacing the heterogeneity of the other' (Derrida 1994: 29). This is what the quasi-concept '*seriasure*' endeavours to relay.

Accordingly, *seriasure* is the singularity neither of exception nor of extremity, since it is no longer the singularity of an integral event, of a given 'one.' It portends disintegration and differentiation, and its diasporic disjuncture frustrates, interminably, the desire for totalization and finalization. *Seriasure* cannot be related to a norm, nor even gathered up within a distribution, since it follows a certain torque in the event itself, which disjoins both the norm and its concomitant distribution. Whilst a point of exception and extremity can always be integrated into an interpretative frame, *seriasure* never stops evading the grasp of a controlling context. Or rather, the 'controlling' context never stops losing control, opening up, and passing out (of itself). In this regard, *seriasure* relays the 'strength of the weak' and the active force of the virtually zeroed. It is in this sense that the Holocaust will never become fastened down or domesticated. The Holocaust will never cease shaking and soliciting 'our' complacent space–times (*cf.* Lanzmann 1985; Lyotard 1990b; Ohlin 1997). Strictly speaking, then, it is *seriasure* that maintains the singularity of the Holocaust without reduction. And it is this singularity that confounds all experience, all memory, all representation, and all interpretation (*cf.* Friedländer 1992; Hayes 1991; Lang 1990; Lyotard 1990a). It confounds every attempt to present and describe the Holocaust in and of itself, even by means of the most necessary and rigorous work of empirical documentation. Hereinafter, the Holocaust can never be quarantined within a space of mere factuality: *it happened*. It demands other spaces, including those that border on, enfold, and unfold our own. Ones that will shake our all-too-frequent efforts of remembering to forget.

Spacing the Holocaust

Having outlined a broad tropology for each of the three figures of singularity, we now want to set them to work in relation to the spaces and spacing of the Holocaust. Above all, we wish to argue that the Holocaust was not 'an accident of geography.' From the outset, then, one should be wary of the all too common reduction of space and place to the minor roles of particularity, passivity, and contingency, so that a fetishized notion of time can be presented as being vital, dynamic, and active. For this 'accident of geography' rested on an irreducible spatial logic, which continues to reverberate through, and thereby shake up, modernity's current ordering of space–time. In the remainder of this chapter we will therefore consider the disadjusted, enduring, and 'untimely' space–time of the Holocaust through the trifid singularity of exception, extremity, and *seriasure*.

As we have already noted, during the 1970s some German historians began to call for the 'historicization' of the Third Reich, drawing attention away from the Nazi elite and the specificity of the Jewish experience of the Holocaust (*cf.* Friedländer 1987, 1990; Kershaw 1993); a few even called for its 'normalization,' thereby reducing Nazism and the Holocaust to mere *examples* of totalitarianism and genocide (*cf.* Carlton 1990; Evans 1987; 1990; Fein 1990; Rich 1987). Great controversy has surrounded Nolte's (1988) claim that the notion of a Final Solution originated not with the Nazi desire for the complete annihilation of the Jewish *race*, but with the Marxist quest for the total eradication of the bourgeois *class*. This claim fails to recognize the fundamental difference in kind between Nazi annihilation (death pure and simple) and Marxist sublation (conservation through expunction). It also serves to map the spaces of the Holocaust in a highly reductive manner. For example, Nolte claims that the Gulag Archipelago was more of an 'origin' for the Holocaust than was Auschwitz, and that the Nazis may have carried 'out an "Asiatic" deed perhaps only because they regarded themselves and their ilk as potential or real victims of an "Asiatic" deed' (quoted in Joffee 1987: 72). In this way, Nolte turns the Holocaust into a defensive reaction and a pre-emptive strike against the massing forces of 'Judaeo-Bolshevism.'

The lack of sensitivity to the topography and the topology of the Holocaust within the discourses and counter-discourses of historical uniqueness is brought to the fore when the spatiality of the Holocaust is given due consideration (Clarke *et al.* 1996). We want to suggest here that two fundamentally spatial concerns underpinned the Nazis' desire for the production of a New Order: *Lebensraum*—the so-called organic 'living space' required by the German Reich and the 'Aryan Race' (Dickinson 1943); and *Entfernung*—the process of objective and subjective distancing and de-stancing that amounted to 'an effective removal of the Jews from the lifeworld of the German race' (Bauman 1989: 120). Whilst *Lebensraum* is intimately connected to geopolitical concerns for the domination of physical space, it should not be detached from the desire to control *social space* itself—cognitive, moral, and aesthetic—especially by means of *Entfernung*. One must therefore adopt a theorization attentive to the impossibility of separating social and physical space. For example, Mayer (1990: 106) stresses that 'Hitler's geopolitically and racially informed imperative in the east was of one piece with his anti-modernist and anti-Semitic project inside Germany.'

Bauman argues that physical notions of distance derive from a phenomenological reduction of the social experience of space into a series of abstract, impersonal, and seemingly objective categories: 'we grasp physical space intellectually with the help of notions which have been coined originally to "map" the qualitatively diversified relations with other humans' (1993: 145). Such is the conventional (mis)understanding of the notion of *Lebensraum*, which only readmits social spacing at one remove from the original reduction. It is a typically modern representation of space: abstract, homogeneous, and empty (*cf.* Lefebvre 1991). The way in which such an abstract space then maps onto other social spaces thus demands sustained attention, not least with respect to the Final Solution. For it is here that one can discern the full implications of the Nazis' desire for a 'purification of space' (Sibley 1988).

Perhaps the greatest contribution Bauman (1989, 1993) makes to the understanding of the relationship between modernity and the Holocaust is his recognition of how the swathing geopolitical ambition of the Nazis was fundamentally at odds with the world of the stranger, personified for the Nazis by their reductive stereotyping of 'the Jews.' The stranger foreshadows the reversal of the traditional, positive correlation between proximity in physical space and intimacy in social space, an overturning of the distance-decay of affect. Yet estrangement in place is one of the defining features of modernity, as is the onset and spread of the strangely familiar, with all of the associated 'risks' of ambivalence and indiscernibility. Only in a world of universal strangeness would the figure of the stranger, who straddles the clean-cut boundaries differentiating this from that, no longer be conceivable: although the fate (and fatality) of modernity resides in its repeated attempts to reinforce the cuts and therefore maintain the discernibility of strangers, aliens, drifters, etcetera (but let us not get sucked into globalization, space–time compression, and postmodernity just here (*cf.* Baudrillard 1990b; Clarke 1997; Clarke and Doel 1994; Doel and Clarke forthcoming)). In other words, the stranger does not enter into the clean-cut spaces of modernity from the outside; modernity cuts out the figure of the stranger from the inside, and attaches it onto various groups of people. 'Strangers are the products of the same social spacing which aims at assimilating and domesticating the life-world,' they are, 'simultaneously, the anchor and bane of existence' (Bauman 1993: 160). This uncomfortably ambivalent feature enabled those in the position of defining the Nazis' social space to adopt a mechanism for 'projecting its inner incongruity upon a selected social target (that is, focusing the ambivalence which saturates the whole of social space on a selected sector of that space)' (Bauman 1993: 106)—the Jews.

In such a context, Bauman (1989), like Lyotard (1990a), writes of real Jews, in contradistinction to the Nazis' proteophobic projection of the boundary-straddling, 'slimy,' conceptual 'Jews:' a dehumanized, depersonalized, and demonized category through which the Nazis rationalized and legitimated their desire to exterminate ambivalence from modernity, to exterminate the Jews, along with a whole host of others. Bauman suggests that the reductive categorization of the Jews initially takes on the role that it does in relation to Christianity. 'The self-identity of Christianity was, in fact, estrangement of the Jews. It was born of rejection *by the Jews*' (Bauman 1989: 38). The later secularization and modernization of society inherited this heterophobic projection of an ambivalent and 'slimy' role onto the Jews. But, as we noted above, modernity itself needed and produced such ambivalent, boundary-straddling strangers, and it is precisely the contradiction between modernism and anti-modernism within the Third Reich that engendered such a vehement, exterminatory strain of anti-Semitism. In particular, and when contrasted with the geopolitical ambitions of the Nazi state, the diasporic character of the Jewish people presented itself as a wholly Other conception of social space, one fundamentally at odds not only with the pan-Germanism of the Third Reich (Barnavi and Eliav-Feldon 1992: vi–ix), but also with the wider currents of (nation-)statism and imperialism. 'The conceptual Jew carried a message; alternative to this order here and now is not another order, but chaos and devastation' (Bauman 1989: 39, emphasis removed). Or again: 'The World tightly packed with nations and nation-states abhorred the non-national void. Jews were in such a void: they were such a void' (*ibid.*: 53, emphasis removed). The mostly indifferent and ineffectual response by the Allies to the plight of the Jews within Nazi-controlled territory accords with such an image of thought. Yet as we shall now indicate, somewhat elliptically, the 'void' is a complicated and overdetermined image of thought.

That the Jews were figured as a void by the Nazis provides one of the most persuasive explanations of the conditions of possibility of the Holocaust: they *ought* not to exist. However, this void was neither simply *empty*—in the manner of a 'blank space' in the fabric of the *socius* (a tear, rent, gap, seam, hole, 'dark continent,' etcetera)—nor *neutral*—in the sense of a 'dark body' without intensity or affect. The void was not figured as an inert and inconsequential region of space–time, whose alleged incompossibility with the space–time of 'nation-states' could simply be ignored and bypassed, or else sutured and overcome. (Sadly, such an indifference with regard to 'dark bodies' is how the majority seem to figure the ethico-political topology of the *socius*, unless the actual or potential clash of incompossibles is all too palpable, especially with regard to the striae of race, class, nationality, ethnicity, gender, religion, and sexuality; but also with regard to the so-called forces of globalization, which are supposedly over-running the regulatory apparatuses of the state: deterritorialized money, media, knowledge, capital, desire, etcetera. Furthermore, these striae are poorly served by essentially quantitative and invariably aporetic notions of (1) *opposition* (the facing off of two givens); (2) *continua* marked by extremes; and (3) *discretion* (the extraordinary use of this word alludes to the discreetness of discreteness, such as the 'standard deviations' of a 'normalized distribution of deviancy:' for example, consider the discretion of a typical, 'majoritarian' distribution of sexuality: heterosexual, homosexual, bisexual, transsexual, asexual, celibate, etcetera). Instead, these striae express qualitatively different habits, modes, and expressions of being-in-the-world: they are transformers and differentiators, rather than duplicators and integraters; lines of flight, rather than axiomatic strictures. If the striae coordinate, it is not to triangulate and pin down; it is to give a body a sense and a direction—that is to say, coordination is a play of force upon force. Striae do not identify (*it is an* . . . x, y, z): they *move* a body. So, 'masculinity' and 'femininity' do not walk hand-in-hand, in a sort of stand-off against other striae. A certain masculinity may share a greater proximity with a certain sexuality or ethnicity—or not. Or else a certain masculinity may enter into a zone of proximity and indiscernibility with another masculinity, or with a certain animal, or a certain form of coloration, or a certain tune . . . . In short, the striae are less like a net or a network and more like a skein, the ravelling and unravelling of which brings us back to the reciprocity of amassing intensities and affects, on the one hand, and the deepening of nihilism and vacuity, on the other.) Now, since the void into which the Nazis interned the Jews was not figured as passive, there was to be no 'letting be.' Why did the Nazis insist that the Jews, although null and void, were nevertheless a strange attractor? How could 'nothing' express a force on something? Or again: why did this void hold such a fatal attraction for the Nazis? Precisely because it was a (w)hole that was *open*: the void is the inside *of* the outside—and vice versa (chaos, chaosmosis on the sk(e)in of the Earth). But also because the void was figured in the extreme: as a *vacuum* and a *black hole*. The quantitative difference between the one image of thought and the other—such as that between 'emptiness' and 'fullness' or between 'zero intensity' and 'infinite intensity'—is secondary to the essential element: the differential relation, in the sense that both a vacuum and a black hole have the quality and power to affect what surrounds them; both exercise a force on the rest of the universe. Figured as a vacuum or a black hole, the Jews threatened to draw in all of those social spaces that surrounded them—either by one body acting directly on another, or indirectly, by each body curving the region of social space in which it is embedded. And this drawing of each into the neighbourhood of the other would of course extend beyond the immediate vicinity and event horizon. One could call such a double movement a 'becoming other,' a 'reciprocal deterritorialization,' or simply 'a life.' And the Nazis loathed them all. Their own fatal attraction was for a death drive—not just for the Jews (amongst others), but also for themselves. For only in death can one remain eternally the same, untouched by any other. By turning their backs on the *socius*, on the social relations of becoming (through) a life, the Nazis sought to produce first an autistic, and then a dead, *socius*. But even autism and death escape this desire for the eternal return of the same . . . . But let us return, once more, to the void.

Since everything effects everything else, one would expect the superior force of numbers of the majority to easily overcome the scant numerical force of the minority—but not if the minority can manage to achieve sufficient vacuity or density, or have that enhancement thrust upon them. Such is the role of the void, which has a varied characterization: vermin, vice, cancer, weed, etcetera. Either the *force* of the void is augmented as its density/nihilism increases, thereby visibly affecting more and more of social space; or else the void itself begins to circulate across the face of the Earth, like a vagrant, vagabond, nomad, or tourist; or else these augmenting and migrating 'strange' and 'dark attractors' begin to *proliferate* across the face of the Earth (by cloning or contagion), deterritorializing the striated, sedentary, and arboreal spaces of the state apparatuses, and reterritorializing the resulting flows onto a new Earth and a new people: an 'anarcho-social' body without organizations, servicing the smooth, nomadic, and rhizomatic spaces of the schizoid desiring-machines. (The brevity of this chapter precludes further development of these distinctions, but a certain unzipping of them can be found in Deleuze and Guattari 1984, 1988, 1994; Doel 1996, forthcoming; Perez 1990.) And for the Nazis, these apparent forces of destabilization *vis-à-vis* the molar (equals 'moral') order, which have been inflated by the Nazis themselves, ought not to exist. It is as if they wanted a Newtonian universe for the *socius*: for an eternal return of the same; and for a regularized, normalized, and above all stabilized machine in perpetual motion (automatism, autism). It is as if they wanted to stop the *socius* from becoming relativized, and from coming alive: to stop the curvature of space–time; and to stop the difference-producing repetition, differentiation, and dynamism of the universe. The figuration of what ought not to exist was much more than an arbitrary image of thought: much more than this or that disparaging and dehumanizing identification in order to naturalize and legitimate mass murder and genocide. But arbitrary or not, for the Nazis the Jews must not exist. 'A crime is met with punishment; a vice can only be exterminated' (Arendt 1951: 87). Or again: 'Cancer, vermin or weed cannot repent. They have not sinned, they just lived according to their nature' (Bauman 1989: 72). For the Nazis, then, the Jews served as the focus for the ambivalence that saturated the social spaces of modernity: they were projected as the 'dark' and 'strange attractors' *par excellence*. And the exterminatory purging of this ambivalence allows Bauman to depict the Holocaust as being singular in the extreme. In such circumstances, it should be clear why so many moral geographies were faced with a smooth cognitive space with little place for resistance or hiding.

On a related, although elliptical, tack, it was the Nazis' desire for *Lebensraum* through *Entfernung* that marks the singularity of the place of the Jews in Nazi ideology and practice. For the Nazis, *Entfernung* represented a *sine qua non* for the maintenance and development of *Lebensraum*. This is why Berenbaum (1993: 105) is so wrong to insist that 'the Holocaust served no political or territorial purpose;' 'the Jews posed no territorial threat to the Nazis. Their murder led to no geopolitical benefit, yielded no territorial gain.' This claim entirely misses the spatial underpinnings of the Final Solution. It cannot see beyond the conflation of physical and social space. For the Nazis, by contrast, physical space literally *amounted to nothing* unless it conformed to a very particular configuration of cognitive, moral, and aesthetic codes. Arendt (1977: 217) refers to a secret speech by Hitler to members of the German High Command, delivered in November 1937: 'Hitler had pointed out that he rejected all notions of conquering foreign nations, that what he demanded was an "empty space" [*volkloser Raum*] in the East for the settlement of Germans. His audience . . . knew quite well that no such "empty space" existed.' Only on this basis is it possible to grasp why the Nazis were prepared to risk conquered territory (physical space) for the production of *Lebensraum* through *Entfernung* (social space). More precisely, *Lebensraum* was thus produced through a deterritorialization of physical space onto a two-fold socio-spatial reterritorialization. First, physical space was turned into a blank page, ready for an inaugural inscription by the Aryan race; and second, the Jews were not only expelled from the face of the Earth, they were also forced into a 'social space of silence' (Olsson 1991; see also Carroll 1990; Lyotard 1988, 1990a). In short, it was precisely the disappearance of the Jews without trace that made *Raum* formation for the Nazis possible. In this way, the despotic deterritorialization and two-fold reterritorialization of the Earth (its 'Germanification') came to be seen as a primordial inscription upon the blank page of a more or less isotropic plane: *Deterritorialization*—'In regions of total planning freedom (*Plannungsfreiheit*), 4.5 million of a total population of 10.2 million were to be removed and the rest resettled within the imposed territorial order' (Cosgrove 1994: 19); *Rural reterritorialization*—'a key feature of the new order imposed across former Poland was the imposition of a planted landscape of fields, farms and hedges, reflecting national Socialist belief in *Blud und Boden* [Blood and Soil], that the nation is born out of the soil' (*ibid.*: 18); and *Urban reterritorialization*—'plans were made to build completely new German cities in the East, with approximately 15,000–20,000 population, at important railway and highway junctions. Such cities were to be surrounded by a 5–10 mile broad belt of German villages' (Kamenetsky, quoted in Rössler 1989: 422). It is not for nothing, then, that Lanzmann's *Shoah* is at pains to demonstrate the impossibility of a purified, empty social space, for even the landscape itself bears the traces of mass murder.

All in all, then, the separation of the social spaces of the Holocaust from the territorial ambition of the *Lebensraum* policy entirely misses their imbrication as an ordering that created and destroyed its own internally engendered waste. For as Douglas (1970: 48) reminds us, 'Dirt is the by-product of a systematic ordering and classification of matter . . . the reaction which condemns any object or idea likely to confuse or contradict cherished classifications.' On this basis, a sensitivity to the extremity of the Holocaust's social spacing can allow one to discern the violence shadowing the creative destruction of ambivalence and strangeness in the apparently more mundane social spaces of 'ordinary' or 'comfortable' modernity. However, the discourses of extremity are dogged by conceiving of a state of affairs as a selection from an array of pre-existing possibilities. Such a (mis)conception haunts the presentation of the Holocaust by both Arendt and Bauman, a haunting that risks dispersing the Holocaust into the ethereal realms of the always already still more of the same. And yet, the more one unfolds the imbrication of normality and extremity, of ordinary modernity and exterminatory modernity, the more one becomes embroiled in a labyrinthine *seriasure*. However, whilst the singularity of extremity gathers up the singularity of exception, since it takes the disavowed and expelled into itself, the singularity of *seriasure* releases the grip of such a holding context.

In an attempt to disrupt the discourses that present the Holocaust as simply past, Barham (1992) has employed the notion of a 'counter-time' that interminably interrupts what is taken to be present; the two temporalities are coterminous but disjoined, in a non-Euclidean parallelism through which the separate series continually criss-cross and interlace with one another (parallel and therefore incompossible events come into contact in one and the same curved *socius*). For Barham, as for Levi (1963), this disjunctive synthesis is exemplified in the bodily comportment of the survivors, for whom the 'recapture of forgotten habits—how to use a toothbrush, toilet paper, a knife and fork—and forgotten tastes and smells—blossom, the sweet scent of rain in spring—finds itself vulnerable to the "counter-time" of Auschwitz, where the rain stinks of diarrhea and the winds carry the odour of burning flesh' (Barham 1992: 40). Barham calls this a 'temporal dislocation' because the survivor who experiences such an intrusion of counter-time is no longer (simply) living in the present. Accordingly, and in contrast to the commonsense view that 'we' are distant from the Auschwitz universe in cognitive, moral, and aesthetic space, Barham wants to emphasize the survivors' (almost) absolute proximity to Auschwitz. Indeed, it is precisely in this context that Barham, following Blanchot, speaks of the 'died event' and 'humiliated memory,' in contradistinction to the sublating movement of the 'lived event' and 'anguished memory:' 'The person may indeed survive but the terms of existence available to her are those of living under threat of a death or dying which has already taken place' (*ibid.*: 41). Consequently, it is important 'to understand the *finality* of the "died event" and to recognize it as a determinate ending, no matter how many new beginnings may come after it. . . . The Holocaust experience "murdered part of the future even for those who survived it"' (*ibid.*: 41–42). However, since this finality traverses the body and its comportment—the social body and body politic, no less than an individual's body—the died event and anguished memory should not be confused with all of those forms of remembering in order never to experience 'it' again (remembering to forget, and forgetting to remember).

The distinction between a unilinear, diachronic temporality of 'normal' everyday life on the one hand, and the intrusion of a dislocating, synchronic counter-time on the other, is underscored by Barham's reliance upon Langer's (1991: 175) problematic distinction between Auschwitz as a 'story' and Auschwitz as a 'plot:' 'Auschwitz as story enables us to pass through and beyond the place, horrible as it may be, while Auschwitz as plot stops the chronological clock and fixes the moment permanently in memory and imagination, immune to the vicissitudes of time.' Hence Langer's suggestion that her everyday life as a survivor is less a living-on 'after' Auschwitz than a living 'beside' Auschwitz—although this warped and twisted t(r)opology is more like a Möbius (s)trip than a simple parallelism: the ins(l)ide is (on) the outside, and the outside is (on) the inside (*cf.* Doel, forthcoming). It is in this sense that Barham claims that whilst Auschwitz represents a rupture in the linear and progressive unfolding of the (meta)narratives of (post)modernity—to the extent 'that the actuality of what took place there cannot be transfigured or absorbed, become "history or past time"' (Barham 1992: 52)—we must nevertheless 'respect the longings of narrative to commence the story anew.' Consequently, the Holocaust becomes (again) an infinitely fissile break-point for the disadjusted space–times of our (post)modernity.

However, this remarkable and mostly compelling presentation is troubling for several reasons. First, who are 'we'? Second, the bifid splitting of the explanatory framework ensures that the terms for situating and comprehending the Holocaust are never put into question: 'we' can still distinguish between the story and the plot, between our time and counter-time, between diachrony and synchrony, between the died event and the lived event, between humiliated memory and anguished memory, between Auschwitz and (post)modernity, and between the survivors and the collectivity. Third, to the extent that it is the intrusion of the plot into the story that dislocates space–time, opening a *seriasure* in the socio-spatial order of things, this dislocation can only be attributed to, and experienced by, the survivors of the Holocaust: it is not 'our' counter-time; it is not part of 'our' (post)modernity. As is so often the case, the flaws of being-in-the-world are condensed onto an Other, since binary logic can attempt to purify only one term (the supposed self-sufficiency of truth, goodness, speech, use value, etcetera) by wasting its Other (the supposed derivative duplic-ity of error, evil, writing, exchange value, etcetera). Fourth, Barham's and Langer's accounts suggest that, in spite of the rupture produced through the *seriasure* of counter-time, the progressive narrative of (post)modernity continues unabated to stutter and stammer along its diachronic path. After Auschwitz, and beside Auschwitz, the (hi)story moves on; 'we' 'pass through and beyond' into fresh cognitive, moral, and aesthetic spaces: 'We shall never know whether Nazism, the concentration camps or Hiroshima were intelligible or not: we are no longer part of the same mental universe' (Baudrillard 1993: 91). Finally, and perhaps most worrisome, despite the enormous heterogeneity of Holocaust experiences, spatial practices, memories, and memorials (Milton 1991; Young 1993, 1994), Barham's theorization collapses all of this down to a single imper-ative. Our obligation to the fact of the survivors' counter-time extends only to a passive memorialization, remembering (not) to forget. Like the survivors of the Holocaust, 'we cannot put Auschwitz behind us, but must live "beside" it,' sug-gests Barham (1992: 52). 'Auschwitz just is that place,' he continues, noting that 'For Freud, a memorial once established was "just there" and to be taken for granted.' But not only does this repress the *politics* of memory (Charlesworth 1994; Milton 1991; Young 1993, 1994), it also implies that 'we' should simply 'let Auschwitz be,' whilst 'our' (post)modernity—without ever suturing or sublating the wounds of the Holocaust—merely passes by and moves on. Clearly, this would entail an impoverished sense of foreclosed responsibility. For one would be called upon to abandon the (died) event and (humiliated) memory to the stutter of an inert, impassive, and amassing factuality: *It hap-pened. It happened. It happened.* And yet, what if it is precisely *our* (post)moder-nity, (lived) events, and (anguished) memories that are untimely, disjointed, and disadjusted by the Holocaust? What if *we* are all survivors? (Doel and Clarke, forthcoming; Kuggelmass 1994; Massumi 1993).

Conclusion

As we have repeatedly emphasized throughout the course of this chapter, every attempt to represent, comprehend, and explain the Holocaust is accompanied by profound difficulties. Indeed, the Holocaust obliges us to become sensitive to the *fact* of the forgotten, a sensitization that we have argued can most effectively be relayed through the threefold figure of singularity: exception, extremity, and *seriasure*. This is not to reject empirical documentation and the testimony of the victims, perpetrators, and bystanders—far from it—but to affirm the necessity of an ethico-political act of *suppletion*. For the supplement bears witness to a remainder that was never part of any original, insofar as it participates without belonging, completing an irreducible lack, gap, or disjuncture *in* the original itself, and not merely *around* it. And whilst each of these figures of singularity can be seen to be embedded within particular discourses—historical, sociological, and philosophical—each is nevertheless oriented around a particular configuration of social, physical, moral, and aesthetic *space*.

Specifically, we have sought to demonstrate how the Nazis' quest for *Lebensraum* was necessarily, and not simply contingently, related to the desire for *Entfernung*. Thus, any consideration of the Holocaust that does not adequately take account of its *spacing*, in every sense of this term—and by that we mean not only the heterogeneous spaces in which the Holocaust took place, but also the spatial practices of its implementation, and the conceptual spacing of its discursive and libidinal economies—is seriously impoverished. But by the same token, *any* account of spacing that does not take into consideration the yearning for purification is similarly impoverished. Indeed, any approach that fails to draw out the link between the spatial practices of Hitlerite ideology, the drive for *Lebensraum*, the thirst for *Entfernung*, and the quest for a Final Solution, will seriously misconstrue the importance of the trimerous concept of singularity—the refolding of normality, exceptionality, and extremity into the ethico-political spacing of *seriasure*—for a rigorous engagement with the diastematic and diasporic spaces of the Holocaust. Thus, our contribution to the understanding of the Holocaust is an attempt to maintain and radicalize the disarticulation, disadjustment, dissimilation, and disjuncture in space–time, and in the very event-ness of the event. For we will never be finished with reading and re-reading the diasporic signs and spaces of the Holocaust, since even in the wake of the most ruthless attempt to terminate the interminable, to finalize a Final Solution, to suture disjuncture, and to compel the Other to vanish from the face of the Earth, there will always be traces, and traces of traces (Derrida 1991b; Lanzmann 1985). The Holocaust trembles. After this singular event nothing will ever truly add up without remainder. Its spacing demands other calculi of identity and difference.

References

Arendt, H. (1951) *The Origins of Totalitarianism*. New York: Harcourt Brace.

Arendt, H. (1977) *Eichmann in Jerusalem: A Report on the Banality of Evil*. Harmondsworth: Penguin.

Baldwin, P. (ed.) (1990) *Reworking the Past: Hitler, the Holocaust, and the Historians' Debate*. Boston: Beacon.

Barham, P. (1992) 'The next village': modernity, memory and the Holocaust. *History of the Human Sciences*, 5(3), 39–56.

Barnavi, E. and Eliav-Feldon, M. (1992) *A Historical Atlas of the Jewish People: From the Time of the Patriarchs to the Present*. London: Hutchinson.

Baudrillard, J. (1990a) *Seduction*. London: Macmillan.

Baudrillard, J. (1990b) *Fatal Strategies*. London: Pluto.

Baudrillard, J. (1993) *The Transparency of Evil: Essays on Extreme Phenomena*. London: Verso.

Baudrillard, J. (1996) *The Perfect Crime*. London: Verso.

Bauman, Z. (1989) *Modernity and the Holocaust*. Cambridge: Polity.

Bauman, Z. (1993) *Postmodern Ethics*. Oxford: Blackwell.

Berenbaum, M. (1993) *The World Must Know: The History of the Holocaust as Told in the United States Holocaust Memorial Museum*. Boston: Little, Brown & Co.

Bergen, D. L. (1994) The Nazi conception of 'Volkdeutsche' and the exacerbation of Anti-Semitism in Eastern Europe, 1939–45. *Journal of Contemporary History*, 29, 569–582.

Bosworth, R. J. B. (1993) *Explaining Auschwitz and Hiroshima: History Writing and the Second World War, 1945–1990*. London: Routledge.

Braun, R. (1994) The Holocaust and Problems of Historical Representation. *History and Theory*, 33(2), 172–197.

Browning, C. R. (1991) *Fateful Months: Essays on the Emergence of the Final Solution*. New York: Holmes & Meier.

Browning, C. R. (1992) *The Path to Genocide: Essays on Launching the Final Solution*. Cambridge: Cambridge University Press.

Burleigh, M. (1991) Racism as social policy: the Nazi 'euthanasia' programme, 1939–1945. *Ethnic and Racial Studies*, 14(4), 453–473.

Carlton, E. (1990) Race, massacre and genocide: an exercise in definitions. *International Journal of Sociology and Social Policy*, 10(2), 80–93.

Carroll, D. (1990) Foreword: the memory of devastation and the responsibilities of thought: 'And let's not talk about that', in *Heidegger and 'the jews'* (J.-F. Lyotard), Minneapolis: University of Minnesota Press, vii-xxix.

Charlesworth, A. (1992) Towards a geography of the Shoah. *Journal of Historical Geography*, 18(4), 464–469.

Charlesworth, A. (1994) Contesting places of memory: the case of Auschwitz. *Environment and Planning D: Society and Space*, 12, 579–593.

Clarke, D. B. (1997) Consumption and the city, modern and postmodern. *International Journal of Urban and Regional Research*, 21(2), 218–237.

Clarke, D. B. and Doel, M. A. (1994) Transpolitical geography. *Geoforum*, 25(4), 505–524.

Clarke, D. B., Doel, M. A. and McDonough F. X. (1996) Holocaust topologies: singularity, politics, space. *Political Geography*, 15(6/7), 457–489.

Cosgrove, D. (1994) Terrains of power. *Times Higher Education Supplement* 11 March, 18–19.

Deleuze, G. (1994) *Difference and Repetition*. London: Athlone.

Deleuze, G. and Guattari, F. (1984) *Anti-Oedipus: Capitalism and Schizophrenia*. London: Athlone.

Deleuze, G. and Guattari, F. (1988) *A Thousand Plateaus: Capitalism and Schizophrenia*. London: Athlone.

Deleuze, G. and Guattari, F. (1994) *What is Philosophy?* London: Verso.

Derrida, J. (1976) *Of Grammatology*. Baltimore, Md.: Johns Hopkins University Press.

Derrida, J. (1978) *Writing and Difference*. Chicago: Chicago University Press.

Derrida, J. (1991a) *Between the Blinds: A Derrida Reader*, P. Kamuf (ed.). Hemel Hempstead: Harvester Wheatsheaf.

Derrida, J. (1991b) *Cinders*. Lincoln: University of Nebraska Press.

Derrida, J. (1994) *Specters of Marx: The State of the Debt, the Work of Mourning, & the New International*. London: Routledge.

Dickinson, R. E. (1943) *The German Lebensraum*. Harmondsworth: Penguin.

Doel, M. A. (1994) Deconstruction on the move: from libidinal economy to liminal materialism. *Environment and Planning A*, 26(7), 1041–1059.

Doel, M. A. (1996) A hundred thousand lines of flight: a machinic introduction to the nomad thought and scrumpled geography of Gilles Deleuze and Félix Guattari. *Environment and Planning D: Society and Space*, 14, 421–439.

Doel, M. A. (forthcoming) *Poststructuralist Geography: The Harsh Law of Space*. New York: Guilford.

Doel, M. A. and Clarke, D. B. (forthcoming) Transpolitical urbanism: suburban anomaly and ambient fear. *Space & Culture*, 1(2).

Douglas, M. (1970) *Purity and Danger: An Analysis of Concepts of Pollution and Taboo*. Harmondsworth: Penguin.

Eley, G. (1988) Nazism, politics and the image of the past: thoughts on the West German *Historikerstreit* 1986–1987. *Past and Present*, 121, 171–208.

Evans, R. J. (1987) The new nationalism and the old history: perspectives on the West German *Historikerstreit*. *Journal of Modern History*, 59, 761–797.

Evans, R. J. (1990) *In Hitler's Shadow: West German Historians and the Attempt to Escape from the Nazi Past*. London: Tauris.

Fein, H. (1990) Genocide: a sociological perspective. *Current Sociology*, 38, 1–7.

Friedländer, S. (1987) West Germany and the burden of the past: the ongoing debate. *Jerusalem Quarterly*, 42, 3–18.

Friedländer, S. (1990) Some reflections on the historicization of National Socialism, in *Reworking the Past: Hitler, the Holocaust, and the Historians' Debate*, P. Baldwin (ed.), Boston: Beacon, 88–101.

Friedländer, S. (ed.) (1992) *Probing the Limits of Representation: Nazism and the 'Final Solution.'* London: Harvard University Press.

Gellately, R. (1990) *The Gestapo and German Success Enforcing Racial Policy 1933–1945*. Oxford: Oxford University Press.

Habermas, J. (1987) *The Philosophical Discourse of Modernity: Twelve Lectures*. Oxford: Blackwell.

Habermas, J. (1989) *The New Conservatism: Cultural Criticism and the Historians' Debate*. Cambridge: Polity.

Halberstam, J. (1988) From Kant to Auschwitz. *Social Theory and Practice*, 14(1), 41–54.

Hayes, P. (ed.) (1991) *Lessons and Legacies: The Meaning of the Holocaust in a Changing World*. Evanston: Northwestern University Press.

Herbert, U. (1993) Labour and extermination: economic interest and the primacy of *Weltanschauung* in National Socialism. *Past & Present*, 138, 144–195.

Herf, J. (1984) *Reactionary Modernism: Technology, Culture, and Politics in Weimar and the Third Reich*. Cambridge: Cambridge University Press.

Hillgruber, A. (1986) *Zweierlei Untergang: Die Zerschlagung des Deutschen Reiches und das Ende des Europäischen Judentums*. Berlin: Siedler.

Joffee, J. (1987) The battle of the historians: a report from Germany. *Encounter*, 69, 72–77.

Kansteiner, W. (1994) From exception to exemplum: the new approach to Nazism and the 'Final Solution.' *History and Theory* 33(2), 145–171.

Kershaw, I. (1993) *The Nazi Dictatorship: Problems and Perspectives of Interpretation*. London: Edward Arnold.

Kugelmass, J. (1994) Why we go to Poland: Holocaust tourism as secular ritual, in *The Art of Memory: Holocaust Memorials in History*, J. E. Young (ed.). New York: Prestel-Verlag, 175–183.

Lang, B. (1990) *Act and Idea in the Nazi Genocide*. London: University of Chicago Press.

Langer, L. (1991) *Holocaust Testimonies: The Ruins of Memory*. New Haven, Conn.: Yale University Press.

Lanzmann, C. (1985) *Shoah: An Oral History of the Holocaust*. New York: Pantheon.

Lefebvre, H. (1991) *The Production of Space*. Oxford: Blackwell.

Levi, P. (1963) *The Drowned and the Saved*. London: Abacus.

Lyotard, J.-F. (1988) *The Differend: Phrases in Dispute*. Manchester: Manchester University Press.

Lyotard, J.-F. (1990a) *Heidegger and 'the jews.'* Minneapolis: University of Minnesota Press.

Lyotard, J.-F. (1990b) *Pacific Wall*. Venice: Lapis Press.

Lyotard, J.-F. (1990c) *Duchamp's TRANS/formers*. Venice: Lapis Press.

Maier, C. S. (1986) Immoral equivalencies. *The New Republic* (December), 36–41.

Mason, T. (1993) *Social Policy in the Third Reich*. Milton Keynes: Open University Press.

Massumi, B. (ed.) (1993) *The Politics of Everyday Fear*. Minneapolis: University of Minnesota Press.

Mayer, A. (1990) *Why did the Heavens not Darken?* London: Verso.

Milchman, A. and Rosenberg, A. (1992) Hannah Arendt and the etiology of the desk killer: the Holocaust as portent. *History of European Ideas*, 14(2), 213–226.

Milton, S. (1991) *In Fitting Memory: The Art and Politics of Holocaust Memorials*. Detroit: Wayne State University.

Nolte, E. (1985) Between myth and revisionism? The Third Reich in the perspective of the 1980s, in *Aspects of the Third Reich*, H. W. Koch (ed.) London: Macmillan, 17–38.

Nolte, E. (1988) A past that will not pass away (a speed it was possible to write, but not at present). *Yad Vashem Studies*, 19, 65–73.

Ohlin, M. (1997) Lanzmann's *Shoah* and the topography of the Holocaust film. *Representations*, 57, 1–23.

Olsson, G. (1991) *Lines of Power/Limits of Language*. Minneapolis: University of Minnesota Press.

Perez, R. (1990) *On An(archy) and Schizoanalysis*. New York: Autonomedia.

Peukert, D. J. K. (1994) The genesis of the 'Final Solution' from the spirit of science, in *Nazism and German Society, 1933–1945*, D. F. Crew (ed.), London: Routledge, 274–299.

Rich, H. (1987) Genocide as a sociopolitical process. *Canadian Journal of Sociology*, 12, 393–340.

Rössler, M. (1989) Applied geography and area research in Nazi society: central place theory and planning, 1933 to 1945. *Environment and Planning D: Society and Space*, 7, 419–431.

Sabini, J. R. and Silver, M. (1980) Destroying the innocent with a clear conscience: a sociopsychology of the Holocaust, in *Survivors, Victims and Perpetrators: Essays on the Nazi Holocaust*, J. E. Dinsdale (ed.), Washington, DC: Hemisphere Publishing Co., 329–358.

Sibley, D. (1988) The purification of space. *Environment and Planning D: Society and Space*, 6, 409–421.

Steiner, G. (1987) The long life of a metaphor: an approach to 'the Shoah.' *Encounter*, 68, 55–61.

Taylor, P. J. (1996) What's modern about the modern world-system? Introducing ordinary modernity through world hegemony. *Review of International Political Economy,* 3(2), 260–286

Weinberg, D. and Sherwin, B. L. (1979) The Holocaust: historical overview, in *Encountering the Holocaust: An Interdisciplinary Survey,* B. L. Sherwin and S. G. Ament (eds). Chicago: Impact, 12–22.

Wytwycky, B. (1980) *The Other Holocaust: Many Circles of Hell.* Washington DC: The Novak Report on the New Ethnicity.

Young, J. E. (1993) *The Texture of Memory: Holocaust Memorials and Meaning.* London: Yale University Press.

Young, J. E. (ed.) (1994) *The Art of Memory: Holocaust Memorials in History.* New York: Prestel-Verlag.

3

FOURTEEN NOTES ON THE VERY CONCEPT OF THE COLD WAR

Anders Stephanson

> The well-known is such because it is well-known, not known.
>
> G.W.F. Hegel

> Ideas are displayed, without intention, in the act of naming, and they have to be renewed in philosophical contemplation.
>
> Walter Benjamin

Introduction

It was exasperation with an omnipresent cliché that occasioned these notes: 'Now that the cold war is over, etc., etc.' Every article on international affairs seemed to begin with it, often followed by reference to that other well-known fact, 'globalization.' The formula became reified punditry, something akin to advertising language. It is not hard, of course, to see why the epoch assumed such a self-evident aura. Doubtless the postwar 'period' was dominated geopolitically by the USA–USSR relationship; doubtless too, therefore, something did come to a resounding end with the Soviet collapse. But the effect of this seamless periodization is to conceal qualitative shifts in the nature of the relationship. We forget, for example, that Richard Nixon announced the end of the Cold War in Moscow in the early 1970s. Typically, moreover, the obvious end is retrospectively inscribed in the beginning and in the whole nature of the period so as to allow its history to be rewritten as an 'explanation' of the obvious. Meanwhile, other possible periodizations are barred or simply subsumed, periodizations, say, in terms of 'decolonization,' 'the economic rise of Japan and Germany,' or 'the universalization of the European model of the nation-state.'

Historians, discomfited by this flattening out of the historical 'real,' tend to modify the image, not by re-examining the nature of the concept itself, but by adding ancillary aspects designed to make the epoch 'fuller,' more 'realistic,' more 'accurate.' One way of achieving this reality effect is to reintroduce on stage the supporting actors of olden times – Europeans for example – as part of an extended main cast. Another is to widen the stage itself to include new actors and aspects (above all, 'marginals' and 'culture'). A third option, perhaps the preferred one, is to focus on the archival findings made possible by the 'end,' sources located in the East and so ensuring by default that reinvestigation of the real will henceforth be framed in terms of Soviet (and Chinese) pathologies. All these efforts are of some intrinsic value. If nothing else, they often generate important empirical findings. But they are also duplicitous, unwittingly, in attenuating the very historical specificity of the Cold War that they were intended originally to attain. They fill in the blanks. Yet the picture to be completed always seems to expand and indeed always will. There is no final or pristine cold war in the archives, or anywhere else for that matter, waiting to be discovered or uncovered.

On the contrary, my wager is that if the 'Cold War' is to have any explanatory value, then its periodization can be achieved only through the exactly opposite process – one of rigorous, relentless narrowing through conceptual inquiry. Periodization, which is what historians do, should always be explicitly theorized and the question here is ultimately whether the very concept of the Cold War can be produced, if indeed it 'has' a concept. Otherwise it is perhaps better left on the heap of everyday banalities.

The general starting point for this genealogical exercise is Lippmann's critique of Kennan's X-Article, launched in a series of articles promptly collected into a short book in late 1947. I used it in an earlier attempt to periodize the Cold War because Lippmann not only introduced the term but also provided, in my view, the historical key to its concept – though the Cold War is nowhere mentioned except in the title. The publicist spotted in Kennan's argument a certain gesture of diplomatic refusal *vis-à-vis* the USSR. This posture became constitutive (I argued) of the Cold War when the USA, to Kennan's own chagrin, institutionalized it under the formula 'no negotiation unless from a position of strength.' The USA and the USSR were at the geopolitical center until 1990, but the era divided into two distinct historical moments. The year 1963 marked the crucial shift. After the Sino–Soviet split and the Cuban Missile Crisis, the relationship became non-antagonistic in the sense that each side recognized the other and 'coexistence' became the new ordering principle. From that vantage point, the very brevity and shallowness of Ronald Reagan's Cold War posturing in the early 1980s showed that the historical underpinnings in fact no longer obtained.

Reading Lippmann, then, produced a diagnosis and a criterion for periodization. However, a proper conceptual determination was still lacking. Such a reflection on the conditions of possibility for talking about something called the Cold War must situate it more distinctly within the very opposition that ultimately framed it: war and peace. One should consider, at the very least, what kind of surrender (or peace) the Cold War presupposed and embodied; and that in turn requires a derivation of our notions of war and peace.

My procedure is essayistic as opposed to 'definitional.' Organized historically around a set of proper names, these are provisional notes, schematic and incomplete. After a sketchy (but not brief) history of the conceptual pair of war and peace, I make a few basic claims in the final notes. Chiefly, I argue that (1) the Cold War was a US project, and (2) its nature or logic was laid out (unintentionally) by Franklin D. Roosevelt during 1939–1941 and epitomized later in the notion of 'unconditional surrender.' I am assuming that 'the project' here is analytically distinct from the related questions of 'origins' and 'justifiability,' or, for that matter, 'causes.' Moreover, once the polarity (itself a problematic metaphor) between the USA and the USSR has been put into question and determined, it will be seen not as an essence but a kind of situation around which other processes, relations and antagonisms evolve and revolve, none of them homologous with it, much less 'the same.'

1. Bond and Leamas

Let us begin, however, with James Bond in 'Nigger Heaven.' That, at any rate, was Ian Fleming's title of the chapter in *Live and Let Die* (1954) wherein Bond goes to Harlem. Fleming's villain is a Smersh (i.e. Soviet) agent known as Mr Big. Mr Big hails from Haiti but now presides over a gangster empire in the ghetto, whence he also serves his murderous Moscow controllers. By means of his Haitian voodoo art, the charismatic Mr Big dupes naive blacks into becoming an army of footsoldier spies, highly efficient because, as servants, they are everywhere throughout (white) society but also 'invisible.' Bond is surprised when M tells him of this exotic master criminal, since blacks, to 007, seem 'pretty law-abiding chaps,' unless of course 'they've drunk too much.' M, however, soon sets him right. For Mr Big, it turns out, has 'a good dose of French blood' mixed in with his Haitian black, thus perhaps explaining his capabilities in the realm of sinister organization.

When, during the very height of detente in 1972–1973, Fleming's book was turned into a film, Mr Big appears in an altogether different frame. He is now the head of a Harlem-based drug operation, whose object it is to flood the US market with enormous amounts of heroin, produced on a Caribbean island that he happens to control in another capacity and identity. Bond is called in to crush this threat to Western civilization that now, of course, has nothing at all to do with any Soviet Smersh. By then, in fact, the Bond movies had been devoted for several years to combating supranational and non-Soviet threats. The spirit (in Hegel's sense) of the Cold War seemed dead as a doornail.

This moment was marked brilliantly by Le Carré's great novel *The Spy Who Came In From the Cold* (1963). With its remarkable series of inversions of basic Cold War binaries and gloomy greyness, the novel made a mockery of the technicolor allure of the contemporary James Bond fantasies. Alec Leamas, existentially weary and manipulated by his British spymasters with the utmost cynicism, was an altogether more historically representative figure of the spirit of 1963 than Fleming's cold warrior. Bond (or his producers) took political notice and turned to comic book adventure in the name of the world. Fleming died around that time, but earlier he had apparently conjured up for his literary admirer John

F. Kennedy a way to kill Fidel Castro, whose beard was to be sprinkled with radioactive material. The joke was taken seriously.

2. Lippmann (I), Baruch and Swope

The Cold War forms a whole semantic field of meaning, whose emergent boundaries may be traced initially to its entry into public usage. Who actually coined it is disputable. Walter Lippmann's book contra Mr X made 'the Cold War' an expression, but others could and did claim authorship. Bernard Baruch, for example, deployed the term in April 1947. Yet his speech was not about the Cold War but about the danger of inflation and the imperative, as this Wall Street financier would have it, for American workers to put in longer hours. He borrowed the term (as he freely acknowledged) from his speechwriting friend Herbert Bayard Swope, who in turn said he had first thought of it in the context of Hitler and 'the phoney war' – a phrase Swope did not like – in 1939–1940. Lippmann countered that he had picked it up from French sources in the 1930s, 'la guerre froide' and 'la guerre blanche' being synonymous expressions for a state of war without overt war. French lexigraphers dispute his account (and, for what it is worth, my own quick perusal of old French dictionaries and encyclopedias revealed no such terms).

All of which is well known and not very interesting except insofar as it points to the 1930s and the proliferation of intermediate 'states' between declared war and peace, more about which later. It remains that Lippmann put it into general usage as a historical and political term. Yet two other preceeding uses are of interest here, one by Lippmann's contemporary George Orwell, the other by Don Juan Manuel in early fourteenth-century Castile (though, on closer inspection, this turns out not quite to be so).

3. Orwell and Burnham

Right after war, in October 1945, George Orwell talked about a 'cold war' in the British Labour journal *Tribune*. His article ('You and the Atom Bomb') argued that the Bomb would become the preserve of a few Great Powers and thus relatively weaken the already weak, opening up a dystopian 'prospect of two or three monstrous super-states, each possessed of a weapon by which millions of people can be wiped out in a few seconds, dividing the world between them.' Probably, these states would then 'make a tacit agreement never to use the atomic bombs against one another.' Hence, Orwell surmised, 'we may be heading not for general breakdown but for an epoch as horribly stable as the slave empires of antiquity.' Three such states, predicted Orwell, would emerge: the USA, the USSR and China/East Asia, the last still being only potential. Each would be 'at once *unconquerable* and in a permanent state of "cold war" with its neighbours.' The Bomb, then, would perhaps 'put an end to large-scale wars at the cost of prolonging indefinitely a "peace that is no peace."'

A 'peace that is no peace,' 'tacit agreement,' geopolitical division of the world in an oppressive order of atomic Great Powers, 'as horribly stable' as the old slave empires: Orwell's scenario illustrates some salient characteristics of what was indeed to come (the question is exactly when). However, his use of 'cold war' passed unnoticed. The ensuing debate in the *Tribune* concerned weapons technology and how it might relate to bigger and smaller powers. Orwell went on, nevertheless, to redeploy the image of three globally hegemonic super-states in his relentlessly bleak classic *1984* (appearing in 1948, hence the inverted title), wherein 'Oceania,' 'Eastasia' and 'Eurasia' fight meaningless, peripheral wars in the name of meaningless propagandistic slogans, every piece of news being manipulated and liable momentarily to be changed into its direct opposite.

Orwell took the geopolitical contours of this dreadful image from James Burnham's famous work of 1941, *The Managerial Revolution*. Burnham, at that earlier moment, had seen another tripartite division (one, curiously enough, with renewed meaning today): Japan, Germany and the United States. None of them would be able ultimately to conquer any of the other two, even in combination. A standstill would follow. Burnham's main point, from our perspective, was that there would be more and diffuse conflicts, but 'since war and peace are no longer declared, it may be hard to know when this struggle is over and the next one begins.' Here again, then, we get an embryonic idea of the Cold War as a condition outside the 'normal' polarity of peace and war. In his next contemplation (1947) within the genre of 'whither the world,' Burnham argued that this war/no-war, what he otherwise referred to as World War III, had actually begun in Greece in April 1944. In the best of cases, however, it 'might end its life in its beginning, like a new bud late-frosted.' Burnham, too, was evidently evoking wintry images in 1947.

By that time, well on his way from Trotskyist renegade to arch cold-warrior (not in itself a very original route), Burnham was now advocating an American world empire. What he had in mind was 'a state, not necessarily world-wide in literal extent but world-dominating in political power, set up at least in part through coercion' and extending 'to wherever the imperial power is decisive, not for everything or nearly everything, but for the crucial issues upon which political survival depends.' He believed that this empire, founded on the atomic monopoly and featuring strong interventionism, could be combined with democracy at home, at the core. If the USA, an adolescent world power, failed in this imperial endeavour – by necessity an offensive one – the Soviet Union would succeed in its stead. An American empire, nevertheless, was already in the making, even if it was not being called that. Burnham's suggestion for an alternative and more congenial name – 'the policy of democratic world order' – has a certain contemporary resonance.

The door was thus open for Burnham's ferocious attack in 1953 on the containment policy for its apparent lack of properly offensive qualities. Kennan's position, Burnham argued, was 'pale and abstract.' In a situation where there was 'no clear line between war and peace' but 'only different forms and stages of the continuous struggle for survival and dominance in the developing world sys-tem of the future,' the United States needed active warfare, political warfare. Containment was merely holding the line, a recipe for defeat.

4. Don Juan Manuel and the Muslims

A more ancient lineage goes back to Don Juan Manuel in early fourteenth-century Spain, or more accurately Castile. Juan Manuel was the grandson of the powerful Castilian king Ferdinand III, a major figure in the reconquest of the Iberian peninsula from the Muslims; and Juan Manuel himself, aside from being one of the first prose writers in Spanish, was part of the same military, political, cultural, and ideological struggle. He (and Iberians in general) had ample reason to ponder the nature of warfare between Christians and Muslims. Some modern writers (notably Fred Halliday) have seen Juan Manuel's analyis of the inconclu-sive, irregular skirmishes and raids with fluctuating frontiers and the context of incommensurate religious world views as analogous to that of the Cold War. Indeed, they claim that Don Juan Manuel was the first to use the term. Halliday cites no authority but my guess is that his source is Luis Garcia-Arias, a noted geopolitical thinker in Franco's Spain during the 1950s and 1960s.

Don Juan Manuel's authorship, alas, happens to be something less than that. It would be more accurate to say that a nineteenth-century editor in Madrid coined 'the cold war' through mistaken transcription of Juan Manuel's work. The pas-sage in *Libro de Los Estados* that Garcia-Arias was referring to in the 1950s actually speaks (in the Spanish of the 1320s) of 'la guera tivia.' *Tivia* (in modern Spanish 'tibia') means tepid or lukewarm, something metaphorically very different, of course, from what should have been 'la guerra fria.' Garcia-Arias, however, was relying on the 1860 (Gayala) transcription and edition of Juan Manuel's book; and this version does indeed say 'la guerra fria.' In a footnote, Gayala says that the fourteenth-century original seems to be 'avia' (a microfiche transcript I found renders it 'la g<u>'rra (avia) [tivia]'); but, as this makes no sense to him, he goes on to exercise a certain editorial privilege by substituting the more sensible 'fria' instead. Thus, then, the first known use of 'the cold war.'

The passage indicates the difficulties of 'the cold war' as a metaphor. Its antonym is presumably hot war, real war, rising temperature. But rising tempera-ture could also mean 'thaw,' an improvement, a lessening of the risk of real war. To the extent, however, that 'cold' also connotes frigidity in the sense of someone unresponsive, it is indeed quite suggestive. From that angle, Juan Manuel's image of the lukewarm war is actually not without relevance. While real war ('muy fuerte et muy caliente') has real results – death or peace – 'la guera tivia' confers upon its respective parties neither peace nor honor. In short, it is not recognizable as a fully fledged, proper war between equal enemies. Inconclusive, it seems not have real peace as its object.

Don Juan Manuel had a good deal of military respect for the Islamic fighters he had to contend with but was in the end too much of a Christian feudal lord to be able to see them as the kind of enemies that were one's equals (he himself had more than a few of the latter in Castile and Aragon). Christian attitudes towards Islam – and Islam was in every way a fundamental problem – actually underwent several changes during the Middle Ages. At no point, however, did this concern result in any real knowledge, for Islam could not be situated within the dominant intra-Christian division between orthodoxy ('the right opinion') and heresy. Islam was a strange bird, monotheistic and Abrahamic, yet also profoundly different. It represented, in short, a thorny problem of classification.

The prevailing Christian view held that the Muslim Saracens of Spain stemmed from Ishmael, Abraham's son by his Egyptian wife Hagar. Thus they were outside the original covenant. Christians, by contrast, descended from Isaac, Abraham's son by Sarah. Isaac prefigured Christ and so by extension also the medieval Church, while Ishmael had been expelled into the desert; and Saracens, of course, were men of the desert. Much analytical effort was thus expended trying to explain away the apparent phonetic paradox of 'Saracens' and 'Sarah'. Such were the preoccupations of the Church intellectuals, policy theorists of the medieval world.

Two monotheistic and universalist religions cannot, if the respective communities understand themselves as vehicles for salvation, truly recognize one another as geopolitical equals. The third Abrahamic religion, Judaism, sees salvation in terms of exclusion of the outside, a war for the preservation of the inside. Christianity and Islam, on the other hand, are marked by expansionary notions. Space over time will become unitary, the outside therefore conceived as a space eventually to be conquered. The question is really only what sort of relation of 'non-recognition' one will maintain with it. Unlike Christianity, however, Islam was territorialized from the very beginning and involved in military conflict. Non-recognition between the world or abode of Islam ('submission') and the world of war would not necessarily take the form of open war. On the contrary, one could engage in temporary treaties – truces – because the world was by definition temporary anyway. Thus *dar al-suhl*, the abode of treaty where one might for reasons of stalemate conclude agreements.

Islam also differentiated between various kinds of 'non-recognizable' enemies: Jews and Christians could be accepted as second-class citizens, while no compromise was possible with atheists and polytheists. *Jihad* ('strife' or 'struggle' in the path of God, 'holy war' actually being a Greek term) could take different forms. Truces, as they tend to do when prolonged, became coexistence. Opinions differ on the matter, but 'real' peace treaties were arguably concluded with Christian powers in the sixteenth or seventeenth century. Not until the end of the Crimean War in 1856, however, did Christian Europe, on its part, fully recognize the Ottomans as part of the 'family of nations' proper. Which brings us to the roots of the Christian view of war and peace.

5. Augustine and Aquinas

Writing at the tail end of the Roman Empire and firmly convinced that the end of time was near, Augustine put forth a set of terms about peace and war that would travel authoritatively down through the centuries, as time in fact did not come to an end. By then, of course, Christianity had become the state religion, territorialized and ready to persecute pagan and other dissent by violent means. Most of Augustine's view on the subject was taken from classical Roman authors, mainly Cicero, with some crucial Christian (indeed 'Augustinian') elements added. World order was, in principle, about peace and justice, *pax* and *iustitia*, connoting a tranquil order of rest where everything would be in its proper, paradisically natural place. But life on Earth after the Fall was inherently tainted by sin, by definition merely temporal. Eternal peace, real peace, *pax aeterna*, could only occur after the Second Coming. Actually existing peace on Earth, meanwhile, was nothing but a *pax temporalis*, a sort of simulacrum of the real thing (Augustine's neo-Platonic leanings are at work here).

Within that shadowy context of imperfection, however, it remained that Christians desired just peace while the heathen wanted an iniquitous one, a perverse peace of domination and subservience, a peace that is 'not worthy even of the name of peace.' Nevertheless, good and bad alike seek *some sort of peace*. Even *pax falsa*, wicked peace, as opposed to *pax vera*, is thus peace of a kind. War, then, is derived and defined in terms of its goal, peace.

By the High Middle Ages, this Augustinan framework had been modified (along with his radical distinction between the earthly and godly domains), so that *pax temporalis* could quite well be imagined in the here and now as *pax vera*. Thus Thomas Aquinas, writing in the thirteenth century, distinguishes *pax vera* from *pax apparens*, the peace of power and injustice. Unreal peace had become, significantly, the province, not of humankind and earthly existence as such, but spatially of the heathen outside. Inside *res publica Christiana*, peace (following, notably, the arguments of that old teleological heathen Aristotle) was indeed the very condition that made it possible for human beings to be human. Aquinas, however, did allow for agreements ('concord') with heathens outside the normative community of Christians, technically signifying an existence 'alongside' without violence.

6. Hobbes and Grotius

Medieval peace, then, is understood as the 'natural' condition, marked by *iustitia* (meaning both 'right order' and justice), *caritas, tranquillitas, securitas*. War is disturbance, upsetting the right order of justice and hence 'unnatural.' Private feud and public war are not clearly distinguished. The medieval order disintegrates in due course and individual states begin to emerge, themselves eventually falling into confessional civil wars in the early modern period. A situation of extreme insecurity ensues. These intra-Christian conflicts actually included absolute negations along the lines of a cold war. Consider Cromwell's position on Spain in the seventeenth century: peace with France was possible but not with with 'papist' Spain, for 'the pope,' as he put it, 'maintains peace only as long as he wishes.' (This, incidentally, corresponds precisely to Kennan's Cold War argument about the Soviet Union.)

Hobbes, Cromwell's contemporary, is the theorist who breaks most decisively with the medieval conception by making war and insecurity the natural state, thus requiring all reasonable human beings to create an unlimited, absolute sovereignty, an artifical man, so as to prevent nature from having its way. Only thus could one make possible commodious living for everyone. Legitimacy, then, is for him solely a matter of *securitas pacis*. Justice has disappeared, or, rather, it is transformed into law and order. Authority makes peace, says Hobbes; truth does not. Truth, on the contrary, is associated with religious claims and so, in his view, with the very fanaticism that had initiated the devastating civil wars.

Yet in the seventeenth century the imposing word of Hobbes was not everywhere the word of polite society. Theorists of 'natural rights' offered less radical alternatives, the most noteworthy here being that of Grotius. Hobbes, interestingly, had expanded the notion of war beyond 'actual fighting' to mean 'the known disposition thereto during all the time there is no assurance to the contrary.' The room for real peace in the sense of full security in external relations seemed correspondingly slender. Grotius, by contrast, maintains the conventional distinction based on the presence or absence of open fighting. He also insists (following Roman models) that war must properly be declared; but he takes one step further and, crucially, turns it into a distinct condition, a state of affairs. War-making as such is conceived of as taking place in a 'theatre,' an external space of confrontation.

In the course of his argument, Grotius invokes the authority of Cicero to the effect that between war and peace there can be no intermediate. But the Ciceronian passage he refers to (which appears in the Eighth Philippic) is actually talking about civil war. Cicero is concerned to classify it as a real war as opposed to a mere 'tumultus.' Grotius transfers this to the whole range of emergent rules and regulations designed to control war in post-confessional Europe, the process whereby war was banished to the outside of the state and, conversely, the inside became an inviolable, absolute sovereignty. War, from then on, is seen as a legitimate property and defining aspect of that sovereignty. It becomes a political means, governed by certain explicit conventions that Grotius does more than anyone to codify as *ius gentium*, international law.

All in all, it is a dehistoricized, seemingly timeless order he construes. One of its founding pillars is indeed the razor-sharp distinction between war and peace. War is conducted for limited aims and does not, in principle, entail the liquidation of the enemy. On the contrary, the enemy is an equal – a just enemy – a conception which becomes the precondition for the edifice of Grotian juridico-political principles. Correct forms of hostility cannot be based on confessional or ideological difference, provided one is a member of the European family of nations. This, in turn, permits theorization of land appropriations from indigenous peoples in the newly 'discovered' lands across the ocean, appropriations that were a precondition for the emergence of the European state system of limited war in the first place.

The medieval *res publica Christiana* is so recast into 'Europe' precisely through the emergence of this new system of regulated war, condensed at the Peace of Utrecht 1713 in the principle of balance of power. Diversity and proper balance, the absence of a single dominant or universal monarchy, is henceforth a central part of the very definition of 'Europe.'

7. Rousseau, Paine, Kant, *et al.*

Europe, in Rousseau's words, is thus 'no mere fanciful collection of peoples with only a name in common, as in Asia and Africa': it is a 'real society' with common 'religion, manners, customs and even laws.' As was his wont, however, Rousseau sets up this idyllic extreme only to demolish it with a paradox or discrepancy. For this 'resplendent sanctuary of science and art' is congenitally given to all manner of bloody carnage. 'So much humanity, in principle,' he says, 'so much cruelty in deed.' Europe is in fact nothing more in the end than a 'pretended brotherhood' where nations are 'in a state of war' with one another and treaties 'represent passing truces rather than true peace.' Hence the need for a system, rationally imposed, of collective security.

Though he is not very original on the topic, Rousseau represents an important shift, the shift by the eighteenth-century Enlightenment towards a critique of war. Morality aside, for these thinkers the worst thing about war is that it is stupid. Balance and natural diversity are good, but war in the name of balance is bad and silly. Yet the condemnation of intra-European war as irrational entailed an interesting corollary. For the Enlightenment also reinvents the principle of just war as civil war against the ruling order. What is reasonable is just, and human beings can determine what is reasonable and thus just. If the ruler/state fails to conform to this reasonable truth or suppresses the right to express it, he/it is illegitimate, whereupon one has the right to inflict violence in response. Crucially, then, the inside (sovereignty itself) is no longer beyond dispute. It is subject to moral reasoning, questions of good and evil, good and bad, absolute notions of right and wrong, now in the name of reason, which is to say, a secularized version of God.

This could then be translated into a notion of international peace as intrinsically linked to the nature of domestic society and its political régime. Thomas Paine, for example, typically assumed that monarchy meant war and republics peace. Europe he considered 'too thickly planted with Kingdoms to be long at peace,' though the republics that did exist there were 'all (and we may say always) in peace.' Republican régimes, being natural, would reasonably negotiate any conflict; monarchies, by contrast, would go to war. Kant developed a concept along the same lines but less radical. Perpetual (real) peace would come only with generalized republican government, a system he conceived as the opposite of 'despotism.' Not to be equated with any democracy, a republican régime is an aggregation of free and equal citizens under a single law coupled with representative government. As a collectivity, these free citizens would naturally not consent to war, because of all the 'calamities' it would impose upon themselves. A 'subject' (despotic) form of state could, by contrast, engage in war arbitrarily at any time, 'as if it were a kind of pleasure party.'

Kant's view has similarities to Augustine's but is ultimately weaker; for everlasting peace does not imply 'tranquillity of order' modelled on heavenly repose, merely rational resolution of extant conflicts. This quintessentially bourgeois outlook found a contemporary counterpoint in Adam Smith's (and Paine's) British understanding of open commerce as a symbol of, and means to, peace. Maximum trade could, henceforth, be contrasted with the mercantilism of rigidly demarcated states, whose very existence was conducive to war. War, in short, was inherently irrational. Bentham, utilitarian *par excellence*, called it 'mischief upon the largest scale.' Kant, writing in a different idiom, put it more pompously: 'from the throne of its moral legislative power, reason absolutely condemns war as a means of determining the right and makes seeking the state of peace a matter of unmitigated duty' (for which practical purpose, Kant opined, one would then need a federation or league of nations). Thus, unlike Paine and the French Revolutionaries, Kant did not envisage any war to implement this new rationality, any war to end all wars. But this was always a possible alternative understanding of the project.

The shift to domestic derivation was an epochal one. In various ways, it would later influence both US and Soviet self-conceptions of what it is to lead the upward progression of objective history.

8. Hegel and Clausewitz

The American and French Revolutions implemented, in differing ways, this new philosophy of enlightened right. But the radical thrust of the French failed and the European system managed to regroup in the nineteenth century into a semblance of its old balance. Parallel, therefore, to British models of rationality and commercial peace (under British commercial hegemony of course), the geopolitical notion of war as a rational means of policy survived. In some quarters, it even transmogrified into a sort of 'bellicism' – war as a good thing for the fiber of society, war seemingly being inscribed anyway in the very struggle for survival that defined life as such. Since the body politic was now understood as a people rather than merely as dynastic possession, war could serve to fuse the multitude into a collectivity. Hegel, against his eighteenth-century predecessors, was so able to deny that war was any 'absolute evil,' though the traditional proviso still obtained that states would 'reciprocally recognize each other as states' even during war. Legitimate war, then, would issue in a settlement, peace. The object of war was victory, but victory was not liquidation. War, in short, was not the 'total' one of the French revolutionary period but the limited version of yester-year, professional and regulated according to 'civilized' rules, modelled on the old personal duel.

Hegel's thinking on identity and difference pertained to the European-centered 'family' of nation-states. Thus he argued conventionally (as did most Americans) that international law pertained to peoples recognized as equals but not to nomads or Amerindians. In a different key, he also maintained that 'religious views may entail an opposition at a higher level between one people and its neighbours and so preclude the general identity which is requisite for recognition.' The reference was to Jews and Muslims but has clear implications for our context.

Perhaps Hegel's original contribution was indeed his phenomenological problematic of recognition and acknowledgement ('anerkennen' meaning both). A much simplified version would run something like this. The self-conscious subject confronts the other as object, demanding recognition without according recognition in return. The other acts similarly and a struggle for death ensues. But such unilateral premises render the situation contradictory. To gain acknowledgement as the universal subject necessitates someone who acknowledges it. Killing the other obviously destroys that possibility; but even if he relents and submits, my victory is hollow and trivial precisely because his recognition will not, insofar as he is subjugated, be a full or proper one in which I can see my value as self affirmed and mirrored. To accord reciprocity, on the other hand, would mean relinquishing my claim to universality.

We will want to return to this contradiction later. Let us now turn instead to Clausewitz, Hegel's contemporary. These two extraordinary thinkers travelled in the same circles in the late 1820s, but their relation, if any, is undocumented and the question of 'intellectual influence' an open one. Both, at any rate, died prematurely in the cholera epidemic of November 1831.

What is of central interest here in the military theorist is his key concept of 'polarity.' Clausewitz railed, rightly, against the empty technicism and metaphoric language of authoritative writings on war in his day. But he himself actually deployed a whole range of metaphors taken from seventeenth- and eighteenth-century science, chiefly but not exclusively physics: friction, mass, force, gravity, evaporation, vacuum, refraction, equilibrium, electrical charge and so on. Polarity appears to have been borrowed from electromagnetics. By the early nineteenth century, however, it had also become an ontological commonplace in German thought: the idea, in other words, that life and things really consist of an interlocking unity of 'attraction and repulsion.' Kant had systematized this notion (on dubious grounds he thought it Newtonian). Through Herder and others it reached Hegel, for whom, famously, it became not only an interactive principle but a negational one: the identity of the opposition is based on negation of the Other and is thus negational in itself.

Later in the nineteenth century, the use of 'polarity' turned nebulous and metaphorical, signifying entities or forces moving in opposite directions. How the term eventually entered international relations theory, I do not know; but there it seems to mean mere opposition, systematic opposition in a spatial configuration of power centers. Hence the beloved 'bipolarity' and 'multipolarity,' both of which, electromagnetically speaking, would seem absurd: redundant in the first case, oxymoronic in the second. Polarity is presumably by definition a duality. This, in any event, is how Clausewitz uses it. Thus for him (and Hegel) it means a situation where the negative and the positive 'exactly cancel one another out,' more concretely, a confrontation of two sides engaged symmetrically in a battle for victory. You win or you lose. But the point (metaphorical or not) is also that there is continuous interaction ('*Wechselwirkung*'). In a way, then, it is a dialectical opposition. Identity ceases to exist when one pole disappears and the opposition does as well.

Hence Clausewitz sees the struggle for victory through decisive battle as the very nature or essence ('*Wesen*') of war. Bloodletting is to war what cash payment is to commercial transactions. 'Like two incompatible elements, armies must continually destroy one another. Like fire and water they never find themselves in a state of equilibrium, but must keep on interacting until one of them has completely disappeared. Imagine a pair of wrestlers deadlocked and inert for hours on end!' War, then, is about throwing the enemy down, eliminating his will to fight. But this ontological proposition is then, in typical Clausewitzian manner, modified in 'reality.' Because, among other things, defense is inherently stronger than offence, war can lapse into a desultory state and lose much of its basic polarity. Intensity is however also a function of politics, political purpose. The less maximalistic one's political aims, the less intense therefore the polarity.

9. Marx, Engels and Lenin

Clausewitz pondered war in a European frame, war as epitomized in a battle performed in a baroque theater. Yet his own experience of war (beginning at the age of thirteen) was the devastating confrontations with the French, and Clausewitz remained uncertain about the relation of 'total war' to political liquidation. Hegel, on the other hand, stuck to the traditional view that the enemy's internal order was beyond attack. International (i.e. European) law protected 'domestic institutions' in times of war.

Hegel's lineal descendants, Marx and Engels, thought otherwise. Nation-states to them were irrational and bound to be undermined by the globalization of capital. More originally, they also claimed that the whole apparatus of inside and outside, sovereignty in short, served to hide the real nature of the state, namely, class rule. In a way, then, one was always already in a sort of war, a class war, whether openly declared or merely smouldering. Class conflict was a state of affairs, resulting from a certain mode of production; and as long as it remained, there could be no *pax vera*, only *pax apparens*.

Traditional war between states, meanwhile, was ultimately irrational and bad, but one had to contend with it as intrinsic to an unjust order that was to be abolished. What they meant by 'contending' actually changed several times, from the view that capitalist war would provide openings for revolution to the opposite, that it would prevent it. Engels, who wrote professionally about war, ultimately came to think that war indeed would mean a world war and be a disaster.

Lenin carried this Marxist–Enlightenment critique to its fullest expression. The state was for him too an instrument of class rule and so, therefore, was the international state system of capitalist rapacity. Against this, he set the legitimacy of class war, his Marxist reformulation of the just civil war so abhorred by Hobbes and every conservative statesman ever since. Capitalism, then, was war, struggle to the death. No Augustinian peace could be envisaged until the world had become socialist and rational.

This was merely reworking Marx's concept of legitimate class war as an always existing structural condition of antagonism. Going beyond Marx, however, Lenin also militarized party politics, eventually rendering it positively Clausewitzian. Class war is class politics by other means. The Party thus becomes the equivalent of the State and politics a matter of battles, alliances, strategy and tactics, all organized around the pivotal notion of a single 'main enemy.' This construes an absolute enemy precisely in the sense of the Clausewitzian battle; but, crucially, it is also an enemy bereft of legitimacy.

Had that concept prevailed in the ensuing century in the form of global class warfare, the question of the cold war would have corresponded much better to the picture of historiographical traditionalism. But the revolution was territorialized in a single, if huge, land mass. By 1923, the international civil war had failed everywhere except in Lenin's native land (and perhaps he was right in thinking that it would then fail there as well). His vision was followed not by Trotsky's internationalism but Stalin's Fortress USSR. At no time was Trotsky's notorious formula at Brest Litovsk – 'neither war nor peace' – in the basic interest of Stalin's fortress. Lenin's view did survive, however, in different and reinvigorated form in the figure of Mao, theorist of protracted civil war and invasion of the enemy's social order; but that is another story.

10. Stalin

The anomaly of a revolutionary socialist state amidst capitalist ones was resolved by Stalin from the mid-1920s onwards. He did this first by territorializing Marx's and Lenin's always existing contradiction between capital and labour into one between the Soviet Union and the outside. The historic interests of progress henceforth were lodged on Soviet territory – or, more precisely, in the Kremlin and the class interests it represented. (A corresponding notion of historical chosenness and progress towards true humanity had of course long since reigned in the USA.) Yet the contradiction of capital and labour, conceived by Marx as a single structure of continuing interaction between two antagonists, could be defused, at least potentially, in the Stalinist reworking because of the physical separation from the outside, the severing of real interaction. And separation was indeed Stalin's instinctive strategic aim. Once situated at the geopolitical level, the 'fundamental contradiction' could thus take any number of forms, since interaction (or the dialectic) was no longer *a priori* present in the structure itself, much less its defining feature. The Other was externalized, symbolically present on the inside only as a constituent hostile outside or as foreign agents, usually in the guise of deviationists serving evil forces.

Stalin's second move (which he did not originate but sanctioned) had to do with alliance politics. Put simply, if monopoly capitalism, based on an everslimmer class basis, was inherently stagnating and so tending to resort to war at home and abroad, then it made no sense at all for the USSR, once the massive power of fascism had been understood, to engage in any drastically offensive maneuvers – or for that matter, after 1945, in any cold war. On the contrary, building coalitions of the widest possible kind against the narrowly based monopoly factions would be the marching order. Thus, beyond the tendency to separation and distance, there was nothing *as such* in the Soviet position after 1935 that made impossible stable, non-revolutionary relations with capitalist powers: the 'main enemy' was not intrinsically linked to bourgeois states or their nature. Contradiction had to do with a transnational class, or more accurately, a small fraction of that class. Whereas the political embodiment of progress was always to be found in the Kremlin, the embodiment of reaction could be found in a variety of places. Its precise location was contingent, subject to decision in typically voluntarist Stalinist manner. Hence the constant postwar reference to the mysterious influence of 'reactionary circles' on Western state policy when the latter turned hostile to Moscow's position.

It is wrongheaded, therefore, to pose the historical question in terms of putative contrast between a realist Stalin and a communist Stalin (or between realpolitik and ideology). There was every reason for Stalin to maintain stable, if distant, relations with the major capitalist powers in the name of a common anti-fascist legacy. This, on his view, was prudent Marxist–Leninist geopolitics, though it did not turn out that way.

Stalin lost the political contest over anti-fascist legitimacy after the war, partly because of the massive Western superiority, partly because, as a crude reductionist, he had a very limited understanding of how the West in general, and the USA in particular, actually operated. Thus in 1949 he found himself faced with precisely the kind of scenario he most devoutly must have wished to avoid, a huge and powerful USA-led coalition directed against his régime in a cold war. From his standpoint, it is in fact difficult to imagine a more disastrous turn of events short of all-out war.

11. Wilson and Roosevelt (I)

It is time, then, to turn west again. The American experiment in Enlightenment politics was allowed to expand, largely undisturbed, in the name of reason and light across the continental expanse. The USA, embodying right, could thus by definition not wage unjust war. Dispute arose in domestic politics about the perversion of this original and universal right, about its concrete meaning. Could it, for example, include slavery? That question had to be settled by a massive civil war. Yet the self-conception of universal right certainly survived.

Fast-forward now to the end of World War I and Woodrow Wilson. Wilson, of course, spoke famously in the name of humanity and rendered all enemies by nature therefore inhumane and/or criminal. In doing this he was, as he himself said, merely expressing American traditions, which were also those of humanity at large. War to end all wars (once it had been decided upon) was perfectly legitimate, just as the radical Enlightenment had said all along; and the opposition to such an obviously legitimate aim had to be be eliminated forthwith, or at least not allowed participation in the new normative community of the world.

The resort to war as an 'analog' is a common one in American history: war against depression, war against drugs, war against poverty, war against a whole range of ills, amounting to nothing less than a homespun sort of metaphorical bellicism. Wilson fused, in traditional American ways, this secular concept of reasonable conduct with a thoroughly Protestant notion of election and mission into a fully fledged ideology of US exceptionalism. His project of a new international order of law, discussion and economic openness met with not much more success than Lenin's alternative. Both projects were ultimately defeated by fascism, the future spectre of which the two originators, in a weird way, had each sensed.

The time, then, has come to make my more radical claim about FDR and also to tie these remarks together. For the matrix or logic of the American Cold War project after the war was established by Roosevelt during 1939–1941 in his attempt, in my view generally justified, to prepare the United States for (and perhaps steer it towards) the 'inevitable' open war.

As alluded to at the outset, the background here was, *pace* Grotius, the proliferating 'intermediate' states of war in the 1930s. Thus the Japanese war against China was called an 'incident'; the Italian Fascists invented the term 'notwar making' to describe their intervention in the Spanish Civil War; and Hitler expanded his territory by means of threats and bullying that never had to become open war. The whole set of conventional distinctions and institutions pertaining to war and peace (declaration, rules of conduct, rights of noncombatants, neutrality, in short, international law) that had emerged from Hobbes and Grotius onwards and been most extensively codified in the Hague Convention of 1907 seemed increasingly meaningless. Not accidentally, it was now, too, that the Roosevelt administration came officially to class the naval war of 1798 against France as a 'quasi-war'; FDR referred to this term in 1939 precisely with regard to the difficulties in retaining the old distinctions of peace and war. The intense legalism that had marked the 1920s – not a return to just war doctrines of the medieval type but the institutionalization of legal procedure as international norm – had the paradoxical effect in the following decade of actually increasing the space for war as non-war.

FDR noted this. Already in his Quarantine speech in October 1937, he was referring to 'times of so-called peace.' Once war had broken out in 1939, he developed the notion into an explicit dismissal of any possible settlement. Hitler's unabashed lawlessness (until Pearl Harbor, notably, Roosevelt spoke almost exclusively about Hitler) rendered any sort of agreement useless: 'Live at peace with Hitler? The only peace possible with Hitler is the peace that comes from complete surrender. How can one speak of a negotiated peace in this war when a peace treaty would be as binding upon the Nazis as the bond of gangsters and outlaws?' It would be hard to guess that the United States was technically 'at peace' with Germany at this point (July 1941).

As FDR saw matters, it was in fact inherently impossible to deal with such a dictator: 'normal practices of diplomacy . . . are of no possible use in dealing with international outlaws.' Out of this notion came the notion of 'unconditional surrender,' enunciated at Casablanca in 1943 but actually present from the beginning in Roosevelt's outlook. Symptomatically, he took the formula (he said) from U(nconditional).S(urrender). Grant and the Civil War, a kind of conflict that could indeed only be fought to that end. Henry Stimson expressed this logic in a more unequivocal and radical fashion when he projected onto the whole world Lincoln's famous dictum that no nation can survive half slave and half free: from now on it was the the world itself that had to be either free or slave. This was a prescription for limitless war, indeed the reinvention of war as civil war on a global scale in the name of total victory and the principle of universal right. The idea followed in the spirit of Wilson's earlier view that the only really secure world would have to be one in accordance with US principles (i.e. those of 'humanity'). Moreover, it was curiously analogous with Lenin's notion of international class war.

To be free as opposed to slave was then given substantial political meaning in the Atlantic Charter and the Four Freedoms. FDR referred to the latter freedoms in the following context: 'we know we ourselves shall never be wholly safe at home unless other governments recognize such freedoms.' His point was not only that the US would remain 'insecure' as long as other governments failed in this respect but also that such failure justified action to rectify it. Very quickly, indeed, FDR's vision took on a globalist tinge: 'An attack today begins as soon as any base has been occupied from which our security is threatened. That base may be thousands of miles from our own shores. The American Government must, of necessity, decide at which point any threat of attack against this hemisphere has begun; and to make their stand when that point has been reached' (July 1941).

The interventionism conjured up here contradicted another Rooseveltian notion, the idea that the Western Hemisphere now under threat on a global scale had achieved peace of a permanent kind, not of the balance of power variety but a genuine *pax vera*, stemming from generalized respect for one's neighbors. To live in the New World was to live in peace and without fear of invasion, where the vice of covetousness had been conquered. As long, therefore, as one allowed no internal dissension (FDR was referring here to Trojan Horses and alien conspiracies), the Western Hemisphere had reached the end of history, some virtually Augustinian 'tranquillity of order.'

Any peace with lawless aggressors, on the other hand, was a mere *pax falsa*, merely 'another armistice' as FDR said. Having formulated a maximalist notion of 'peace' and simultaneously divested all non-Western spaces of the traditional distinction between war and peace, Roosevelt had really declared that the United States was always already in a state of quasi-war and would so remain until, negatively, the last dictator had been eliminated and, positively, the Four Freedoms had been everywhere secured. However, this situation was not any American doing; it was the Nazis who, argued FDR, had deliberately declared themselves civilizationally against the American way of life.

This outlook can be condensed into three propositions: (1) everything that is not *pax vera*, a true peace, is by definition war, whatever the actual current relations; (2) there can be no true peace with power X because of certain qualities Y in the domestic make-up of that power; and (3) whoever is not my explicit friend (friendship being a question of identity with a set of universals) is my explicit enemy. We have here, in its essentials, what would become the matrix of the Cold War as a US project.

12. Roosevelt (II) and Truman

Roosevelt, being Roosevelt, adjusted his frame politically after 22 June 1941 and began *de facto* to differentiate between dictators, pragmatically deciding that the actual dynamics may well require that one such dictatorial régime had the potential, if treated as a would-be member of the world of peace, eventually to become a real one. So while the matrix remained in place, the juggling began. Roosevelt, in other words, had to engage in a game of simulation, hoping that the world of events and realities would come to approximate the world of shadows. Hence the playing for time, hence the exclusive focus on things military, hence the avoidance of fundamental political problems and contradictions.

Regrettably, this did not work. One of the reasons was precisely differences in the Soviet and US conception of the wartime coalition. The former was negative, logically so in accordance with the theory of anti-fascist class and state alliances: common interests from bourgeoisie to working class dictate that monopoly capitalism in its most reactionary and warmongering form be prevented. 'Winning' would not completely remove these avatars of stagnant monopoly capitalism, but the postwar coalition of anti-fascist forces could, if properly managed and maintained, keep them under wraps. The preconditions for a *pax vera* would then doubtless come at some indefinable future date, once the historic example of rationally planned production had been demonstrated near and afar. This is why, even after 1947, the official line of opposition to the USA took place defensively under the name of 'national independence,' not socialist revolution. The American stance, by contrast, was positive. It was not only a matter of preventing something from recurring but of achieving, in principle, *pax vera* in the here and now. Once, on closer inspection, it turned out that the Soviet Union did not fit positively the bill of a true friend, it could logically only be a true enemy, not an equal enemy of the duellist kind but an absolute enemy with whom there could be no real peace, only a peace, in Augustinian terms, 'not worthy even of the name of peace.'

By that time, the New Deal elements in the original Rooseveltian peace ('freedom from want,' etc.) had been compressed by Truman into an entirely abstract notion of freedom, defined positively as that which resides in the United States with spiritual environs and negatively as that which is the opposite of totalitarianism. Deciding who was a true friend from then on was comparatively easy. One should add here, because of our present-day obfuscations, that freedom was not yet directly equated with capitalism, at least not unblushingly. The collective memory of depression and war was still fresh enough to make one liable to think that behind 'free enterprise' lay enterprise but nothing automatically free.

The totalitarian Other, then, was constitutively present in symbolic form as a constant threat to universal freedom and concretely as evil foreign bodies. Given concrete historical sanction through the lessons of Munich, Roosevelt's original matrix was thus recast and redeployed in a project of unprecedented global scope, military, political, ideological, and economic.

NSC 68 epitomized this new and transformed negation. The enemy is said here to expound 'a new fanatic faith, antithetical to our own,' to seek 'absolute authority over the rest of the world' by 'violent or non-violent methods in accordance with the dictates of expediency,' to have initiated a 'cold war' by its very nature. All the basic Rooseveltian themes of the implacable enemy, infiltration and subversion, civilizational negation, worldwide struggle and infinite strategic needs are present in NSC 68. 'The cold war,' says NSC 68, 'is in fact a real war.' Only by 'frustrating' totalitarian designs will the Free World eventually 'force the Kremlin' to change its aim of domination and 'negotiate acceptable agreements.' This agenda of global frustration of evil design would then of course have to be a lot more active than is implied by the simple concept of containment.

The authors also express concern not to make the existing 'diplomatic freeze' into a prolonged period; but the potential openings they have in mind are in fact defined *a priori* as modifications and retractions of the Soviet position, deemed altogether too powerful and aggressive at present. Negotiation, in the end, must thus always be deferred until a proper relation of strength has been achieved and the Soviet Union can been 'forced' into the realm of the 'acceptable.' Success on that score would, given the essential nature of Soviet expansionism, eventually spell the end of the Soviet Union itself. To ensure this scenario, then, only one policy was rational for the United States: massive expansion of the war machine.

13. Lippmann (II), Kennan, Fanon

All of which brings us full circle back to that famous non-debate between Lippmann and Kennan in 1947–1948. For in the end I want to argue that Kennan's early postwar position did express (let us leave aside his 'intentions') the logic of the US-induced Cold War perfectly and that Lippmann saw this instantly and quite rightly hit him hard. Built into Kennan's notion of containment was a deliberate moment of diplomatic refusal, a period of recharging the Western batteries and rearranging the power configuration. To Kennan's dismay, as mentioned, that temporary recharge became a perpetual and indeed accelerating recharge, coupled with endless deferment of diplomacy. Kennan himself actually imagined the future along the lines of a metaphor he took from Molotov at the time, the image of a long-term fencing match, a game of thrusts and parries, back and forth, not a lethal exercise but within the range of measures short of war, eventually resulting in some new and perhaps more favorable situation. In that sense, there would be real dialogical interaction, though conflictual.

What ensued instead was something like the frozen dialectic of Franz Fanon's colonial world, the absolute spatial separation between settler and native in which there is no real *Wechselwirkung*, no interaction in Hegel's and Clausewitz's sense and therefore no mediation either, only potential annihilation. 'The zone where the natives live is not complementary to the zone inhabited by the settlers. The two zones are opposed, but not in the name of any higher unity. Obedient to the rules of pure Aristotelian logic, they both follow the principle of reciprocal exclusivity. No conciliation is possible, for of the two terms, one is superfluous.' Kennan's original formulation, however, was duplicitous in the actual movement to freeze things. His vacillation between the idea of Moscow as nefarious power professionals and Moscow as nefarious fanatics locked in the shadow world of Plato's cave meant that he could never offer a real rationale why Washington should risk dealing with them. And Washington never really did, much to Moscow's surprise. Kennan himself was surprised. At one point, he was even baffled to discover that it was the Soviet Union that was behaving like a 'traditional' great power, while the United States was unorthodox. So, in that perspective, one should take seriously the ensuing Soviet conception, however self-serving, of the Cold War as the Western policy of 'strength' and non-negotiation as opposed to Moscow's line of 'peace' and reduction in tension. One need not embrace the Soviet position to see that the Cold War as embodied in the American stance was utterly against Stalin's interests, that he would have liked precisely what he said he wanted: negotiations, deals and reduction in tension, coupled with relative isolation, above all, recognition as an equal. Instead the USSR became a pariah.

One can object that this was exactly what the régime deserved because it had impinged unduly on the security interests of others and/or because it remained wedded to revolutionary long-term goals. Here I side with Lippmann.

Lippmann saw in containment the danger that diplomatic dialog, normal relations, probing negotiation, and resolution of issues of mutual interest would pretty much cease. We do not have to like them, he said, just negotiate their and our own withdrawal from Central Europe and a certain normality and independence will return. In the end, it would take half a century, an arms race of unimaginable waste, and a collapse to achieve that 'normality.'

14. Politics, polarity, and space

For the United States, communism was the equivalent of war and the communist HQ lay in Moscow. There could be no real peace, consequently, with the Soviet Union, indeed no real peace in the world as such, unless the Soviet Union ceased being the Soviet Union and communism ended. For the Soviet Union, on the other hand, there could be peace with the United States but not until the influence of 'reactionary, warmongering monopoly capitalism' had been neutralized and the régime assumed a more 'normal' bourgeois character. The index of such normality and obverse decrease in reaction was of course the precise extent to which it responded to the Soviet-led overtures for inter-state peace in the world. Such a theoretical procedure would have been unthinkable to Marx, but it was one plausible codification of the unexpected necessity of a post-Leninist geopolitics. This, then, was the structural 'difference' or discrepancy that gave rise to the Cold War as a situation and provided its laws of motion.

The master signifier around which the struggle initially came to be articulated was World War II, or more precisely, what it had meant to negate fascism in that war. No one could question that act in itself; it was a universal Right. But to claim the same role now and, obversely, to cast the former ally and present enemy in the role of fascism, was not mere repetition. It was a new constitution of the Other and a new affirmation of the self as the negation of that which was thus being excluded.

What now remains is a brief elucidation of the ensuing 'epoch' itself. I need to raise, in particular, the question of Clausewitzian polarity and Hegelian recognition. If all politics is in some sense about polarization, dualistic configurations of friend and enemy, the problem is still if this particular polarity is a battle to the death, a clash of two wills to complete victory, or another kind of antagonism. A battle to the death the Cold War certainly was, but to a kind of abstract death. Elimination of the enemy's will to fight – victory – meant more than military victory on the battlefield. It meant, in principle, the very liquidation of an enemy whose right to exist, let alone equality, one did not recognize. Liquidation alone could bring real peace. Liquidation is thus the 'truth' of the Cold War. In that sense, civil war is the real analog. Yet all of this is 'in principle.' For the more important fact is of course that the Cold War was never a 'real' war. The authors of NSC 68 got it wrong. As Raymond Aron says somewhere, the leaders of the USA and the USSR always made every effort to avoid real war. Only for very brief moments (Berlin, Korea) did it even approximate to

Roosevelt's concept of 'quasi-war.' NATO and Warsaw Pact powers never once went to war with one another. Soon, indeed, the Cold War took the impossible Clausewitzian form of deadlocked wrestlers rather than armies continually destroying one another. Anything but lukewarm, it nevertheless brought 'neither peace nor honor' to its antagonists. Had the struggle escalated into open war or one side 'capitulated' early on, there would have been nothing much to ponder. So it is the deadlock that warrants exploration: a struggle to the death that is at the same time Orwell's 'tacit agreement.'

The terminological problem here has already been noted. 'The Cold War' is tricky because it is both metaphor and not metaphor. Its meaning hovers uncertainly between war and warlike. Absolute hostility, the antithesis of peace, is coupled with the absence of real war. 'Interaction' freezes, or is reduced to ideological and political monologues, the polarity marked by immobility and frigidity. In a way, then, it is the very reverse of a Clausewitzian understanding of war: the political purposes are total, maximalistic, intensely polar, but unlimited enmity is not reflected in real fighting. The defining, decisive battle never comes. Unlike the escalating intermediate forms of war leading up to World War II, this one freezes at the center. Spatial demarcation and immobility mark the polar 'axis.' The Cold War both produces a space and is produced by it. Perhaps, then, the original magnetic metaphor is better than Clausewitz's appropriation would have it: in the very middle a neutralized nullity between poles locked in the equilibrium of attracting opposites.

One might thus reformulate the matter as a paradox: the Cold War is warlike in every sense except the military. Its truth is 'war for unconditional surrender' but the reality is the kind of war one has when war itself is impossible. It is war as an ideological, political, and economic claim to universality, taking place not in the two-dimensional space of traditional battles but mediated through other realms when not, as universality, actually eliminating space altogether. The militarization of the respective inside and the attendant strategic games are an interaction of continuing mutual destruction endlessly deferred. Real war, meanwhile, is displaced beyond the militarized heartlands onto the 'periphery,' articulated in regional and local conflict, which often had little to do with the polarity as such. Thus the Cold War appears in spaces of the third kind as militarization and death, as crushing effects, but these are not exactly the same thing. If the term designates a certain antagonism between the USA and the USSR, the specificity of these other conflicts and processes can be preserved and grasped. The Cold War was not everything that happened between 1947 (or any other year) and 1990.

If one sees the relationship accordingly as a conflictual mixture where both sides are utterly opposed but also always realize the impossibility of open war, then the real driving force of the Cold War is the contradictory unity of non-war and non-recognition, where the latter dominant is not only warlike but the 'higher' kind of lack of equality that Hegel is referring to in situations of normative incommensurability. The end of the Cold War, in my analysis, will then come when both sides recognize each other explicitly as legitimate antagonists, when 'they recognize themselves as mutually recognizing one another,' when they recognize that conflict can never be resolved by means of war, when China goes its own way, when the Cuban Missile Crisis is over and the Test Ban Treaty is signed, when deterrence replaces liquidation as the master signifier and new dominant, when Leamas dies his defiant death in Berlin because both sides have essentially become the same, and when Bond goes on to fight villains of a new kind.

Bibliographical note

Aside from obvious classical texts (e.g. Saint Augustine's *City of God*, Hegel's *Phenomenology of Spirit, The Science of Logic, The Philosophy of Right*, etc.), I have necessarily borrowed from many secondary sources. First among these are Wilhelm Janssen's two extensive entries 'Krieg' and 'Friede' in *Geschichtliche Grundbegriffe* (eds, O. Brunner, W. Conze, R. Koselleck). The work of Reinhart Koselleck hovers over the entire exercise, and through him, a bit more distantly, so does that of Carl Schmitt (especially the latter's 'Die Geschichtliche Struktur des Heutigen Welt-Gegensatzes von Ost und West,' in *Freundschaftliche Begegnungen. Festschrift für Ernst Jünger zum 60. Geburtstag* (Frankfurt: Klerstermann, 1955) and 'Die Ordnung der Welt nach dem Zweiten Weltkrieg' in *Schmittiana-II* (Brussels: Piet Tommissen, 1990). Don Juan Manuel's *Libro de Los Estados* is available in a reliable Oxford edition in Spanish (1974). On Christians in the Iberian frame, see Elena Lourie, 'A Society organized for War: Medieval Spain', in *Past & Present*, 1966, 54–76. For Luis Garcia-Arias, see his 'El Conceptio de Guerra y la Denominada 'Guerra Fria' in *idem, La Guerra Moderna y la Organizacion Internacional* (Madrid, Instituto de Estudios Politicos, 1962) [I do not pretend, by the way, to any expertise in Spanish: my delineation was done with the help of dictionaries, French, an Italian restaurant owner in New York, a waiter from El Salvador and my colleague Edward Malefakis]. My own periodization (where the Cold War ends in 1963) was originally presented in a historiographical essay, 'The United States,' in David Reynolds (ed.) *The Origins of the Cold War in Europe: International Perspectives* (New Haven, Conn.: Yale University Press, 1994). On the historico-juridical question of war, see Frederick H. Russell's excellent *The Just War in the Middle Ages* (Cambridge: Cambridge University Press, 1975); Fritz Grob, *The Relativity of War and Peace: a Study in Law, History, and Politics* (New Haven, Conn.: Yale University Press, 1949); Josef L. Kunz, 'Bellum Justum and Bellum Legale' [1951] in *The Changing Law of Nations* (Columbus: Ohio State University Press, 1968); John Kelsay, James Turner Johnson (eds) *Just War and Jihad: Historical and Theoretical Perspectives on War and Peace in Western and Islamic Traditions* (New York: Greenwood, 1991); Majid Khadduri, *War and Peace in the Law of Islam* (Baltimore, Md: Johns Hopkins University Press, 1955); R.W. Southern, *Western Views of Islam in the Middle Ages* (Cambridge, Mass.: Harvard University

Press, 1962). Barry Gowan's 'Gravity, Polarity and Dialectical Method' in M. J. Perry (ed.) *Hegel and Newtonianism* (Dordrecht: Kluwer, 1993) was very help-ful. George Orwell's article 'You and the Atom Bomb' appeared in *Tribune* on 19 October 1945. See also Alfred Allan Lewis's chatty *Man of the World: Herbert Bayard Swope: A Charmed Life of Pulitzer Prizes, Poker and Politics* (Indianapolis: Bobbs-Merrill) and William Safire, *Safire's Political Dictionary* (New York: Random House, 1978) [though he gets only part of the story and some of it wrong]. On FDR, see Raymond G. O'Connor, *Diplomacy for Victory: FDR and Unconditional Surrender* (New York: Norton, 1971) and Anne Armstrong, *Unconditional Surrender: The Impact of the Casablanca Policy on World War II* (New Brunswick: Rutgers University Press, 1961); and of course Warren Kimball, *The Juggler: Franklin Roosevelt as Wartime Statesman* (Princeton, NJ, 1994). On Marx, see Amanda Peralta ... *med andra medel: Från Clausewitz till Guevara – krig, revolution och politik i en marxistisk idetradition* (Gothenburg: Daidalos, 1990). Raymond Aron's *Clausewitz* is a bit disappointing, considering Aron's philosophical credentials, but it says some useful things about polarity. Fanon's remarks are to be found in *The Wretched of the Earth* (New York: Grove Weidentfeld, 1968), 38–39; having read the passage many times without noting anything much, I was brought back to it through Ato Skyi-Otu's *Fanon's Dialectic of Experience* (Cambridge, Mass.: Harvard University Press, 1996).

4

THE OCCULTED GEOPOLITICS OF NATION AND CULTURE

Situating political culture within the construction of geopolitical ontologies

Carlo J. Bonura Jr

It is best to begin a journey of exploration with a map.
(Putnam 1993: 83)

For contemporary comparative political science, Gabriel Almond and Sidney Verba's *The Civic Culture: Political Attitudes and Democracy in Five Nations* has become a classic text in the study of political culture and democracy. Through its use of empirical methods, *The Civic Culture* specifies those cultural qualities necessary for the development of a stable democracy, including political equality and participation, trust, and cooperation. Ronald Inglehart, a major figure in the current efforts to re-establish political culture as a significant field in political science, suggests that the primary contribution made by Almond and Verba consists in their 'providing a well-developed theory of political culture based on cross-national empirical data . . . and moving away from the realm of literary impressions to that of testable propositions' (1988: 1204). In support of Inglehart's claim, *The Civic Culture* did, in fact, present one of the first large-scale attempts to employ quantifiable data in the examination of the relationship between certain cultural forms and democratic stability.

However, to maintain that Almond and Verba successfully moved such analysis away from 'literary impressions,' requires a closer look at the logic of their methods. In the midst of their explanation of the methods employed in *The Civic Culture*, Almond and Verba interrupt the flow of their argument to provide the following geographic elaboration of their project:

It is as if a [political] system were a large map on the wall of a darkened room, and all we know of it is what is revealed by one thousand separate pinpoints of light. These points of light (our interviews) illuminate the spots on the map that they touch. But they light up only a small

86

part of the map and leave the areas between the dots completely dark. We want to say something, not merely about the points that are illuminated, but about the entire map itself.

(1989: 42)

The end goal of their work, therefore, is to 'illuminate the territory between' each interviewee (*ibid.*: 42).

The inclusion of this 'metaphor' to clarify the concept of statistical 'representativeness' demonstrates an underlying visualization of their research dependent upon a textual representation of the nation-state: the map. This geographical representation of the political system through a map calls on its viewers to imagine the system's borders even though they cannot see them. To 'say something . . . about the entire map itself' is to incite an understanding of the political system as a bound geographical entity. The 'yet-to-be-illuminated territory' between interviewees is assumed to be the continuous cultural space found within the 'system,' sovereignly opposed to those 'other' discontinuous spaces found outside of it. The textuality of the map in this conceptualization of representativeness achieves geographically bound 'impressions' in its viewers in ways that are necessary and prior to the 'testable propositions' so enthusiastically celebrated by Inglehart. In Gearóid Ó Tuathail's word's, these geographic impressions and the role they play in the meaningfulness of Almond and Verba's comparisons reflects their map's location as 'already in a place other than where it claims to be' (1994: 542). Although throughout the field of comparative politics there lies a general acceptance of the meaningfulness and representativeness of international boundaries in differentiating national differences, it is in the context of political culture that the concept of culture becomes configured with reference to the territoriality of sovereign states. Cultural identities become bounded through their particular position located within the territory of the modern nation-state. After situating political culture in the context of new scholarship in critical geopolitics, I will explore how this relationship between geographic knowledge, cultural identity, and political boundaries provides the central mechanisms of identity necessary to discourses and practices of geopolitics.

Situating political culture within discussions of critical geopolitics

At first glance, comparative political culture[1] is an unlikely field in which to inquire critically into the workings of geopolitics. Suprisingly, however, the study of political culture presents a rich site through which to examine the construction of the ontological identity of the nation-state (Agnew and Corbridge 1995) and the reiteration of this identity within geopolitical discourses. From this perspective, the function of political culture in international relations and its formal study in comparative politics demonstrates how 'spatial practices and representations of space are dialectically interwoven' (*ibid.*: 47). Analyses of political culture provide seemingly objective and statistical representations of the cultural and political forms within a nation-state that directly shape the 'reading' of these states by theorists in an effort to explain a certain 'outcome,' as well as by geopoliticians who depend upon the reading of political culture for insights into a nation or leadership's 'character.' As such, this chapter will focus on disassembling the 'territorial trap,' in Agnew and Corbridge's words, upon which political culture rests and on explicating the geopolitics of political culture, demonstrating how the study of political culture both informs and is informed by geopolitics (*ibid.*: 79). These two topics lie at the core of any serious attempt to apply critical geopolitical methods to problems of international relations. Suggesting the relevance of political culture to critical geopolitics, therefore, simultaneously demands a critique of the unquestioned status of sovereignty within the study of political culture and outlines new ways in which to approach the creation and contestation of national and transnational identities within international relations.

In his recent work, *Critical Geopolitics*, Gearóid Ó Tuathail examines particular 'elite' political cultures of international politics that have informed certain traditions of geopolitics. He does so by presenting a series of mini-intellectual biographies designed to develop a historical context for geopolitics as well as document the influential careers of some of its most important 'geopoliticians.' The formal study of national political culture in relationship to geopolitics, however, has received little attention. Political culture accounts for a significant amount of research and resources in comparative political science often unrelated to the debates and theorization of geopolitics and international relations. The subsequent demarcation of research topics and questions reflects one of the most important contrasts between comparative politics and international relations surrounding each field's treatment of sovereignty and the state. In the study of international relations, theoretical and practical questions of sovereignty are of primary importance. To the contrary, the sovereignty of those nation-states compared in comparative political science is always assumed to be unproblematically stable, intact, and positioned beyond the boundaries of the field itself. Thomas Biersteker and Cynthia Weber, in *State Sovereignty as Social Construct*, have raised serious questions concerning the different ways in which sovereignty is produced and socially constructed. They provide a definition of sovereignty helpful for rethinking disciplinary boundaries between the study of international relations and comparative politics:

> The modern state system is not based on some timeless principle of sovereignty, but on the production of a normative conception that links authority, territory, population (society, nation) and [the] recognition [of sovereignty] in a unique way and in a particular place (the state).
>
> (1996: 3)

A full analysis of sovereignty and geopolitics would necessitate an additional inquiry into constructions of political culture. Discussions of political culture are for the most part devoid of critical, or at least theoretical, treatments of sovereignty that allow for a complex account of the socially constructed relationship between state, sovereignty, and society. The state, in comparative political science, appears to have fixed those geographical and discursive boundaries that guarantee the division of the domestic from the international. Biersteker's and Weber's definition assists in rethinking this absence of sovereignty and its effects in comparative analyses. Missing, with regard to comparative analyses of the modern nation-state, are inquiries into the importance of 'territory' in the scripting of narratives of national political cultures. This chapter will architect a starting point from which to begin questioning the epistemological function of territory in the 'production of a normative conception' of the modern state.

Ó Tuathail (1994: 534) specifies one of the primary tasks of critical geopolitics as calling into question 'the delimitation of the relationship between geography and politics to essential identities and domains.' This production of essentialized identities and spaces through the conjoining of geography and politics produces narratives of geopolitical and national identity that inform 'the strategies by which maps of global politics are produced [at] governmental sites' in the everyday activities of geopoliticians (*ibid.*: 535). Those narratives of nation and identity found within political culture can be understood as geopolitical in so far as they posit and inscribe the bound domestic spaces crucial to geopolitical discourses. Richard Ashley has described the importance of these narratives in that '"International politics" and the prospect of war are invoked [in modern statecraft] primarily in opposition to a construct of "domestic society," conceived as a [self-]identical social *whole* that is the very embodiment of a "reasonable humanity," a "civilization," a "nation," a coherent "modern community of sovereign men"' (Ashley 1989: 303). Hence, political culture becomes the very set of activities against which the international is defined. It is the condition of the sovereign domestic sphere that is observed and replicated through each comparison. Moreover, the ongoing practices of 'geopoliticians,' international media, activists, and academic scholarship all participate in the 'knowledgeable practices of statecraft that functions to produce the effects of modern domestic societies – social identities consisting of populations subordinate to a rational [political] center' (*ibid.*: 304). Taking seriously the conceptualization of society and culture found in comparative political culture allows for the interrogation of a site of the continuous production of sovereignty. To do otherwise would be to accept Almond's and Verba's occulted geographies as real representations of the places they claim to compare.

This assessment of comparative political culture will critically analyze generalizable summaries of national political cultures that 'arrest questioning and suggest essentialist explanations' of political organization and behavior (Ó Tuathail 1994: 539). It will be 'especially attentive to the historical emergence, bounding, conquest, and administration of social *spaces*,' recognizing as incomplete those conceptualizations of national culture that 'accord to moral claims, traditional institutions, or deep interpretations the status of a fixed and homogeneous essence . . . or an ultimate origin of international political life' (Ashley 1987: 411). The ambiguous status of territory within these explanations calls attention to the place of geopolitics in the imagining of political culture. Incorporating critiques of comparative political culture into 'critical geopolitics' necessarily denies the disciplinary mandate of political science to relegate the study of the international to the field of international relations and that of the domestic sphere to comparative politics. In challenging these disciplinary boundaries and demonstrating the dependence of political analysis upon a 'geocultural knowledge' it becomes possible to reconsider central assumptions of political culture and the very foundations of its comparisons.

Revisiting the renaissance of political culture

This critique of the inherent spatialities and geopolitics of comparative political culture takes place during, in Almond's words, 'the return of political culture' into mainstream political science (1993: xii). Inglehart himself has termed this return 'the renaissance of political culture.' Both Inglehart's own *Culture Shift* (1980) and Robert Putnam's *Making Democracy Work* (1993) comprise the core of this reconsideration of the merits of work done on political culture in the 1950s and 1960s. Prior to uncovering the geo(graphical)/political assumptions at the foundations of political culture it is important to revisit the characterization and operationalization of culture found in its study. Focusing particularly on the explanation of culture makes clear how the 'essentializing explanations' of political culture script narratives of the ontological identity of international relations' 'actors.'

Sidney Verba has defined political culture in terms of 'the system of empirical beliefs, expressive symbols, and values which defines the situation in which political action takes place' (Pye and Verba 1965: 513). Examinations of political culture locate these beliefs, symbols, and values as important variables in the processes involved in establishing national identity, political and economic development, modernization, democratization, authoritarianism, postindustrial society, and the distinction between modern and traditional societies; although in varying ways, political culture throughout these themes is viewed as an important element for the explanation of a society's economic and political development or 'backwardness.' As such, it is recognized as a core component of the modern nation-state. Reflecting this claim, research on political culture begins with the assumption that 'in any particular community there is a limited and distinct political culture which gives meaning, predictability, and form to the political process' (*ibid.*: 7). Political culture, according to this view, makes up the central feature of modern political systems, without which political activity and community would be impossible.

According to this approach to political culture, the study of patterns of beliefs and values is crucial to understanding the political qualities and outcomes found in any given society. Necessary for the success of a comparative analysis based on culture is identifying 'the empirical [and] . . . fundamental beliefs about the nature of political systems and about the nature of other political actors' (*ibid.*: 518). These beliefs attain their status as 'empirical' because they reflect 'primitive' political beliefs 'so implicit and generally taken for granted that each individual holds them and believes all other individuals hold them' (*ibid.*: 518). In other words, the beliefs at the foundations of political culture emerge from an 'essential' political identity. This is not to demand that political cultures be considered as homogeneous in the orientations that they encapsulate. Literature within comparative political culture in fact recognizes that a lack of 'integration' and 'consistency' may be present in a political culture (*ibid.*: 520). Rather, heterogeneity within political culture is recognized only if it remains analytically contained within the political and geographical boundaries of the nation-state.

Political culture in Brian Girvin's analysis resonates with this move simultaneously to privilege national approaches while recognizing the potential for cultural heterogeneity. In his view, political culture makes up 'a shared pattern of beliefs within which there may be many subcultures but a common source of values which inform those beliefs' (1989: 34). This common source capable of withstanding the possible heterogeneity of subcultures reflects an 'irreducible core' of 'national identity which has established the basis for differentiation and enmity between groups that do not share such an identity' (*ibid.*: 34). Such a characterization of national identity underscores the importance of critical geopolitics in uncovering the central ontological function of political culture in the social construction of sovereignty. The 'irreducible' ontological 'core' of national identity advanced in the study of political culture provides the basis for sovereign difference and political enmity at the center of geopolitics. One crucial element in the formulation of security discourses rests in the coupling of political identity to the sovereign nation-state. Security depends upon spatialized notions of sovereignty in that it allows for a 'rigid separation between those people within the territorial space pursuing "universal" values (politics) and those outside practising different, and nominally inferior, values' (Agnew and Corbridge 1995: 87). Scholarship on political culture works to achieve this rigid differentiation signified by a nation's sovereign boundaries as well as to formulate international priorities concerning the modernity and democraticness of individual states' political culture.

Girvin specifies the foundation for this understanding of political culture in the political and cultural events of nineteenth-century Western Europe at which point 'industrialism, nationalism, and mass-democracy blend within the modern state structure' (1989: 35). After its historical 'transmission' to the non-European world, a process narrated but not explored by Girvin, national political culture becomes the privileged macro-analytic category for the comparison of 'religious belief, political symbols and forms of interpersonal behavior' in the politics of any nation-state (*ibid.*: 36). Remarkably, these assumptions concerning political culture as the ultimate sign of national identity approximate Ernest Gellner's definition of the nation in *Nations and Nationalism*. In an effort to explore the nexus of political community and culture in the nation that arose in the nineteenth century, Gellner maintains that:

> nations can indeed be defined in terms both of will and of culture, and indeed in terms of the convergence of them both with political units. In these conditions, men will to be politically united with all those, and only those, who share their culture. Polities then will to extend their boundaries to the limits of their cultures, and to protect and impose their culture with the boundaries of their power.
>
> (1983: 55)

Gellner imagines the nation as a homogeneous cultural space that extends its political culture to the sovereign edges of this space. National identity, in this way, gains its analytic uniqueness from an understanding of politics as 'so deeply rooted in the native genius of each nation that the continuity of separate political traditions constantly resists the leveling forces at work in the social and economic spheres of modern life' (Pye and Verba 1965: 4). Regardless of the social and economic transformations, such as modernization, that a society might encounter, the national political culture derived from its 'native genius' still retains its strength to shape the political and social formations within that society. Narratives of this 'native genius,' therefore, posit the ontological identity of the nation through the concept of political culture.

Similarly, Almond and Verba identify political culture as 'the particular distribution of patterns of orientation toward political objects among the members of the nation' (1989: 13). This portrayal stresses the capacity of quantitative analysis to describe and explain the cultural realities of politics within the nation-state. The 'distribution' referred to here is spatial, but not in the geographic sense earlier alluded to in the still unilluminated map. This reference to *statistical* distributions is reaffirmed over thirty years later by Inglehart (1988: 1207) as he submits that 'through statistical procedures it is possible to distinguish between the underlying cultural component and short-term disturbances' of national political culture. The spatial distributions of quantitative analyses, therefore, are produced with the spatial limits of the nation-state already assumed in their outcomes. These distributions, responsible for delineating new national 'cultural components' from their 'disturbances,' work to consolidate cultural and sub-national differences under the category of the nation-state. The task of critical geopolitics, therefore, in the face of this renaissance of political culture, lies in a disassembling of 'this relay between a citizenry and *polis* grounded in the ontological principle of identity' (Spanos 1996: 152).[2] The examination of this 'indissoluble relay,' as Spanos claims, is central to a rethinking of the 'unthought ontological conditions of global power relations precipitated by the end of the cold war' (*ibid.*: 172).

Geographic knowledge and the production of bound cultural spaces

Paramount to this project of recognizing the ontological conditions of geopolitics is an exploration of geographical and sovereign boundaries in the production of national identity. In this relationship, however, geography does not simply complement the nation-state by unproblematically *describing* its boundaries and essentially relating to its national culture. Rather, it will be argued in these next sections that the invocation of geography in discussions of political culture actually *inscribes* bound cultural spaces necessary for the function of the nation-state as a unit of analysis and a meaningful political form. In this way, geography ensures the knowability of culture by aligning ontological cultural limits with the geographical limits of the nation-state.

This reassessment acknowledges the incorporation of geography into comparative political culture not as a part of a larger systematic methodological framework but as the deployment of a certain kind of knowledge required for the production of continuous political and cultural fields. To foreground geography's epistemological function in the study of political culture is to acknowledge that 'space itself becomes a kind of neutral grid on which cultural difference, historical memory, and societal organization are inscribed. It is in this way that space functions as a central organizing principle in the social sciences at the same time that it disappears from analytical purview' (Gupta and Ferguson 1991: 7). This conjunction of geography, nation-state, and ontology, therefore, should not be taken as an objective or organic one in which political forms naturally adhere to their territorial referents. The rejection of an organic relationship between space and modern political forms resituates geography as a theoretically powerful site for the production of knowledge. Highlighting this epistemological function, Klause-John Dodds and James Sidaway, relying on the work of Michel Foucault, position this argument in terms of geopolitics:

> The concepts of power, knowledge, and geopolitics are thereby bound together in a provocative way. What is suggested is that forms of power/knowledge operate geopolitically: a certain spatialisation of knowledge, a demarcation of a field of knowledge, and the establishment of subjects, objects, rituals, and boundaries by which a field (and the world) is to be known.
>
> (1994: 516)

The utility of geography in the study of political culture, therefore, allows for the formation and dissemination of the effects of power through the production and reification of a spatially bound nation-state and its *statistically* aggregated subject.

National political cultures and identities, continuous national spaces (discontinuous with regards to other national spaces), and national publics appear in the study of political culture by denying a critical analysis of their very constitution. These bound cultural entities appear in support of the theorization of a nation-state that naturally and successfully 'represents' the culture within its domestic sphere. 'The distinctiveness of societies, nations, and cultures,' in these accounts, 'is based upon a seemingly unproblematic division of space, on the fact that they occupy "naturally" discontinuous space' (Gupta and Ferguson 1991: 6). In explicating this foundational conceptualization of space, Ashley refers to a Cartesian 'spatialization' so 'crucial to the modern demarcation of the field of "international politics."' Moreover, he recognizes the centrality of 'bound' cultural and political identities with regards to the 'question of space':

> the Cartesian practice . . . imposes the expectation that there shall be an absolute boundary between 'inside' and 'outside,' where the former term is privileged. The *inside* is taken to be the space of identity and continuity – the privileged space of the Self. . . . [It] is a sharply bounded identity – an identity that is hierarchically ordered, that has a unique center of decision presiding over a coherent Self and that is demarcated from and in opposition to an external space of difference and change.
>
> (1989: 290)

This spatial logic of what becomes national identity, dependent upon both geographical and discursive boundaries and inscriptions, allows for the 'very possibility of rational political subjectivity' (*ibid.*: 290). Within the narratives of political culture this subjectivity is apparent in the shape of 'national identity.' This becomes especially clear in Inglehart's recent work. In 'The Renaissance of Political Culture' he maintains that 'the study of political culture is based on the implicit assumption that autonomous and reasonable enduring cross-cultural differences exist and that they can have important political consequences' (1988: 1205). The theoretical justification for such 'cross-cultural autonomy' is never made clear, except that the only manifestations of 'autonomous cultural components' are under the signs of nation-states compared in his analysis. As a result, the 'important political consequences' found in Inglehart's study do not involve enduring cross-cultural empirical differences but rather the very constitution of national identities through his own methods of analysis.

Akhil Gupta and James Ferguson (1991: 12) refer to this kind of lack of theoretical accuracy in Inglehart's treatment of culture as 'national naturalism,' or 'an association of citizens of states and their territories . . . [that] presents associations of people and place as solid, commonsensical, and agreed-upon, when they are in fact contested, uncertain, and in flux.' Linking the capacity of comparative political culture to comparisons based on the presence of bound cultural entities, Verba (Pye and Verba 1965: 519) remains committed to the proposition that 'in the perspective of cross-national comparisons . . . one can find rather sharp differences among different political cultures in terms of general beliefs, a fact that makes them a useful explanatory tool.' The containment and representation of bound cultural differences, in this view, becomes the foundation for the utility of political culture in the broader 'explanatory' project of comparative political science. Placing its explanatory capacity in the untheorized geographical 'autonomy' of the nation-state, however, results in 'reified and naturalized national representations' (Gupta and Ferguson 1991: 12). The necessity of these representations to the 'essentializing explanations' of comparative politics demonstrates the constitutive capacity of geographical knowledge in the maintenance of Spanos's relay of citizen and polis. One point at which the production of these representations becomes apparent exists in the relationship between the individual, geography, and national identity posited in comparative political culture.

Constituting national subjects through the geographic knowledge of political culture

Situating the study of national identity in political culture, Verba has claimed that 'the creation of a national identity among the members of a nation is the cultural equivalent of the drawing of the boundaries of the nation' (Pye and Verba 1965: 530). Culture and geography play a similar function in the production of the nation-state. National identity in Verba's analysis is the most important 'political belief.' To emphasize this point he extends his analogy further by suggesting that 'just as nations may have unsettled or ambiguous boundaries, so may the sense of identity of the members of that system be unsettled and ambiguous' (*ibid.*: 530). The bound whole space of the nation-state's geography, therefore, becomes the physical template for the individual's identification with the nation. Moreover, this conjoining of geography and 'identity' demarcates the cultural spaces and activities of a national subject. Individuals *become* the nation in a manner that makes Inglehart's 'naturalisms' possible.

Additionally, individuals as a unit in comparative political science play a fundamental role in the identification of sovereign cultural difference across political cultures. They provide crucial data most often collected through surveys as to the 'empirical' realities of a political culture. As such, 'political culture assumes that each individual must, in his own historical context, learn and incorporate into his own personality the knowledge and feelings about the politics of his people and his community' (Pye and Verba 1965: 7) In this cognitive exploration of culture the political subject is one who is representative of a people or a community. The possibility never arises that some who reside in a particular community might not identify with that community's nation. To know which nation an individual identifies with, as Verba's analysis makes clear, is to locate that individual within that nation's cultural and geographical boundaries. The category of 'great dimensions,' the nation-state, becomes coterminous with the busy activities of that political culture which the nation-state claims to represent.

In these claims, the constitution of the individual acquires a fixity in the same manner as the boundaries of national political culture. Through geographical representational practices, individuals appear in bound cultural settings already certain of their identity as members of a particular political culture. Cautioning against such ontological assumptions, Foucault urges:

> The individual not to be conceived as a sort of elementary nucleus, a primitive atom.... In fact, it is already one of the prime effects of power that certain bodies, certain gestures, certain discourses, certain desires, come to be identified and constituted as individuals. The individual is not the *vis-à-vis* of power; it is one of its prime effects. The individual is an effect of power, and at the same time, or precisely to the extent to which it is that effect, it is the element of its articulation.
>
> (1980: 98)

The 'explanatory' project of comparative political culture, therefore, relies upon this production of the individual whose opinions make possible its quantitative aggregates. The invocation of the geography of the nation-state in Verba's claims produces the effect of the national citizen-subject.

In a critical move away from understandings of cultural and political spaces that work to reaffirm hegemonic constitutions of individuals as unproblematically integrated into national political culture, Gupta and Ferguson (1991: 20) argue that 'physical location and physical territory, for so long the only grid on which cultural difference could be mapped, need to be replaced by multiple grids that enable us to see that [cultural] connection and contiguity vary considerably by factors such as class, gender, race, and sexuality.' The mechanism for describing such 'connection and contiguity' in comparative political culture is the 'representative' sample of the statistical survey. Respondent's answers to questions on their 'satisfaction' with institutional performance or their lifestyle[3] are assumed to reflect these differences through the rules of quantitative inquiry. Gupta and Ferguson (*ibid.*) continue that this reconceptualization of cultural connection and contiguity allows for an analysis of 'those in different locations in the field of power.' By recognizing every individual response as equal to the next, quantitative approaches make this kind of analysis impossible. Seeing the sign 'United States' in Inglehart's charts, the viewer, in a similar manner to the viewer of Almond's and Verba's map, can assume the presence of national subjects devoid of the particularities of class, gender, race, and sexuality. These approaches impose a field power in which, contrary to the outcomes imagined by Gupta and Ferguson, differences in location and privilege evaporate under the aggregated sign of the nation-state.

The recent work of Michael Shapiro also raises significant questions of method and the subjects of analysis in the study of geopolitics, and particularly war. Shapiro outlines two theoretical approaches to the study of war and its history: that of the strategic and of the ethnographic. This dualism, developed early in his text, reflects the primary methodological concerns at the core of *Violent*

Cartographies. Strategic approaches toward war 'seek to deepen identity attachments by politicizing boundaries and locating dangers outside of them, while [ethnographic approaches] seek to attenuate identity commitments by reflecting on the boundary practices and history-making narratives through which they are shaped' (Shapiro 1997: 138). Yet, there is an intermediate category, perhaps one of many, that intercedes at a crucial point to reaffirm the sovereign claims to boundaries in strategic studies and deny the political and methodological legitimacy of ethnography. This category marks the place of the academic discourse of political culture. Through the inscription of a bound domestic sphere, political culture works to elide the difference and uncertainty of political and national identities at the moment of their methodological interrogation through ethnography. It functions as a seemingly neutral category in the effort merely to describe domestic elements of a state's politics. Political culture appears unimbricated in the ontological claims, to borrow Shapiro's phrasing, of geopolitics and statecraft. More importantly, it does so through an epistemological assumption of the capacity of geographical boundaries to represent the difference located in the juxtapostioning of the international and the domestic. As such, it generates the necessary 'ideological inscriptions that assign a certain identity and coherence to [geopolitics] as part of an argument about its nature and relationship to state and society' (Ó Tuathail 1996: 142). Its spatial implications not only amount to the production of a concealed geography that forcibly aligns the administrative boundaries of the state to a 'national' culture. It also fosters a kind of geopolitics in its recognition of the sovereign state as the proper container and representative category for the analysis of cultural difference. Yet, in order to fully destabilize the geopolitical configuration of dangerously discontinuous national identities and war, the category of political culture cannot be ignored.

Geological/cultural approaches and the geography of 'sovereign jurisdiction'

As Verba's identity-based nexus of culture and geography demonstrates, geography's incorporation into explanations of political culture articulates a seemingly organic relationship between the national and those spaces and cultures that it claims to represent. Under this organic code, representation is taken for granted as unproblematically guaranteed by geographical boundaries. Larry Diamond's suggestion that political culture be understood 'geologically' advances a similar kind of organic thesis. Diamond offers a reorientation toward political culture, asserting that it 'is better conceived not purely as the legacy of the communal past but as a geological structure with sedimentary deposits from many historical ages and events' (1993: 412). The nation-state as the unit of analysis rests upon these 'deposits,' which in turn can be identified and mapped. Additionally, geopolitics culminates in the 'layering' and 'erosion' of national identities. According to Diamond, the advantages of this geological analysis lie in its emphasis on recognizing the heterogeneous foundations of political culture.

This cultural heterogeneity, however, is still geographically contained. Geographies of states, regions, and entire continents at times can be shown to have been 'influenced by similar processes of geological layering' (*ibid.*: 412). In doing so, Diamond invokes the continental impact of colonialism on Africa as an example of the cultural geology of geopolitics. 'Geological' evidence and the heterogeneity that it symbolizes, however, never call into question the very capacity of a state's geography to represent the 'layers' beneath its territory. R. B. J. Walker's work, in contrast, considers cultural approaches like Diamond's that attempt to encapsulate both the temporal and spatial dimensions of political culture as still dependent upon the geography of the nation-state. 'Spatially delimited identities,' according to Walker, require 'accounts of the modern state as institution, container of all cultural meaning, and site of sovereign jurisdiction over territory, property and abstract space, and consequently over history, possibility and abstract time' (1993: 162). Diamond's geological analysis articulates this 'sovereign jurisdiction' over space and time perfectly. The sedimentation of political culture in locatable 'deposits' reflects its historical and temporal qualities, while its manifestation in deposits crystallizes its spatiality. As such, the 'sovereign jurisdiction' of the state becomes reaffirmed. The end result of Diamond's analysis is to posit the organic geological foundations for sovereign cultural differences as nations weather the history of geopolitics.

The visible effects of this assumption of the state's sovereign jurisdiction are largely apparent throughout comparative political culture. Statistically based analyses, as in Inglehart's *Culture Shift*, depend upon the presence of long-range patterns in values and behavior so as to suggest the possible effects of political culture on outcomes in political life. In other words, these studies attempt to identify the history of a political culture that is distributed representatively across a space delimited by the choice of the unit of analysis, predominantly the nation-state. Perhaps this 'silent' theoretical maneuver, in Walker's words, can best be described in Inglehart's peculiar term 'syndrome.' A 'cultural syndrome' refers to a long-range distribution of beliefs and values that reflect a change in the political culture of a nation (1988: 1215). To return to Verba's 'native genius' of the nation-state, the term syndrome is an organic metaphor that works to invoke the geography of the nation-state. In quantitative analyses, the distribution of 'empirical beliefs and values' over time must always reflect their distribution over a continuous space. The geographic connotations that arise out of the organic metaphor of a 'cultural syndrome' highlight geography's function as that which allows the state 'as a category of analysis . . . [to be] treated as the silent condition guaranteeing all other categories' (Walker 1993: 176).

Under Walker's critique Inglehart's untheorized assumptions about culturally autonomous spaces make evident the importance of the 'principle' of state sovereignty to the conceptualization of space and political identity. 'Though relatively silent in contemporary social and political analysis,' he maintains 'the principle of state sovereignty has become indispensable to our understanding of what a state, nation or political identity can be' (*ibid.*: 164). Contrary to this suggestion of the silent dominance of the state in current thinking on 'sovereign' identities, Theda Skocpol, through her contributions to statist literature, has worked to 'bring the state back in' to social and political comparative analysis. By way of transition in her article 'Bringing the State Back In,' Skocpol refers to her treatment of the state as providing a 'conceptual frame of reference . . . informing future research on states and social structures across diverse topical problems and geocultural areas of the world' (1985: 8). Although Skocpol does not engage and has often critiqued comparative political culture, her use of the phrase 'geocultural area' in fact brings together the nexus of geography and culture with reference to a bound space: the singular 'area' of the modern state. In this sense, the study of political culture is primarily a 'geocultural' inquiry.

Most clearly found within area studies (Emmerson 1995) or in Samuel Huntington's *Clash of Civilizations*, the geocultural is a prominent topic in examinations of political culture and recent geopolitical transformations. This being the case, it is important to distinguish the geocultural knowledge produced in comparative political culture from that of geopolitics. Political culture, like the conjoining of international relations and geography as explicated in Gearóid Ó Tuathail's *Critical Geopolitics*, on the one hand aids in the production of geopolitical knowledge fundamental to imperial projects and imaginaries. On the other hand, it generates a geocultural knowledge just as epistemologically violent and necessarily complicit in the establishment of post-World War II and post-colonial international proscriptions for economic and political development. In this way, the introduction of the state into comparative analysis, whether as a unit of analysis or as part of a larger methodological project (as in Skocpol's case), still produces a silence, in Walker's terms, of its a/effects. Acknowledging Diamond's geological narratives as an essentially geocultural knowledge foregrounds the function of the state and its sovereign jurisdiction in supplying the very terms for imagining space and time pivotal to the practice of geopolitics. In the concluding two sections, I will use the critical insights developed up to this point to scrutinize closer an example of geocultural knowledge in the work of Ronald Inglehart and political culture's role in the practice of geopolitics more generally.

Geocultural analysis and Inglehart's 'cultural geography of the world'

Both in *Culture Shift* and 'Changing Values, Economic Development and Political Change,' Ronald Inglehart situates his studies within a geocultural frame that utilizes the epistemological function of the state (his unit of analysis) as the 'condition guaranteeing all other categories,' as theorized previously by Walker. Inglehart works to uncover the primary cultural values found within the 'familiar syndrome of industrialization, occupational specialization, bureaucratization, centralization, rising education levels and a configuration of beliefs and values closely linked with high rates of economic growth' (1995: 379). In doing

99

so, the emerging 'cultural clusters' from this statistical analysis of the World Values Surveys, covering 'the values and beliefs of the publics of 43 societies representing 70 per cent of the world's population,' present a 'cultural geography of the world' (*ibid.*: 393).

Although, Inglehart's task makes claims toward studying global political culture, cross-cultural differences quickly becomes interchangeable with cross-national ones. The cultural clusters on the charts provided in his research attempt to 'map' their distribution as reflections of 'cultural space' (*ibid.*: 393). Contrary to this contention, however, these clusters are made up of the statistical aggregations of surveys done within nation-states. The spatial distributions in both studies refer to the 'preferences' of national political cultures. Gupta and Ferguson (1991: 6) comment on this slippage between nation-states, cultures, and geography as 'it is so taken for granted that each country embodies its own distinctive culture and society that the terms "society" and "culture" are routinely simply appended to the names of nation-states.'

This theoretical and practical running together of nation and culture is the very premise of geocultural knowledge that allows it to be so effective in generating continuous cultural spaces and erasing disruptive cultural differences. 'If culture is read through the principle of state sovereignty,' according to Walker, 'it can only refer to the diversity of national cultures. If culture is read through a geometry of territorial exclusions, through a metaphysics of identity here and a non-identity there, it can only refer to an absence of community, a relativity of values and a clash of different ways of life' (1993: 181). But the representativeness of statistical methods refutes this relativity by reproducing the territoriality of the nation-state in the limits and error margins of its sample. The claim to a statistically representative sample mirrors the precise claim made by the sovereign state over the cultural and political meaning of the domestic sphere. It fixes the spatial identity of the state through the sample's parameters so as to recognize the history of the patterns of such a constructed space.

The actual geographies produced in Inglehart's works reflect this eliding of complex cultural difference described by Walker. In 'Changing Values, Economic Development and Political Change,' Inglehart engages in a complex discussion of the cultural outcomes of spatial distributions, claiming that 'geographically contiguous countries tend to have similar cultures' (1995: 393). His opening example lies in a detailed discussion of the 'compact cluster' of 'the Nordic countries.' Other more difficult to sustain examples found in *Culture Shift* concerning the determinacy of these clusters refer to the juxtaposition of 'the Islamic world' to 'the Confucian-influenced zone of East Asia' and, 'Protestant Europe' to the 'Confucian cultural area' (1980: 61). The capacities of the boundaries of these clusters to 'represent' the presence of at times radical cultural difference within these clusters are never seriously discussed. The clusters appear as continuous spaces comprising 'geocultural areas.' As such, they depend upon the epistemological function of the nation-state as a primary unit of analysis for their geographical limits.

The geopolitics of political culture

By emphasizing these questions of territory and culture in comparative politics it becomes possible to discern the dependence of political culture on certain 'normative conceptualizations,' in Simon Dalby's words, of state and sovereignty that make possible the very act of comparison. Moreover, these normative conceptualizations facilitate political culture's status as a principal resource in geopolitical discourse. Of central importance to the study of political culture is the 'ideological representation of identity and difference' through the 'division of space into "our" place and "their" place ... distinguishing "us" from "them," the same from "the other"' (Dalby 1991: 274). The conclusion of Almond's and Verba's *Civic Culture* that the political cultures of Britain and the United States, in contrast to those of Germany, Italy, and Mexico, are more 'supportive of stable democratic processes' underscores this contention (1989: vii, 37). The geopolitics of this particular case is two-fold: (1) it reaffirms the truth and legitimacy of US and British democracy at the height of the Cold War and, (2) through its comparisons of 'sovereign identities' it positions the nation-state as *the* privileged actor of the modern state system. Through a focus solely on the domestic sphere political culture provides narratives of the 'sovereign' political and social identities of domestic spaces necessary to the very substance of geopolitics and international politics.

Geopolitical discourse during the Cold War provides one such example of the way these identities are formed. Agnew and Corbridge identify the construction of specific binary oppositions within the 'ideological geopolitics' of the Cold War. These discourses included efforts to create 'a homogenization of global space into "friendly" and "threatening" blocs in which universal models of capitalism-liberal democracy and communism reigned free of geographical contingency' (Agnew and Corbridge 1995: 65). With the development of this global division came a 'semiotics of the "three worlds"' in which the 'Western first world' became not only the referent of economic and political development but also the world leader in that development (*ibid.*: 71). The 'third world,' on the other hand, represented a 'geographical zone not yet committed to a particular path to modernity' but nonetheless still a zone that the superpowers focused their attention on so as to 'recruit candidates for their respective models of political economy' (*ibid.*: 71). Within US academic discourse and foreign policy, political culture as an object of inquiry has played a significant role in the formulation of these zones, oppositions, divisions, and blocs. Modernization theory and its relationship to questions of the quality of a state's political culture and democratization created the very standards that fueled the global arrangement of political economy and *culture*. The positioning of the USA and England as the primary referents for democratization and citizen participation conform to the geopolitical demands of these two countries. The world leadership of the USA by the time of the publishing of *The Civic Culture* is confirmed through the international study of political culture. This appears in a similar fashion in the current renaissance of political culture as the central position of the West, and its post-materialist values in these studies reaffirm geopolitical efforts to steer the mass transformations of political structure and culture that have taken place in the last ten years. Most interesting is the fact that the study of political culture in the USA fell to its lowest point of interest toward the end of and in the wake of the Vietnam War, when US world leadership had been directly challenged and the cause of global democratization became less certain.

Political culture, however, plays a more direct part in the creation and scripting of secure and 'threatening' geopolitical spaces. Its conceptualizations of national identity are not only necessary for developing broad geopolitical categories but also for the essentialized differences that make geopolitical conflict a possibility. In exploring identity as a mechanism of war, Shapiro submits that 'analyses of global violence are most often constructed within a statecentric, geostrategic cartography, which organizes the interpretation of enmities on the basis of an individual and collective national subject and on cross-boundary antagonism' (1997: 175). The formal study of political culture makes possible the construction of 'collective national subjectivities' and the significance of their boundaries. Within academic analyses, the comparisons at the basis of political culture focus on the quality of modernization, democracy, and participation with regard to a predetermined referent, normally that of the USA. Yet, the geopolitics of these comparisons exceeds the 'empiricism' upon which they are founded. Instead, these comparisons provide an opportunity to recognize the (political) cultural differences between the geopolitical referent and aid recipients, neighbors (i.e. Mexico, Cuba, or Eastern Europe), or those states located outside the European–North American traditions of statecraft and democracy. Through political culture, 'the interpretation of enmities' and cultural difference become conjoined. Inglehart's 'discovery' of certain cultural clusters that are geographically bound leads to an understanding of geopolitical proximity in which neighbors that share enough core political values (i.e. Western democracies) will act accordingly and avoid military conflict with one of the same. They are less likely to engage in wars against each other, the argument goes, because of a joint recognition of cultural sameness. This emphasis on the coalescence of enmity and cultural difference is even more pronounced in Huntington's *Clash of Civilizations*. As the future arenas for regional and possibly global conflict take shape, they are marked by peaceful zones of cultural similarity, the building blocks of which remain sovereign states, and contestuous fault lines of cultural difference. The comparisons of political culture, therefore, generate the very narratives of national subjectivities and cultural difference so important for statist analyses of 'global violence.'

Finally, the study of civic culture has also been a dominant element in new scholarship on societal transformations in the former Soviet Union and Eastern Europe, comprising a political and cultural geography of 'transition.' The great mobilization of academic, political, and economic resources in response to the political transformations of Eastern European countries and the former Soviet

Union demonstrates how the future quality of political cultures often becomes the object of discourses of international politics. Continual discussions within US foreign policy concerning international aid programs to 'foster democracy,' especially in areas that have come to represent political instability, depend upon certain expectations and conceptualizations of political culture. That these understandings are at times vague or in other cases simply reiterate basic arguments within the study of political culture only stresses the importance of questioning how this category has become such a constitutive part of the imaginary of international relations. Speaking specifically to the challenges faced by states formed after the break-up of the Soviet Union, Daniel Franklin and Michael Baun argue (1995: 8) that 'it is often assumed that homogeneity and consensus can be achieved through the adjustment of state boundaries and physical relocation or the disfranchisement of minority groups.' Although they acknowledge the inherent violence in such activities, they continue that the end goal of these processes 'is possible to achieve . . . under a certain set of special conditions. For example, if a people is geographically concentrated, such as in the old Czechoslovakia or many of the republics of the former Soviet Union, peaceful nation building is possible upon the foundation of the former state' (*ibid.*). Such possibilities raise questions about the role of political-cultural knowledge in the international processes of engineering geographic boundaries designed to contain culture differences. By acknowledging that 'the formal recognition of ethnic boundaries within a state can be divisive,' however, Franklin and Baun outline one of the greatest challenges faced by governments during 'transition.' The task of comparative political culture in this case is to produce geographies of cultural difference in relationship to both the transitional politics of constitutional development and international security.

Conclusion

In a chapter of *Culture Shift* entitled 'Cultural Change and the Atlantic Alliance,' Ronald Inglehart attempts to identify how support for or against NATO bases and priorities is linked to the European public's trust of different geopolitical actors. Based primarily on Euro-barometer surveys throughout the 1980s, Inglehart's study identifies a particular ambivalence toward the USA and an overall, but not static, lack of trust toward China and the former Soviet Union. His thesis inquires into how post-materialist values (consisting of a deprioritization of concerns over 'security' and ideology) are shaping a decline in support for, in particular, European cooperation with the USA over security matters and, in general, NATO itself. Ingelhart (1980: 400) opens his analysis with the claim that 'trust and distrust of given nationalities seems to be part of a stable cognitive map.' The subject of this assertion, possibly a public, person, nation, or leadership, is unclear. In fact, throughout the analysis of the Euro-barometer questions of 'man' and the 'state' bleed together in their undifferentiated assessments of the possibility of 'war' and of their support of the

Atlantic Alliance. Instead, to reconfigure Inglehart's claim, interpretations of trust and distrust of given nationalities in the study of political culture seems to be part of a stable geopolitical map in which the trust of a political culture is transformed into geopolitical discourses as the distrust of enemies and the wariness of Europe's dependence on a powerful ally. This chapter has demonstrated how the mapping of the geopolitical and the political culture are 'interwoven' both ontologically and politically. Tracing this imbrication serves as one possible method for dissecting those powerful 'relays' that allow for the construction of essentialized ontological identities so necessary for geopolitical discourse and conflict. With this gesture, the 'stable' geographies found within political culture reiterate the need for scholars of critical geopolitics to interrogate a multitude of disciplinary and popular sites in their analyses of territory, sovereignty, and ontology.

Acknowledgements

I would like to thank Simon Dalby and Gearóid Ó Tuathail for their helpful comments on an earlier draft of this essay. Also, without the insights and encouragement of Jacqueline Berman and Matt Sparke this essay would not have been possible.

Notes

1 The study of political culture in comparative political science is commonly referred to as a specified study under the broader approach to economic and political development termed 'modernization theory.' As I am focusing solely on political culture in this chapter I have chosen to demarcate this subdivision of modernization theory with the title 'comparative political culture'. This title is not designed to present the literature on political culture as a single bound approach to the topic. I recognize that there is a diversity of approaches within modernization theory toward questions of culture. Rather, this title is meant as a heuristic device both for simplicity and to call attention to the very mechanisms of comparison in political culture: geography and the nation-state.

2 Although Spanos arrives at this notion of a relay in a reading of imperialism and the 'reciprocal relationship between (Roman) citizen and Empire,' the utility of this concept aids in unpacking the 'territorial trap' of the nation-state and political identity.

3 Satisfaction with institutional performance and lifestyle are the two broad topics of questionnaires that supply the data for Inglehart's *Culture Shift* and Putnam's *Making Democracy Work*.

References

Agnew, J. and S. Corbridge (1995) *Mastering Space: Hegemony, Territory, and International Political Economy*. New York: Routledge.

Almond, G. (1993) 'Forward,' in L. Diamond (ed.) *Political Culture and Democracy in Developing Countries*. Boulder, Colo.: Lynne Rienner Publishers.

Almond, G. and S. Verba (1989) *The Civic Culture: Political Attitudes and Democracy in Five Countries*. Newbury Park, Calif.: Sage Publications.

Ashley, R. K. (1987) 'The Geopolitics of Geopolitical Space: Toward a Critical Social Theory of International Relations.' *Alternatives*, 12: 403–434.

Ashley, R. K. (1989) 'Living on Borderlines: Man, Poststructuralism, and War,' in J. Der Derian and M. Shapiro (eds) *International/Intertextual Relations: Postmodern Readings of World Politics*. Toronto: Lexington Books.

Biersteker, T. J. and C. Weber (eds) (1996) *State Sovereignty as Social Construct*. Cambridge: Cambridge University Press.

Dalby, S. (1991) 'Critical Geopolitics: Discourse, Difference, and Dissent.' *Environment and Planning D: Space and Society*, 9: 261–283.

Diamond, L. (ed.) (1993) *Political Culture and Democracy in Developing Countries*. Boulder, Colo.: Lynne Rienner Publishers.

Dodds, K. and J. D. Sidaway (1994) 'Locating Critical Geopolitics.' *Environment and Planning D: Space and Society*, 12: 515–524.

Emmerson, D. (1995) 'Region and Recalcitrance: Rethinking Democracy Through Southeast Asia.' *The Pacific Review*, 8: 223–248.

Foucault, M. (1980) *Power/Knowledge*. New York: Pantheon Books.

Franklin, D. P. and M. J. Baun (eds) (1995) *Political Culture and Constitutionalism*. London: M.E. Sharpe.

Gellner, E. (1983) *Nations and Nationalism*. Ithaca, NY: Cornell University Press.

Girvin, B. (1989) *Contemporary Political Culture: Politics in a Postmodern Age*. London: Sage Publications.

Gupta, A. and J. Ferguson (1991) 'Beyond "Culture"; Space, Identity, and the Politics of Difference.' *Cultural Anthropology*, 7: 6–23.

Huntington, S. (1996) *The Clash of Civilizations and the Remaking of World Order*. New York: Simon & Schuster.

Inglehart, R. (1980) *Culture Shift*. Princeton, NJ: Princeton University Press, 1980.

Inglehart, R. (1988) 'The Renaissance of Political Culture.' *American Political Science Review*, 82: 1203–1230.

Inglehart, R. (1995) 'Changing Values, Economic Development and Political Change.' *International Social Science Journal*, 47: 379–404.

Ó Tuathail, G. (1994) '(Dis)placing Geopolitics: Writing on the Maps of Global Politics.' *Environment and Planning D: Space and Society*, 12: 525–546.

Ó Tuathail, G. (1996) *Critical Geopolitics*. Minneapolis: University of Minnesota Press.

Putnam, R. D. (1993) *Making Democracy Work*. Princeton, NJ: Princeton University Press.

Pye, L. and S. Verba (eds) (1965) *Political Culture and Political Development*. Princeton, NJ: Princeton University Press.

Shapiro, M. (1997) *Violent Cartographies*. Minneapolis: University of Minnesota Press.

Skocpol, T. (1985) 'Bringing the State Back In: Strategies of Analysis in Current Research,' in P.B. Evans, D. Rueschemeyer and T. Skocpol (eds) (1985) *Bringing the State Back In*. New York: Cambridge University Press.

Spanos, W. V. (1996) 'Culture and Colonization: The Imperatives of the Centered Circle,' *Boundary-2*, 23: 135–176.

Walker, R. B. J. (1993) *Inside/Outside: International Relations as Political Theory*. Cambridge: Cambridge University Press.

5

STABILIZING BORDERS

The geopolitics of national identity construction in Turkey

Kim Rygiel

Introduction

In March 1996, one of Turkey's most prominent and internationally acclaimed writers of Kurdish origin, Yasar Kemal, was arrested for writing an article 'The Dark Cloud Over Turkey.' He was charged with engaging in 'separatist propaganda' that threatened 'the indivisibility of the Turkish state' and given a twenty-month suspended sentence. It was the vision of the Turkish nation as a mosaic of different cultures, in particular, which the state considered to be dangerous. Kemal described his vision in the following way:

> For me the world is a garden of culture where a thousand flowers grow. Throughout history all cultures have fed one another, been grafted onto one another, and in the process our world has been enriched. The disappearance of a culture is the loss of a color, a different light, a different source. I am as much on the side of every flower in this thousand flower garden as I am on the side of my own culture. Anatolia has always been a mosaic of flowers, filling the world with flowers and light. I want it to be the same today.
>
> (1995a)

For Kemal, the world is a garden of culture and the heart of present-day Turkey, Anatolia, a place where many cultures have grown. His vision of the Turkish nation as a mosaic of cultures differs substantially from that of the Turkish state, which views cultural difference as a threat to its integrity. To continue with the analogy, for Turkey, the world is not one but many gardens, each harvesting its own particular color of flower. It is a vision, then, that sees the Turkish nation, not as a diverse group of people whose own cultures can flourish freely and intermingle, but rather as a unified group of people who identify themselves first and foremost as Turkish citizens. Thus, according to the state, cultural identities other than Turkish should be expressed in the private realm but regulated in the public. As the state perceives its survival to be dependent upon a homogeneous Turkish nation, it perceives this regulation of difference in the public sphere as necessary for its continued survival.

The above example presents two different visions of a nation. The first, founded upon difference, often clashes with the second, more common, vision, informed by current twentieth-century geopolitical thinking, which, being hostile to notions of difference, uproots cultural difference in multi-ethnic societies by demanding allegiance to a singular national identity within a demarcated territory. Implicit in this second vision is an understanding of identity as something that is primordial and homogeneous throughout the nation. However, as Stuart Hall, David Held and Tony McGrew note, national identities are not primordial but are constructed through systems of cultural representation, which, however, often present themselves as identities that are, in fact, primordial, fixed, essential and homogeneous (1992: 292).[1] Hall *et al.* argue that national identity is never the homogeneous entity that it presents itself as because it is always created out of the intermingling of cultures. Homogeneity is a myth because, in the first place, 'most modern nations consist of disparate cultures which were only unified by a lengthy process of violent conquest – that is, by the forcible suppression of cultural difference' where each conquest 'subjugated conquered peoples and their cultures, customs, languages and traditions and tried to impose a more unified cultural hegemony' (*ibid.*: 296). Furthermore, before an allegiance to a more unified identity can occur, 'these violent beginnings which stand at the origins of modern nations have first to be forgotten' (*ibid.*: 296). Finally, 'nations are always composed of different social classes and gender and ethnic groups' (*ibid.*: 297).

In an effort to contribute to the project of rethinking geopolitics, in this chapter I wish to explore the problematic relationship of national identity and state borders, which informs this second vision and much current geopolitical thinking. I will do so by examining the way in which the state constructs borders, those of both national identity and territory, by using spatial strategies that homogenize identity and space. The article will discuss how, by homogenizing national identity, the state tries to fix or secure its borders, both those which define membership in the nation over which it can claim jurisdiction and the territorial borders that define the space in which it can then claim sovereign rule. It will demonstrate how the state secures its national identity by suppressing or eliminating different identities such as ethnic or gender-based identities that challenge the official representation of the national identity as well as by disciplining space within its borders. As R. B. J. Walker (1993) notes, the reproduction of a national identity requires that the modern state confront, and to some extent eradicate, particularistic identities in order to ensure that more universalistic loyalties, such as to the state, remain the primary loyalty. Walker writes that the 'claims of universality within states' depend upon 'the explicit but often silent recognition that such claims to universality are in fact particularistic, are made on behalf of a particular group of citizens, rather than of people as members of a common community' (*ibid.*: 63). Ethnic nationalist movements within the state recognize the particularity of state claims when claims do not reflect their own cultural-political identity and needs and can be understood, in part, as a response to this tension.

I will present my argument through a case study of Turkey and its relationship to its Kurdish population. I will first discuss how the Turkish state constructs identity by presenting the Turkish nation as a coherent national identity and how, in doing so, it marginalizes other cultural identities such as Kurdish. Kurds who identify themselves first and foremost as Kurdish organize a competing politics using a similar logic to that of the state: a coherent Kurdish ethnic identity must be constructed by the Kurdish nationalist movement in order to represent Kurdish demands. However, this narrative of a unified Kurdish identity hides real and important differences, such as gender. I will include a discussion of gender in my analysis of the construction of national identities as it enables me to demonstrate the diverse ways and reasons women participate in national struggle and, thereby, to show the diversity that exists within a politicized identity, a diversity that is then silenced in official representations of that identity. As Nira Yuval-Davis notes, 'gender divisions often play a central organizing role in specific constructions of ethnicity, marking ethnic boundaries and reproducing ethnic difference' (in Moghadam 1994: 413). Due to gendered roles within society, women engage in, and are affected by, the practices and political struggles of ethnic and nationalist projects in different ways to men and thus become the particular target of state or group action when cultural or ethnic distinctness or purity is desired (see, for example, Meznaric 1994). Paying attention to gender in my analysis draws attention to the ways in which the ethnic collectivity is constructed and to the heterogeneity that exists within the collectivity (see also Yuval-Davis 1993). Moreover, the connection of gender to ethnicity and nationalism is ultimately important because 'wherever women continue to serve as boundary markers between different national, ethnic, and religious collectivities, their emergence as full-fledged citizens will be jeopardized' (Kandiyoti 1991a: 435).

Constructing the Turkish nation

The regulation of Kurdish identity

Turkey did not evolve into being over several centuries but was deliberately created from the ruins of the Ottoman Empire. Pressure to build a modern Turkish state began by the end of the nineteenth century, when the multi-ethnic Ottoman Empire was in decline from both external and internal pressures. European powers desired greater influence in the region; nationalists among the non-Turkish peoples (such as the Greeks and the Armenians) desired independence, and the Turks themselves desired modernization and democratization

(Moghadam 1993: 80). With the collapse, defeat and subsequent division of the Ottoman Empire, President Woodrow Wilson acknowledged the right of non-Turkish minorities to 'autonomous development'. Articles 62–64 of the 1920 Treaty of Sèvres specifically stipulated that the Kurds were to be given, first, 'local autonomy,' with the possibility of 'independence' at a later date (Laizer 1991). However, the subsequent 1923 Treaty of Lausanne, signed by Ataturk, which ended the war of independence against the occupying powers and established the modern secular Turkish state, failed to recognize the Kurdish people as an independent minority with rights. Only three groups were recognized as having special rights. Following the practice of protecting the religious rights of distinct religious minorities, which was established under the millet system[2] of the Ottoman Empire, the Turkish state recognized minority rights for the Greek, Armenian and Jewish communities based upon their religious identities being different from the Turks. The predominantly Muslim Kurds were not given special status, as their difference was one of culture not religion.

The idea of recognizing distinct cultural differences within the public sphere was thought to be, from the beginning, antithetical to the process of state building. Rather, the belief was that the creation of a modern and secular Turkish nation state out of the multi-ethnic Ottoman Empire depended to some degree upon the forced assimilation of cultural differences in order to construct a homogeneous community that could be governed by a centralized state system. This was especially true given the fact that the expression of ethnic difference, such as that expressed by the Greeks and Armenians, was held to be partially responsible for the collapse of the Ottoman Empire. Thus Ataturk believed that the construction of a universal Turkish national identity based upon an inclusive civic nationalism was necessary to achieve unification within the Turkish state. Cultural differences, which the Ottoman Empire had permitted to flourish under its rule, were subsumed under this one new national identity. In other words, the Turkish state relegated the many particular cultural identities within the new Turkish polity to the private sphere and to a personal status. It was thought that the assimilation of difference within the polity was necessary to foster unity among the different cultural groups within Turkey and thereby to strengthen a single allegiance to the Turkish state.

In order to construct and to reproduce a coherent national identity from a culturally heterogeneous population, the Turkish state pursues policies that assimilate, integrate or eradicate difference such as Kurdish particularity within the polity. For example, the Turkish state promoted a '"Turkish only" policy in arts, culture, education and politics' (Entesser 1989: 93). It introduced laws, for example, prohibiting the use of the Kurdish language, such as the prohibition in 1924 of teaching Kurdish in school. In the 1960s, Kurdish children were forced to attend boarding schools, where Kurdish was forbidden (Chaliand 1993: 72). Articles 26 and 28 of the 1982 Constitution prohibit the dissemination of thought in any language that is prohibited by law. While Article 28 was repealed in 1991 by former president Turgut Ozal, the use of Kurdish was permitted only in so far as it was not for political or educational purposes (Fuller 1993: 114). Moreover, many Kurds are now tried under Article 8 of the Anti-Terror Law, which acts as a catch-all law that 'forbids any "propaganda" that threatens the territorial or national indivisibility of the Turkish state' (UNHCR 1994). Despite amendments to Article 8,[3] the Anti-Terror Law gives wide powers to the state to define which activities constitute a threat to the 'indivisibility of the state,' to prosecute and to confiscate property. For example, many Kurds are now tried for using Kurdish under this law when matters discussed are not strictly historical or cultural but rather of a political nature (UNHCR 1993).

In addition to policies that limited the use of Kurdish, official state discourse denied the existence of an independent Kurdish identity outright. For example, Ms Fugan Ok, the head of the Human Rights Department of the Foreign Ministry in the 1980s, stated that 'the Kurds are not a minority, since according to the Lausanne Treaty of 1923 only religious minorities are recognized' (Whitman 1990: 2). In fact, up until a decade ago, official discourse did not use the name Kurds but referred to the Kurds as 'Mountain Turks' (Ignatieff 1993: 136). The existence of an independent Kurdish identity within the Turkish polity was also forcibly denied through common sayings such as 'Happy is he who is a Turk,' which, as Yasar Kemal notes, was written 'on the mountain sides everywhere' in Kurdish eastern Anatolia, and through sayings such as the one that every Turkish child had to repeat at school 'I am a Turk, I am honest, I am hard working' (1995b).

However, the state's relationship to its Kurdish minority is somewhat more complex than in many cases of ethnic conflict where the state pursues a consistent strategy of suppressing the rights of the minority in question. Despite denying that a separate Kurdish identity exists and suppressing Kurdishness within the Turkish polity, on the one hand, the state actively welcomes Turkish citizens of Kurdish origin to participate in public political life on the other. This is because the greater purpose behind the state's treatment of the Kurdish minority is that of regulating difference and thereby homogenizing the polity. For example, several prominent politicians are, in fact, Kurdish, such as former prime minister and president Turgut Ozal and former foreign minister Hikmet Cetin. Moreover, as Merhdad Izady notes, 'it is a fact widely known in Turkey that nearly a fifth of the Turkish members of Parliament are unassimilated Kurds' (1992: 200). In order to pursue a public life, however, the state demands that its citizens leave their ethnic identities in the private sphere. One cannot be, for example, actively Kurdish in the public sphere. Strict adherence to the Turkish identity is required in the public sphere and diversity within the polity is not recognized at the level of state discourse. For example, former member of parliament Serefettin Elci was sentenced in 1981 for having said in parliament 'I am a Kurd. There are Kurds in Turkey' (Whitman 1988: 7). Since the state allows its citizens to express their cultural identity in the private, it views 'its Kurdish problem' as one of terrorism rather than one of ethnicity. It does not see its persecution of the Kurdish nationalist party, the Kurdish Workers Party (PKK), as an issue of ethnic oppression or racism, because it recognizes all individuals as having equal economic and political rights protected under the constitution. The state's logic is explained by the Journalists' Association of Ankara. It argues that 'there is an ethnic reality in Turkey. There is no ethnic problem. Ethnic differences in our country are not a weakness, on the contrary they are a rich cultural heritage' ('PKK Reality': 1). It justifies this argument with the following explanation:

> there is no discrimination between citizens in Turkey. All citizens have the right to vote and be elected, and possess equal political rights within the framework of principles protected by law . . . The 10th Article of the Turkish Constitution states that 'all citizens are equal before the law with no discrimination as to language, race, color, political leanings, philosophy, religion and similar factors' . . . There is only one nation, the members of which are all equal before the law. Every person, irrespective of roots, religion or race, can claim Turkish identity.
> (*ibid.*: 4–5).

According to this logic, it follows that expressing Kurdish identity at the state level threatens the integrity of the nation and all members having equal status before the law regardless of their cultural heritage. This logic is again demonstrated by the fact that many of the laws charging those who use Kurdish to express political views claim that using Kurdish threatens the 'indivisible unity' of the nation (Article 8 of the Anti-Terror Law) or 'national unity' (Article 89). From the perspective of the state, then, the Kurdish problem is regarded as one of terrorism not ethnicity first, because the state sees itself as permitting the expression of ethnic particularity in the private realm and thus not suppressing ethnicity and second, because it believes that it provides equal rights to all its citizens regardless of ethnic difference. However, what the state fails to realize, and what Kurdish nationalists do, is that the very claims to universality that the state makes on behalf of the Turkish citizens actually are particularistic. Recalling R. B. J. Walker's (1993) observation, the problem stems from the fact that claims to universality made by the state in the name of all of its citizens are in fact particularistic, made on behalf of a certain group of Turkish citizens, and which exclude the particularistic nature of Kurdish identity.

The regulation of gender

In addition to cultural difference, such as Kurdish, the Turkish state also regulated gender by making women integral to the process of state building in order to construct a more homogeneous national identity. The founding of a modern secular Turkish state under Ataturk in 1923 made a clear association of women's emancipation with modernization through Westernization. Influenced by the French Enlightenment, Ataturk introduced the Kemalist reforms, which included

'economic development, separation of religion from state affairs, an attack on tradition, Latinization of the alphabet, promotion of European dress . . . and the
replacement of Islamic family law by a secular civil code' (Moghadam 1993: 81).
Given that Islam has very specific rules regulating the conduct of men and
women,[4] Ataturk's secularization of the state and the introduction of civil law
immediately provided women with a different status. However, Ataturk also
actively advocated the emancipation of women beyond these changes. In a 1923
speech he stated:

> A civilization where one sex is supreme can be condemned, there and
> then, as crippled. A people which has decided to go forward and
> progress must realize this as quickly as possible. The failures in our past
> are due to the fact that we remained passive to the fate of women.
>
> (quoted in Moghadam 1993: 82)

Elsewhere, Ataturk expressed similar thoughts stating:

> Our enemies are claiming that Turkey cannot be considered as a civi-
> lized nation because this country consists of two separate parts: men
> and women. Can we close our eyes to one portion of a group, while
> advancing the other, and still bring progress to the whole group? The
> road of progress must be trodden by both sexes together marching arm
> in arm.
>
> (*ibid.*)

Ataturk's comments equate the extension of full citizenship rights to women as
necessary for the creation of a modern Turkey. The extension of rights to women
enables women to participate actively in the nationalist struggle and in doing so,
to commit their allegiance to the Turkish state. Furthermore, Ataturk's comments demonstrate the identification of women as representatives and symbols of
the nation. More specifically, it is the liberated woman who comes to represent
and to define the boundaries of the new modern, secular and Western Turkish
nation. One of the ways that the Turkish state defines the new nation is through
women, who become important in defining its boundaries. The liberated woman
distinguishes the modern Kemalist state from the Ottoman state, which is defined
by contrast as Islamic and backward. As Nira Yuval-Davis and Floya Anthias
(1989) have noted, women frequently come to symbolize the nation in times of
nationalist struggle. They argue that because of women's distinct roles as biological and cultural reproducers of the collectivity, and as symbols of cultural identity
and authenticity, women's dress and conduct become symbolic of group difference. For example, the boundaries of membership into the new Turkish nation
were enforced and reproduced, in part, through the behavior and dress of women
as modern secular women rather than as Islamic women.

Paying attention to gender, however, also disrupts the image of the nation as homogeneous because it draws attention to the particular needs of women, which sometimes challenge the boundaries of the nation. For example, while citizenship rights were extended to women during the state-building process, women's rights were also limited because they were associated with the state-building process. As Sirin Tekeli notes, 'the price to be paid for this "state feminism" was that it was the creation of the myth that Turkish women had full equal rights with men, that they acquired these rights before women in many European nations and that consequently there was no more need for women's organizations' (1992: 140). This led to the closing of many women's organizations in the 1930s, the implication being that women's rights could be attained only through the state. They could not be achieved through a separate feminist movement or separate women's organizations. Tekeli writes:

> Women's rights were seen by Kemal as a strategically important part of the revolutionary process of transition from the theocratic multinational monarchy of the Ottoman Empire into the secular nation-state of the Turkish Republic . . . Women were expected to believe that only the state with reforms from above, could bring a solution to their grievances. Kemalist women believed that education was the key to ending discrimination against women.
>
> (*ibid.*: 142–143, endnote #2)

Extending citizenship rights to women in order to engage women in the nationalist struggle and thereby unify the polity has led some scholars, such as Deniz Kandiyoti (1991a), to conclude that state intervention on behalf of women's rights is somewhat of 'a sham.' Kandiyoti warns against 'the purely instrumental agenda of nationalist politics that mobilize women when they are needed in the labor force or even at the front, only to return them to domesticity or to subordinate roles in the public sphere when the national emergency is over' (*ibid.*: 429). Kandiyoti's argument is demonstrated in Turkey by the fact that despite the state's advocacy of equal rights for women as Turkish citizens at one level, on another level women are marginalized as second-class citizens through the Civil Code, which places women in a dependent position upon the husband as the head of the family. Sirin Tekeli writes that 'Kemalist women tried to protect their long-held beliefs about Kemalist reforms, but finally had to admit that the legal status of women as defined in the Civil Code was not one of equality but of dependency' (1992: 140). Tekeli points to Articles 152 and 159 of the Civil Code as evidence of this dependency.[5] Article 152 states that 'the husband is the head of the family' (*ibid.*). For some feminists, this is the clearest expression of patriarchy and must be abolished. However, others advocated changing Article 159, amended in July 1992, which states that 'the woman requires the husband's permission if she wishes to take up gainful employment outside the home' (*ibid.*). In addition to Articles 152 and 159, Ayse Saktanber lists Article 440 of the Penal Code, which defines adultery differently for men and women

(1994: 124–125). Saktanber argues that with adultery 'the wife is charged with adultery if she engages in a relationship with a man other than her husband on one single occasion, whereas in the husband's case it has to be proved that he actually lived with another woman' (*ibid.*: 125). In other words, the state incorporated women into the nationalist project during the process of state building, on the one hand, and thereby regulated the difference of gender in order to homogenize national identity. On the other hand, gender difference was maintained in the home into the present day through the Civil Code in such a way that women are, in certain respects, treated as second-class citizens.[6]

Another example that illustrates how women's particular needs challenge the boundaries of the nation and disrupt its image as homogeneous is the turban or headscarf debate that occurred in the early 1990s in Turkey. The representation of the nation as modern and secular according to Ataturk, symbolized by the liberated Western woman, is currently being challenged through Islamic women. In the early 1990s, female students clashed with state authorities over the right to cover their heads at university. For Islamic women, the wearing of the turban enabled them to participate in the public sphere at the university while still symbolically adhering to the Islamic notion of women's proper place in the private sphere (Kadioglu 1994). For the Turkish state, however, the association of the turban with Islam and Islam with backwardness informed its refusal of the public display of the turban. The state interprets Islamic headdress as a threat to the secular identity of the Turkish state. The campaign platform of the governing party at that time, the Islamic Party or Refah Partisi, to allow women to wear headdress in government buildings continues to produce outcries from both government and secular citizens. The example of the headscarf illustrates that women both demarcate and challenge the boundaries of the nation. The association of the liberated, secular and Western woman with the Turkish nation is as important today in Turkey as it was in the 1920s. However, today it is also women, Islamic women, who are challenging this definition of the Turkish nation.

Constructing the Kurdish nation

Those involved in the Kurdish nationalist struggle organize their politics around a similar understanding of identity as that used by the Turkish state. The Kurdish identity must be presented to some degree as a coherent unified ethnic identity, and as distinct from Turkish, in order to wage a national struggle for self-determination. However, like all identities according to a more fluid understanding of identity, the Kurdish identity consists of great diversity. As David McDowall notes, the Kurdish population is quite heterogeneous, consisting of several religious and linguistic groups. Although predominantly Sunni Muslim, the Kurdish people include Jews, Christians, Yazidis and sects such as the Alevis and Ahl-e Haqq sect (1992: 11). In addition, the 'existence of substantially different dialects cuts further lines of division across the simplistic idea of a Kurdish nation' (*ibid.*: 11). For example, the Kurdish language consists of two major dialects and several sub-dialects: Kurmanji spoken in Northern Iraq and Turkey and Sorani (or Kurdi) spoken in Iraq and Iran with sub-dialects Kirmanshhi, Leki, Gurani and Zaza (McDowall 1989: 7). Another Kurdish scholar, Martin Van Bruinessen, explains this diversity with an amusing anecdote of a conversation he had with a young man who, in response to the question as to whether he was Kurd or Azeri, replied that he was 'both a Kurd and an Azeri, and . . . a Persian as well' (1992: 45). Van Bruinessen explains the man's response in the following passage:

> His mother spoke only Kurdish well, but with his father he conversed in both Kurdish and Azeri, and sometimes in Persian. So his father was an Azeri and his mother a Kurd, I ventured, glad to have understood his reaction. No, he objected, his father was also a Kurd and an Azeri and a Persian. These terms were for him purely linguistic, not ethnic labels as I defined them . . . My question was of course prompted by my own belief in the objective existence of ethnicity.
>
> (*ibid.*: 45)

Van Bruinessen explains that although many of the people he met of 'mixed ancestry, defined themselves unambiguously as Kurds, Turks or Arabs,' there were others whom he thought 'to be objectively Kurds but who insisted they were Turks – they spoke both languages' and 'there were many people in the region who apparently had no clear-cut ethnic identity' (*ibid.*: 45–46). Van Bruinessen cites two further examples to illustrate the fluidity of Kurdish identity. He notes 'the case of numerous young Turkish Alevis who, in the 1970s, redefined themselves as Kurds out of political sympathies' (*ibid.*: 46) and the case in areas where there have been many Sunni–Alevi conflicts, 'people define themselves primarily as Sunni or Alevi rather than as Turk and Kurd' (*ibid.*: 47). He notes that in these examples 'the major factor in the change of ethnic identity was political' and that the situation determined which of 'a number of partially overlapping identities . . . he or she will emphasize or de-emphasize' (*ibid.*: 46). Van Bruinessen concludes, however, that the 'emergence of Kurdish nationalism as a significant political force compelled many people to opt for an unambiguous ethnic identity' (*ibid.*: 48).

Amidst the closure of pro-Kurdish political parties like the Democratic Labor Party (DEP), closed in June 1994, and the People's Democratic Party (HADEP), closed in June 1996, which tried to represent the demands and needs of Kurdish people peacefully through the political process, the Kurdish Worker's Party (PKK) has increasingly come to represent the needs of Kurdish people, albeit problematically (see Imset 1992; Bulloch and Morris 1992: 186). The PKK is a Marxist-based group, recruiting mainly from rural Anatolia, with a membership consisting largely of workers, peasants and unemployed youth. As journalist Ismet G. Imset has argued, it was during the period when the DEP was closed that 'the Kurdish origin people had little if no political outlet' and when the PKK 'argued that it was now the *representative* of the Kurdish people' (1992: 329).

The relationship of the PKK to the Kurdish people is complex and illustrates the diversity of positions within what the PKK tries to present as a unified Kurdish identity. As an organization it advocates violence and is responsible for the deaths of many of its own people, including teachers, women and children. As one villager bluntly stated, 'We will lose either way . . . we don't like the Turkish soldiers and we fear the PKK . . . both sides kill our people and burn our towns . . . you tell me which I should support' (UNHCR 1994). Despite the use of violence and different goals, however, the PKK still 'represents Kurdishness to these people' because it has been the 'only Kurdish group tough enough to withstand eleven years of Turkish military pressure' (*MEI*, 25 August 1995: 12). In other words, the PKK has increasingly come to represent the Kurdish population living in the southeast simply for the fact that there is no other voice representing them. As Aliza Marcus summarizes, 'Ankara in its zeal to crush Kurdish nationalism, has managed to undermine and destroy non-violent Kurdish movements, in effect helping ensure PKK dominance. Between the state and the guerrillas, Kurds have not had many options for protesting restrictions on their identity' (1994: 19).

The regulation of gender within the Kurdish national movement

One way in which the PKK tries to homogenize identity is by incorporating women into the movement, and thereby, regulating gender difference. Gender is one of the aspects that distinguishes the PKK from other Kurdish political organizations and parties. The PKK's social base is not only open to, but is dependent upon, the participation of what it claims to be thousands of women among its ranks (Hassanpour 1994: 7). The PKK's 5th Congress Resolution attests to this distinctive element of the party, stating that 'The potential of women who make up half of the society in the service of the revolution, and their hidden and suppressed talents and intelligence in creating an entire society based on equality, are the most humane and radical characteristics of our revolution.'[7]

Today, the relationship between women and the PKK's nationalist struggle for liberation resembles that of women to the Turkish state during the state-building process. Just as women's emancipation was intrinsic to the modernization project in Turkey, the liberation of Kurdish women is central to the PKK's struggle. In Turkey, the liberated woman, represented by the image of the woman in European-style dress as distinct from the Islamic woman in headdress, symbolized a new, modern, secular and Western Turkey. Similarly, the liberated Kurdish woman is represented by the image of the woman guerrilla in army fatigues who is readily identifiable from the more traditional rural woman and the KDP

116

woman *peshmerga*, both in more traditional Kurdish dress. The liberated Kurdish woman represents the new Kurdistan freed from 'oppression' and 'exploitation.' In a declaration to the International Kurdish Women's Conference on 8 March 1994, International Women's Day, PKK leader Abdullah Ocalan expressed this association: 'I can confidently say that the revolution in Kurdistan is a woman's revolution. The essence of the liberation of the people of Kurdistan is the liberation of women' (*Resistance* 1995: 50).

Given woman's role as biological and cultural reproducer of the nation, women come to symbolize the Kurdish nation. From a PKK perspective, the oppression of women in Kurdish society is viewed as a microcosm of the oppression of Kurds by Turks. This is illustrated, for example, in one Kurdish publication, which explains:

> Within both society and the family, a woman had neither a right to self-expression, nor any authority. Women have been held within a captive state created over hundreds of years, maintaining outdated traditions and conventions. It can be said that the fate of Kurdish women has been that of Kurdistan. It follows that changing the destiny of Kurdistan will be closely linked with changing the lot of Kurdistan's women.
>
> (*Kurdish Women* 1992: 1)

Within PKK discourse, the oppression of women is also linked to the 'colonization'[8] of Kurdistan and feudal structures that are organized around the family unit under which women are confined (*ibid.*: 1). This linkage enables the PKK to unite three of its goals: the liberation of women and the termination of feudal structures with the Kurdish national liberation struggle. These goals are integral to defining the PKK as a modern party and distinguishing it from other Kurdish political organizations and positions taken by Kurdish people in the southeast.[9] By incorporating women into the Kurdish national struggle, the PKK is able, first, to represent itself as a modern movement that is as distinct from other organizations that maintain feudal ties; and second, to represent itself as a modern political organization that challenges traditional Kurdish structures such as the traditional patriarchal family unit. In this way, women play a role in the process of defining the boundaries of membership in the PKK as distinct from other Kurdish groups and Kurdish subjectivities.

The inclusion of women as active participants in the PKK struggle has enlarged the social base and unified the nationalist movement to a greater extent. As in the case of the Turkish state, the PKK is better able to homogenize its identity by including women in the national liberation struggle. Yet the participation of women also draws attention to the diversity that exists within the collectivity that disrupts the PKK's representation of the Kurdish nation as a unified collective identity. This is because, first, the ways in which women participate in the liberation struggle, and second, their motives for doing so, may differ from those of men. Women play a variety of roles not only as guerrillas, politicians and activists but also as mothers and wives. Women's motive for participating in the liberation struggle may not always be primarily that of waging a nationalist struggle. For example, their participation also enables them to break free from traditional gender roles, develop new skills and attain new rights and, in turn, transform traditional patriarchal relations within the family.

While women's emancipation plays an integral role to national liberation movements, the literature on gender and nationalism[10] suggests that women's concerns, when they differ from those of men or those of the nationalist group, may not be addressed by the ethnic group or nationalist movement. Thus women simultaneously 'participate actively in, and become hostage to' national and ethnic projects (Kandiyoti 1991a: 431). Cynthia Enloe argues, for example, that while 'nationalism has provided millions of women with a space to be international actors,' and 'national consciousness has induced many women to feel confident enough to take part in public organizing and public debate for the first time in their lives,' when women's needs and opinions challenge the nationalist struggle they quickly learn that their needs as women are secondary to those of the nation's survival (1990: 61). The problematic relationship of women to the nationalist struggle has even forced some scholars, like Nahla Abdo, to conclude that 'historically, there has yet to be a national struggle in which the feminist agenda was compatible with the nationalist agenda. National struggles have tended to prioritize the external enemy – the occupying or colonizing force – thereby blurring the internal struggles and contradictions within the nation' (1993: 32).

In the Kurdish case, one example that illustrates some of these assertions is that of the woman guerrilla. According to PKK statistics, women now make up 30 per cent of PKK guerrillas (*Resistance* 1995: 49). Women undertake political and military training in camps with men in the mountains and supposedly participate equally in military ambushes as well as in political debates. According to one member, 'the work is not shared on the basis of sex' (*ibid.*: 34). However, while more women may be joining the guerrillas, their reasons for doing so may be other than patriotic duty. In an interview with journalist Sheri Laizer, one 23-year-old woman guerrilla, Nermin, explained that she decided to join the guerrillas after her father had announced an arranged marriage for her the following year, saying 'I am not ready to be married and have children, even though it's normal for girls my age to start a family' and that 'I didn't want to be married before having lived my own life, as you know. It would not have been much of a life, always feeling captive and uncertain about tomorrow. Here in the mountains we really value our lives' (*ibid.*: 27). While Nermin does include the 'fight for a free Kurdistan' and the desire 'to travel and get to know my country better' among her reasons for joining the struggle, she places her emphasis on the opportunity it provides her to escape marriage and to live a life where she plays a role in the control of her own destiny (*ibid.*: 27). As she explains:

The PKK really attracts me because I can see that this party offers us a different way of life. Even if I die, I will die with honor. What is it to be a housewife in comparison with this, always living under a shadow, never really knowing what life might have been like had we been able to choose our own political future.

(*ibid.*)

As Nermin points out, becoming a guerrilla offers much to women: travel, education, escape from early marriage, or from marriage altogether, a sense of control and decision making over one's own life, and a less subservient role in society, none of which are offered by the traditional role of housewife. It is for these reasons that Laizer concludes:

Would Nermin have been better off as a housewife? I doubted it, simply because for Nermin's generation, the PKK and the social role the Party extends to women is highly desirable. It offers them the possibility of both personal and political achievement, and a worthwhile sense of sacrifice. If there is to be sacrifice then, they believe, it must be a sacrifice for their country and beliefs, not just for another master occupying their house.

(*ibid.*: 28)

Defining state borders

In the above discussion, I have tried to show some of the ways in which the state attempts to construct and represent its national identity as a stable, homogeneous identity in order to fix or secure, first, the borders that define membership in the nation, over which the state can claim jurisdiction; and, second, the territorial borders that fix the space in which the state claims sovereign rule. I have suggested that the state secures its national identity by suppressing or eliminating different identities that challenge the official representation of the national identity. In this section, I wish to argue further that the state also secures its national identity by disciplining space within and beyond its borders.

The internal division of space within a state is used as a means politically to control diversity within a population that might otherwise challenge the construction of the political collective. As David Smith explains, 'the sovereign control of territory by the state ensures that the dominant culture can circulate freely throughout the space ... within its boundaries. It can divide and rule its territory so as to hinder or prevent attempts by subordinate cultures from developing a solid base from which to reproduce their own culture' (1990: 11). Perhaps one of the most explicit examples of how the Turkish state has used space in the past to control diversity within the southeast was a law, passed in 1932 as a response to the Mount Ararat revolt in 1930, that organized mass deportations in order to implement four zones according to the following specifications as described by Gerard Chaliand (1994):

- No 1 zones comprise those regions where it is desired to increase the density of the populations having a Turkish culture.
- No 2 zones comprise those regions where it is desired to establish populations which require assimilation into the Turkish culture.
- No 3 zones comprise those territories where immigrants of Turkish culture may freely establish themselves, without the aid of authorities [the most fertile of the Kurdish regions].
- No 4 zones comprise those territories which it is desired to evacuate, and which are prohibited areas, for medical, cultural, political, strategic and public order reasons [this last zone includes the least accessible Kurdish regions].

The state continues to use similar spatial strategies today to control diversity within its polity. For example, the Turkish state uses physical segregation to separate the predominantly Kurdish area of southeast Anatolia from the rest of Turkey by declaring a state of emergency rule over the southeast. When the content of an area cannot be controlled, the area is evacuated. The following section describes these strategies in greater detail.

Segregating the Kurdish areas of Turkey: emergency rule in the southeast

Martial law was in place in the Kurdish provinces in the southeast from 1979 until 1987, when it was replaced by a state of emergency (Whitman 1988: 24). Emergency rule gives the government additional powers over its Kurdish population, as the southeast has traditionally been the predominant living area of the Kurds. Amnesty International notes that 'the State of Emergency Region Governor controls the armed and police services in the region and can assume control of any functions of the civil administration. The Governor and the forces at his disposal also enjoy a high degree of official immunity from prosecution' (1996: 6). Emergency rule also gives the government power to treat its Kurdish citizens differently from its Turkish citizens. For example, the military has 'command of the regional security forces and full authority to evacuate villages' (Whitman 1988: 24). With the introduction of identity cards, the military was able to control not only the flow of information but also of people entering and leaving the area (*ibid.*: 24). The regional governor also has the authority to censor news and impose internal exile. Emergency rule effectively enables the state to segregate the southeast and isolate the Kurds while also giving power to the state, and the military more specifically, to police the area to enforce obedience to the state. This strategy of segregation, in turn, is part of the larger process of enforcing homogeneity within the nation. The state of emergency, in effect, legitimizes state interference in the private daily lives of the Kurdish people, in that space where, at least theoretically, the expression of Kurdishness is supposed to be permitted.

The government' s creation of the 'village guard' system aids its policing of the area by creating division between the PKK and the Kurdish population in the southeast. The village guards are 'local peasants who are paid monthly stipends and provided with arms' (*ibid.*: 25). Participation in the village guards is supposed to be voluntary, but Kurdish villagers are caught between the PKK and the government. If they serve as village guards they are likely to become the target of PKK violence. Refusal, however, is regarded by the Turkish military as passive support for the PKK, and if villagers support the PKK, even to provide food, shelter or medical assistance, they will be subjected to interrogation, possible torture and imprisonment by the Turkish military (UNHCR 1993; see also Marcus 1994: 18). More insidious is the operation of the 'special team members,' who, speaking Kurdish, 'present themselves in the villages as PKK guerrillas asking for provisions in order to test the loyalties within the village' (UNHCR 1993). This strategy creates an environment of fear and intimidation as well as distrust within the Kurdish population and between villages.

When enforcing discipline in the southeast fails, the state evacuates the area and forces the Kurdish population to move into larger cities and into western Turkey, where assimilation has proven to be more successful. Thus far, the state has evacuated approximately 2,000 villages and 300,000 people directly, while another 2 million people have left for economic reasons, or in pursuit of safer lives (*MEI*, 21 July 1995: 13). As President Ozal said, commenting on the flow of refugees after an attack on the town of Sirnak on 18 August 1992: 'many problems would be solved much more easily if 500,000 people left here and settled in the West' (Pope, *MEI*, 10 Sept. 1992: 10). Evacuation has thus led effectively to the depopulating of Kurdish villagers from the southeast.[11] Evacuation, like emergency rule, is an effective state strategy because, as Allen observes, it 'disrupt[s] expressions of ethnicity in the southeast' (1994: 26). By eradicating Kurdish identity sympathetic to the PKK, and replacing it with a Kurdish identity that is loyal to the state, the state creates an environment where PKK guerrillas cannot survive. Again, the larger assumptions that the conflict and the understanding of state security rest upon is the belief that cultural homogeneity within the nation must be enforced in order to govern over a specific territory. Any difference such as Kurdish identity is perceived to threaten the integrity of the state and the coherence and governance of the national collective.

Enforcing state borders: regulating identity beyond the state

In addition to using spatial strategies to discipline identity within the state, the state also reinforces the boundaries of the polity by policing the physical boundaries of its territory, thereby mediating between members of the state and foreigners. The Kurds are the fourth largest ethnic group in the Middle East, with a homeland divided between Turkey, Iraq, Iran, Syria, Armenia and

Azerbaijan (Hassanpour 1994: 3). As a Kurdish state would entail the break-up of any one or all of these states, most of these states pursue a security policy that includes the manipulation, assimilation and policing of its Kurdish minority to prevent an autonomous self-governing region from developing. States fear that any recognition of autonomy would be a first step in the creation of an independent Kurdish state. McDowall notes that in the case of Turkey, 'Turkey's main problem has been to seal the border areas, since Kurdish activism is contagious, and since groups like the PKK use safe havens in Iranian and Iraqi Kurdistan' (1989: 14). McDowall also observes '[t]here are now, apparently an increasing number of Turkish Kurds who feel a common identity with Kurds beyond Turkey's borders' (1992: 21). He warns that, in fact, without positive measures to address the Kurdish situation within Turkey, pan-Kurdish feeling may strengthen in spite of military measures (*ibid.*: 21).

The implementation of the safe haven in northern Iraq at the end of the 1991 Gulf War in order to protect Iraqi Kurds from further attacks by Saddam Hussein increased Turkey's fear of greater independence for its own Kurdish population. In June 1992, effective autonomy of the region was handed over to the Kurdish regional government, dominated by the PUK and the KDP.[12] The PKK, however, has used the region as a base from which to launch attacks into Turkey. Since the implementation of Operation Provide Comfort, the Iraqi Kurds have faced a double embargo, with UN sanctions against Iraq and Iraqi sanctions against the north. With trade restricted, this double embargo[13] made it necessary for the Iraqi Kurds to negotiate with Ankara in order to secure their own lifeline. In order to establish safe relief supply routes through Turkey (the only available routes into northern Iraq), the Iraqi Kurdish leadership agreed to help police the PKK from the area and to allow Turkey to conduct periodic cross-border raids.[14] On 4 October 1992 the Kurdish regional government 'issued an ultimatum to the PKK: either withdraw from the border passes or be expelled' (Kutschera 1994: 14). Heavy fighting began shortly thereafter between the PKK and Iraqi Kurds and involving extensive Turkish air raids (*ibid.*). This example demonstrates the way in which the Turkish government enforces the territorial borders between itself and Iraq. By fueling conflict between groups of Kurds in Iraq and Turkey it prevents pan-Kurdish solidarity from developing.

In addition to taking advantage of divisions among the Kurds, the Turkish military also enforces the Turkish–Iraqi border by harassing Kurds living along the border and by conducting military invasions or 'hot pursuits' into northern Iraq (*ibid.*),[15] allegedly in pursuit of PKK members in northern Iraq, the most recent being that of 14 May 1997 involving between 25,000 and 50,000 troops (*MEI*: 31 March, 12 May, 21 July 1995, 30 May 1997). To assist with its invasions, Turkey plans to create 'a three-mile buffer zone along the Iraqi side of the border' (*MERIP*, May–June 1992: 27). In addition, with US assistance, Turkey will build 'a network of heat sensors and observation posts . . . to stem the flow of Kurdish rebels from Iraq' (from *Jane's Defence Weekly*, 11 Nov. 1995, noted in AKIN 1995). As it is nearly impossible to distinguish PKK members from other Kurdish villagers seeking refuge, such pursuits enable the Turkish military to harass Turkish Kurds and to create division within the Kurdish community. Iraqi Kurds, annoyed by attacks that affect their own people, agree to aid Turkey in the policing and rounding up of Turkish Kurds in exchange for peace and autonomy. The state hopes that by making life unbearable it will force evacuation of the area, which will prevent cross-border movement and the blurring of Iraqi–Turkish Kurds (McDowall 1989: 14). However, according to McDowall (1992), one possible unintended outcome of such a policy is the disruption of more local community ties, such as tribal and feudal ties. In the past, such ties have made it more difficult for a unified Kurdish nationalist movement to develop not only in Turkey but also in Iran and Iraq. The disruption of local ties, alternatively, may facilitate the growth of pan-Kurdish nationalism in its place.

To summarize, Turkey enforces its borders, first, from the Iraqi safe haven. By creating division within the Kurdish population, the state tries to separate its Kurdish population from Iraqi or Iranian Kurds and thereby strengthen the boundaries defining Turkish identity. Maintaining clear distinctions between the Kurdish groups in each state prevents greater unity within the Kurdish nationalist movement from developing and prevents the possibility of secession of the Turkish Kurds or of demands for a Kurdish state. Borders are also enforced through forced assimilation of the Kurdish identity into the greater Turkish identity by segregating the southeast, evacuation, harassment and/or forced migration. Again, the purpose is to homogenize the population within Turkey to create a unified polity within the state. As the Turkish state's pursuit of the Kurds shows, difference is always the target of state policies aimed at consolidating the nation and presenting an image of a unified nation. Such a practice and logic can only ever lead to greater insecurity and violence for those against whose difference the state has consolidated its own identity.

Towards an alternative politics: multiple identities sharing space

I have argued that the twentieth-century geopolitical practices by which states like Turkey construct their national identity and delineate borders have produced a politics hostile to diversity. State building and the reproduction of national identity depend upon silencing differences such as ethnicity and gender within the polity. They also depend upon envisioning space as something that can be homogenized and bounded, and is ultimately linked to territory. While homogenizing space is an integral part of defining and reproducing a unified national identity, homogenizing national identity also becomes an important strategy for strengthening territorial jurisdiction. This politics is today, more than ever, a dangerous politics around which to organize societies, since globalization brings greater movement and interaction between cultures and peoples.

Students of the social sciences need to rethink geopolitics to find a more peaceful way to live with difference.

In the introduction, two ways of envisioning the nation were presented. The alternative vision of society described by Yasar Kemal, in which multiple identities share space, needs to be given careful consideration. Implementing this alternative vision requires resisting notions of homogeneous identity and space so integral to present forms of state governance. Rethinking space, for example, might begin with Michael Shapiro's observation that 'states, and many nations within states have residual aspects of cultural alterity *within* them. Such aspects of difference cannot be resummoned by redrawing geographical boundaries; they exist as invisible forms of internal otherness. Every practice which strengthens boundaries produces new modes of marginalized difference' (1994: 496). Living more peacefully with difference therefore 'requires relaxing the spatial imperatives of the order' (*ibid.*: 499).

Rethinking identity, on the other hand, might begin by accepting Kemal's vision of the state as a place where a thousand cultures grow. Rather than viewing the state as bounded territory that contains an unproblematic, fixed, homogeneous identity, we might recognize that identities are always unfixed, contested, multiple and hybrid. By hybrid, I refer to the fluid and historically and discursively constructed nature of identity, the observation made by Edward Said, who wrote 'Partly because of empire, all cultures are involved in one another, none is single and pure, all are hybrid, heterogeneous, extraordinarily differentiated, and unmonolithic' (1993: xxv). In other words, rather than pitting identities against one another, we might regard identities as mutually constitutive, the national state identity defined in relation to a range of other ethnic, cultural and gender identities. For Turkey, this means acknowledging that multiple identities do exist and that Kurds may identify themselves both as Turkish citizens and as Kurdish. Moreover, it means seeing Kurdish not as something different and antithetical to Turkish identity but rather as an identity that, while different, is deeply connected to the history of the peoples living in the region and to the history of Turkey. The histories and cultures of the various ethnic groups are too intertwined with one another to be separated by simple notions of distinct bounded communities or policies that try to purify the group, whether on the part of the state or by the PKK.

Notes

1 The key to this discussion is the need to see identity as something that is constructed rather than primordial. A constructivist approach sees identity as something that is produced through social, material and discursive practices and as involving multiple, competing identities, which are fluid and change over time and space. This differs from the primordial approach, which perceives identity as having a fixed essence or unified core; the vision of the modern subject that has informed Western thought since the Enlightenment and the period of modernity. According to the primordial understanding of identity, the identity of an individual, ethnic group, nation, state or any other such category is grounded upon an essence that determines its identity. As Roxanne

Lynn Doty further explains, 'This essence provides for unity within a category and in other categories which are also defined by an essence. Identity and unity lie within. Difference lies without' (1993: 451). The problem with this approach to identity, as those living through and writing on the Bosnian war will attest (see Sorabj 1993; Campbell 1994), is that it understands ethnicity according to the same logic as does the exclusionary politics of ethnic nationalism: namely that difference is perceived to be external and, therefore, a threat to identity. A primordial perspective reinforces policies that seek to eliminate difference from the community by targeting those individuals whose identities challenge a national or ethnic group's desired representation of itself.

2 Suna Kili writes that 'the Ottoman administration attempted to create unity, yet allow for diversity. This policy was implemented through the "millet" system' which (quoting Shaw) she describes as being 'a party of the structure of state as well as of society. Each millet established and maintained its own institutions to care for the functions not carried out by the Ruling Class and state, such as education, religion, justice and social security' (1994: 303).

3 Pressure to abolish Article 8 came about as a result of the customs union between Turkey and the European Union. The European Parliament stated that its granting approval of the customs union was subject to verification of the improvement of Turkey's human rights record, of which modifications to Article 8 were identified as a key component. Amnesty International expressed disappointment with the changes to Article 8, exclaiming 'we can find little to applaud in changes which leave intact legislation under which people can be jailed for expressing non-violent opinions. The changes in the Anti-Terror law neither secure the release of prisoners of conscience currently in custody, nor rule out future prosecutions and prison sentences for people expressing non-violent opinions' (Amnesty International, 30 Oct. 1995). Amendments of Article 8 of the 1991 Anti-Terror Law reduced the maximum sentence from five to three years. Approximately 170 writers and intellectuals have been sentenced and are serving prison sentences under Article 8, while another 5,500 are currently on trial (Lowry 1996: A11).

4 See, for example, Binnaz Toprak (1994) as this applies to women in Turkey.

5 See also Arat (1996).

6 For further reading on the history of women and feminism in Turkey see Tekeli (1986, 1991, 1995).

7 Due to a paucity of literature on Kurdish women and on the question of gender in Kurdish nationalism, much of the information cited in this section comes from literature published by the Kurdistan Information Centre and the Kurdistan Solidarity Committee, organizations that to some extent provide a public relations function for the PKK. It is likely, therefore, that some of the publications from which I have extracted information serve as PKK propaganda. At the very least, the publications uncritically present a favorable bias towards the PKK. While using the information, I have been attentive to this fact and have tried to use the material critically and with reference to the analytical arguments made within the literature on gender and nationalism. It is for this reason that I frequently use quotation marks to draw attention to propagandist phrases that need to be read critically.

8 I have put quotation marks around the term 'colonization' because I believe this to be an example of ahistorical reading of Kurdish nationalism in an attempt to draw lines of continuity from an oppressive present to the past. The term 'colonization' implies some prior existence of Kurds as a self-consciously distinct people who lived in the southeast before the arrival of the Turks, who settled and took the land away from the Kurds. I believe a more historical account to be that prior to the establishment of the Turkish state, Kurds and Turks (in addition to other ethnic groups) coexisted in the southeast, often fighting on the same side of the same Ottoman–Persian battles. Gerard Chaliand, for example, writes that the 'war of independence . . . was widely

supported by the Kurds in the name of Moslem solidarity' (1994: 4). I believe it ahistorical to suggest that these people constructed their identity primarily along ethnic lines. Rather, other factors such as religious, feudal, tribal and power relations played an important role in identity construction. For this reason, it is incorrect to suggest that the Turks 'colonized' Kurdistan, given that Kurds were active participants, for example, with Ataturk in the war of independence and the beginnings of the building of the new Turkish state. Colonization as a term reflects the late twentieth-century politicization of ethnic identity, that is the self-consciousness of belonging to what is perceived today by Kurdish nationalists as a separate politicized ethnic identity. As part of PKK discourse, it also reflects the process involved in Kurdish nationalism to narrate a coherent identity of the Kurds as victims by rooting this identity in a mythical past.

9 For example, members of the PKK refer to themselves as guerrillas rather than as *peshmergas* to distinguish their anti-feudal position from the feudal movement in Iraq, the KDP led by Barzani and members of the Barzani clan (Kutschera 1994: 14).

10 See Yuval-Davis (1991, 1993, 1994); Yuval-Davis and Anthias (1989); Yuval-Davis and Sagal (1992); Kandiyoti (1991a, 1991b); Moghadam (1993); Jayawardena (1986); Abdo (1991, 1993); Sharoni (1993); and Enloe (1990: ch. 3).

11 Gerard Chaliand states that 'Istanbul has thus become the premier Kurdish town, with an estimated immigrant population of approximately 800,000' (1994: 40).

12 The Kurdish regional government had completely dissolved by the summer of 1996 due to intensified PUK and KDP rivalry over an almagamation of issues including customs revenues, land disputes, patronage networks and party politics, personality clashes between the respective leaders, Jalal Talbani and Masud Barzani, and a historical rivalry between the two parties (McDowall 1995). Accusations on both sides alleging Iranian external support for the PUK and Iraqi support for the KDP provided the final catalyst for the PUK–KDP fighting that broke out in August 1996.

13 See also *Middle East Report*, March–April 1995, for further reading on the effects of sanctions and the double embargo.

14 A lucrative black market has also developed involving 'private trading companies with partners in the South, customs officials, Iraqi Kurdish contractors, and Kurdish truck drivers from Turkey' (Natali 1995: 21). This smuggling trade and its extensive network makes it more difficult to end the military violence against the Kurds, since many, including Turkish officials, have an interest in maintaining the status quo. Natali notes, for example, that 'from 1991–1993, Turkish border guards were loosely interpreting border regulations and allowing drivers to keep as large as 4,000 liter tanks underneath their trucks to store fuel exports. During this period, truck drivers could earn up to $5,000 on a good five-day journey from Southeastern Turkey to Saddam's refineries in Mosul or Kirkuk' (1995: 21).

15 It should be noted that the Turkish military capability necessary to conduct these raids is largely funded through the United States as well as France, Germany and the Netherlands. As William Schultz, the Executive Director of Amnesty International, USA, noted in a recent episode of *60 Minutes* entitled 'An American Dilemma,' which aired on 14 January 1996, 'This year Turkey will receive 320 million dollars of US taxpayer money ... which is going to the Turkish Government for the purpose of killing their own citizens.' This makes Turkey the 'third largest recipient of US economic and military aid in the world' after Israel and Egypt, according to Republican congressman John Porter. Turkey's membership in NATO is one of the stated reasons for the continued US military aid to Turkey.

References

Abdo, Nahla (1991) Women of the Intifada: Gender, Class and National Liberation. *Race and Class* 32(4): 19–37.

Abdo, Nahla (1993) Middle East Politics Through Feminist Lenses: Negotiating the Terms of Solidarity. *Alternatives* 18(1): 529–538.

Allen, Thomas B. (1994) Turkey Struggles for Balance. *National Geographic* 185(5): 2–37.

American Kurdish Information Network (AKIN) Kurdish News and Information Sept.–Dec. 1995. Electronic News. Arm the Spirit Home Page.

American Kurdish Information Network (AKIN) February 1996. Electronic News.

Amnesty International (1995) Turkey: Disappointing Legal Changes Will Not Restore Freedom of Expression. 30 October. Electronic News. Amnesty International Home Page.

Amnesty International (1996) *Turkey: No Security Without Human Rights.* New York: Amnesty International.

Arat, Yesim (1996) On Gender and Citizenship in Turkey. *Middle East Report* January–March: 28–31.

Bulloch, John and Morris, Harvey (1992) *No Friends but the Mountains: the Tragic History of the Kurds.* New York, London, Victoria and Toronto: Penguin Books.

Campbell, David (1994) The Deterritorialization of Responsibility: Levinas, Derrida, and Ethics After the End of Philosophy. *Alternatives* 19: 455–484.

Chaliand, Gerard (ed.) (1993) *A People Without A Country – The Kurds and Kurdistan.* New York: Olive Branch Press.

Chaliand, Gerard (1994) *The Kurdish Tragedy.* London and New Jersey: Zed Books.

Doty, Roxanne Lynn (1993) 'The Bounds of Race' in International Relations. *Millennium: Journal of International Studies* 22(3): 443–461.

Enloe, Cynthia (1990) *Bananas, Beaches and Bases: Making Feminist Sense of International Politics.* Berkeley and Los Angeles: University of California Press, 42–64.

Entesser, Nader (1989) The Kurdish Mosaic of Discord. *Third World Quarterly* 2(4): 83–101.

Fuller, Graham E. (1993) The Fate of the Kurds. *Foreign Affairs* 108–121.

Hall, Stuart, David Held and Tony McGrew (eds) (1992), *Modernity and its Futures.* Oxford and Cambridge: The Open University in association with Polity Press and Blackwell Publishers.

Hassanpour, Amir (1994) The Kurdish Experience. *(MERIP) Middle East Report* 24(4): 2–12.

Ignatieff, Michael (1993) *Blood and Belonging: Journeys into the New Nationalism.* Toronto: Viking.

Imset, Ismet K. (1992) *The PKK: A Report on Separatist Violence in Turkey (1973–1992).* Ankara: Turkish Daily News Publications.

Izady, Mehrdad R. (1992) *The Kurds: A Concise Handbook.* Washington, DC: Taylor & Francis.

Jayawardena, Kumari (1986) *Feminism and Nationalism in the Third World.* London: Zed Books, 1–41.

Journalists' Association, Ankara. PKK Reality in Turkey and in the World. Electronic News.

Kadioglu, Asye (1994) Women's Subordination in Turkey: Is Islam Really the Villain? *Middle East Journal* 48(4): 645–660.

Kandiyoti, Deniz (1991a) Identity and its Discontents: Women and the Nation. *Millennium: Journal of International Studies* 20(3): 429–443.

Kandiyoti, Deniz (ed.) (1991b) *Women, Islam and the State*. Basingstoke: MacMillan and Philadelphia: Temple University Press.

Kemal, Yasar (1995a) The Dark Clouds Over Turkey. The Netherlands: Infogroup Schism. Electronic News.

Kemal, Yasar (1995b) Turkey's War of Words. 6 May. Electronic News.

Kili, Suna (1994) The Jews in Turkey: A Question of National or International Identity, in *Nationalism, Ethnicity and Identity*, Russell Francis Farnen (ed.) New Brunswick, NJ: Transaction Publishers, 299–317.

Kreyenbroek, Philip and Stefan Spiel (eds) (1992) *The Kurds: A Contemporary Overview*. London and New York: Routledge, 1–67, 95–114.

Kurdish Militancy in Turkey: The Case of the PKK (1989) *Crossroads: an International Socio-Political Journal* 29: 43–59.

Kurdish Women: The Struggle for National Liberation and Women's Right (1992) London: Kurdistan Information Centre (KIC)/Kurdistan Solidarity Committee (KSC) Publications.

Kurdistan Report #75 (1995) Kurdish Women Want Freedom: Interview with Helen Ates of the Free Women's Union of Kurdistan (YAJK). July–August.

Kutschera, Chris (1994) Mad Dreams of Independence: The Kurds of Turkey. (*MERIP*) *Middle East Report* 24(4): 12–16.

Kutschera, Chris (1995) Kurds in Crisis. *The Middle East*, 25 Nov., 6–10.

Laizer, Sheri (1991) *Into Kurdistan: Frontiers Under Fire*. New Jersey: Zed Books.

Lowry, Reuben (1996) Pro-Kurdish Paper Struggles for Free Speech in Turkey. *Globe and Mail*, 5 Jan., A11.

Marcus, Aliza (1994) City in the War Zone. (*MERIP*) *Middle East Report* 24(4): 16–19.

McDowall, David (1989) *The Kurds: The Minority Rights Group Report No. 23*. London: The Minority Rights Group.

McDowall, David (1992) The Kurdish Question: a Historical Review, in *The Kurds: A Contemporary Overview*, Philip Kreyenbroek and Stefan Spiel (eds) London and New York: Routledge, 10–31.

McDowall, David (1995) *Middle East International*, 28 April, 17.

Meznaric, Silva (1994) Gender as an Ethno-Marker: Rape, War and Identity Politics in the Former Yugoslavia. *Identity Politics and Women: Cultural Reassertions and Feminism in International Perspective*, Valentine M. Moghadam (ed.) Boulder, Colo.: Westview Press, 76–97.

Middle East International (MEI). 10 Sept. 1992: 10; 491–511. 6 January–20 October 1995. 31 March 1995. 21 July 1995: 13. 25 August 1995: 12. 30 May 1997.

(MERIP) Middle East Report: The Kurdish Experience. 24(4): 1994.

(MERIP) Middle East Report: The Iraq Sanctions Dilemma 25(2): 1995.

Moghadam, Valentine M. (1993) *Modernizing Women: Gender and Social Change in the Middle East*. Boulder, Colo., and London: Lynne Rienner Publishers and the United Nations University.

Moghadam, Valentine M. (ed.) (1994) *Identity Politics and Women: Cultural Reassertions and Feminism in International Perspective*. Boulder, Colo.: Westview Press.

Natali, Denise (1995) The Kurdish Experiment in Democracy: 1992–1994. Paper presented at the Middle East Studies Association (MESA) 29th Annual Meeting, Washington, DC, 9 Dec.

PKK 5th Congress Resolution Concerning the Women's Army and the Free Women's Movement. Electronic News.

Resistance: Women in Kurdistan (1995) London: Kurdistan Information Center (KIC)/Kurdistan Solidarity Committee (KSC) Publications, August.

Rouleau, Eric (1996) Turkey: Beyond Ataturk. *Foreign Policy* 103: 70–87.

Sack, Robert David (1986) *Human Territoriality: Its Theory and History.* Cambridge: Cambridge University Press, 1–27.

Said, Edward (1993) *Culture and Imperialism.* New York: Vintage Books.

Saktanber, Ayse (1994) Becoming the 'Other' as a Muslim in Turkey: Turkish Women vs. Islamist Women. *New Perspectives on Turkey* 2: 99–134.

Shapiro, Michael (1994) Moral Geographies and the Ethics of Post-Sovereignty. *Public Culture* 6: 479–502.

Sharoni, S. (1993), Middle East Politics Through Feminist Lenses: Toward Theorizing International Relations from Women's Struggles. *Alternatives* 18(1): 5–28.

Sixty Minutes (1996) An American Dilemma. *Sixty Minutes.* 28(16), 14 Jan.

Smith, David (1990) Introduction: the Sharing and Dividing of Geographical Space, in *Shared Space Divided Space: Essays on Conflict and Territorial Organization,* Michael Chisholm and David Smith (eds) London, Massachusetts, North Sydney and Wellington: Unwin Hyman, 1–21.

Sorabji, C. (1993) Ethnic War in Bosnia? *Radical Philosophy* 63: 33–35.

Tekeli, Sirin (1986) The Rise and Change of the New Women's Movement: Emergence of the Feminist Movement in Turkey, in *The New Women's Movement,* Dahrelup (ed) London: Sage, 179–199.

Tekeli, Sirin (1991) Women in the Changing Political Associations of the 1980s, in *Turkish State, Turkish Society,* A. Finkel and N. Sirman (eds) London: Routledge.

Tekeli, Sirin (1992) Europe, European Feminism, and Women in Turkey. *Women's Studies International Forum* 15(1): 139–143.

Tekeli, Sirin (ed.) (1995) *Women in Modern Turkish Society: A Reader.* London and New Jersey: Zed Books.

Toprak, Binnaz (1994) Women and Fundamentalism: The Case of Turkey, in *Identity Politics and Women: Cultural Reassertions and Feminism in International Perspective,* Valentine M. Moghadam (ed.) Boulder, Colo.: Westview Press, 293–306.

Turkish Daily New (1995) Professor Dogu Ergil: Separatism Has No Appeal For Our Kurdish Citizens. 23 June. Electronic News.

United Nations High Commissioner for Refugees (UNHCR) (1993) The Current Situation of Kurds in Turkey. Center for Documentation on Refugees 18 December. Electronic News.

United Nations High Commissioner for Refugees (UNHCR) (1994) Background Paper on Turkish Asylum Seekers. Geneva: UNHCR), Center for Documentation on Refugees. September. Electronic News.

US Department of State (1994) *Country Reports on Human Rights Practices for 1993,* Washington, DC: US Department of State, 993–1010.

Van Bruinessen, Martin (1992) Kurdish Society, Ethnicity, Nationalism and Refugee Problems, in *The Kurds: A Contemporary Overview,* Philip Kreyenbroek and Stefan Spiel (eds) London and New York: Routledge, 33–67.

Walker, R.B.J. (1993) *Inside/Outside: International Relations as Political Theory.* Cambridge: Cambridge University Press.

Whitman, Lois (1988) and (1990) *Destroying Ethnic Identity: The Kurds of Turkey, A Helsinki Watch Report*. New York and Washington, DC: US Helsinki Watch Committee, March.

Yuval-Davis, Nira (1991) The Citizenship Debate: Women, Ethnic Processes and the State. *Feminist Review* 39: 58–68.

Yuval-Davis, Nira (1993) Gender and Nation. *Ethnic and Racial Studies* 16(4): 621–632.

Yuval-Davis, Nira (1994) Identity Politics and Women's Ethnicity, in *Identity Politics and Women: Cultural Reassertions and Feminism in International Perspective*, Valentine M. Moghadam (ed) Boulder, Colo.: Westview Press, 408–424.

Yuval-Davis, Nira and Floya Anthias (eds) (1989) *Women–Nation–State*. London: MacMillan.

Yuval-Davis, Nira and Gita Saghal (eds) (1992) *Refusing Holy Orders: Women and Fundamentalism in Britain*. London: Virago Press.

6

MANUFACTURING PROVINCES

Theorizing the encounters between governmental and popular 'geographs' in Finland

Jouni Häkli

Introduction

One of the uniting factors in the heterogeneous 'critical geopolitics' approach has been the aspiration to analyze the taken-for-granted constructions on which conventional politics is based. These include, for instance, the imagination of the world as cultural, geo-economic or geopolitical regions, and the conception of global politics as a strategic game played out on a patchwork of distinct territorial units. In seeking alternatives to reality as presented by the dominant players of the global and national politics, the practitioners of critical geopolitics have called into question the very foundation on which relations between states and social groups are forged, political decisions made, hostilities commenced, and treaties negotiated (Dalby 1991: 264–269; Dalby and Ó Tuathail 1996: 452).

In critical geopolitics, an integral part of this foundation, the geographical depictions of the world produced by the intellectuals of statecraft is treated not as a passive mirror of 'that which exists' but rather as a vital constituent of the political world. While activities with the label 'geopolitical' are material in the sense that they always are embedded within governmental routines, global economy, and political contestation, it is recognized that they are also discursive, which means that geopolitical phenomena are dependent on, as well as conducive to, the production of geographical knowledge, statements, and understandings (e.g. Hepple 1992: 136–140; Ó Tuathail 1996: 17–18).

In this connection, the term 'geograph' has been introduced to underline the persistent yet context-dependent nature of geopolitical imagination (Ó Tuathail 1989). According to Dalby (1993: 440–441), geographs are 'particular frequently used descriptions of the world' that come to structure the ways in which different social groups and agents formulate their political opinions and conceptions of events in society. Importantly, different kinds of geographs inform different interpretative communities. Thus, there are dominant 'geographs,' which structure the institutionally established geopolitical reasoning of the intellectuals

131

of statecraft, but in addition to that, there are 'geographs' that circulate and traverse other spheres of social life. The latter can be politically motivated 'dissident' or 'counter-hegemonic' geographs, or they can be part of the lived geographies of everyday life, and thus characterizable as 'popular' or 'vernacular' geographs (*ibid.*: 441; Häkli 1998a).

In this chapter, I explore the history of encounters between governmental and popular geographs giving shape to the Finnish provinces (*maakunta*). The term 'province' derives from the Latin word *provincia*, which means conquered territory. Despite this rather military etymology, in many languages the word province has come to bear the connotations of spontaneity, naturalness and organicism. This was the case at least with the French regionalists, as well as with the provincialism of Josiah Royce in early twentieth-century America (Rabinow 1989: 197–199; Entrikin 1991: 68–71). Therefore, in what follows the term province refers to a region with the organic and spontaneous connotations of the Finnish word *maakunta*.

In focusing on provinces I argue, with Paul Routledge (1996: 509), that political processes at scales other than the global also need to be considered in the critical analysis of geopolitics. Rather than international geopolitical imagination, I set out to analyze a geopolitical discourse connected to a series of reforms of regional administration in Finland, one in which different geographical conceptions of provinces have surfaced. From a governmental point of view the provinces have represented a natural and spontaneous subdivision of the Finnish state's territorial space. However, in popular or vernacular imagination the provinces are rather vague, historically embedded cultural regions with no fixed boundaries or meaning. Whereas for government the 'essence' of provinces lies in the junction of social spontaneity and spatial demarcability, in popular use the provinces have no essence, and the provincial geographs remain open to contestation, contradictions, and contextual readings.

In discussing the distinction between governmental and popular geographs, I am not proposing an oppositional relation between these two, as they in fact constantly resort to similar language and images. Governmental agents appeal to popular ideas and identities to gain support, while people routinely express their cultural identity and distinctiveness in starkly territorial terms (Kaplan 1994; Häkli 1998b). Thus, what differentiates governmental and popular geographs cannot be read out directly from their linguistic or visual expressions. Rather, they should be understood as systems of signification embedded in and differentiated by what institutional and individual agents do, not so much by the particular objects they refer to. In other words, the distinction makes sense as an argument for the contextuality of social life, not as an attempt to account for better or worse access to 'real provinces.'

To claim that governmental discourse makes very different use of provincial geographs compared with provinces as lived social spatiality is, therefore, really no more than to acknowledge the particular character of institutional social action. While institutional practice seeks to define, demarcate, and accomplish universal and universally applicable provinces, no such urge can be found systematically in the myriad contexts of social life. Outside the administrative pressure for conformity, the provinces remain contradictory and contested realities subject to place-bound sentiments and regional identities, but also economic interests, place marketing, and parochialism. This is a crucially important assertion in that it begs the question of who wins and who loses when undefined and ephemeral regions are brought into governmental discourse and action, which, by necessity, has to arrive at an objective and universal definition of provinces. More specifically, a question arises as to how the practitioners of statecraft are related to the 'ordinary people' when the former are engaged in an administrative reform that seeks to territorialize the provincial geographs of the latter.

The notion of 'deep space' points to the different modalities of social action from which governmental and popular geographs emanate. Neil Smith has referred to deep space as 'the relativity of terrestrial space, the space of everyday life in all its scales from the global to the local and the architectural in which . . . different layers of life and social landscape are sedimented onto and into each other.' He sees deep space as 'quintessentially social space; as physical extent fused with social intent' (1990: 160–161). This is the contextual and multilayered social space in which the Finnish provinces exist as symbolic landscapes of the social life subject to continual contestation, negotiation, rereading and rewriting. This is the deep space that in the governmental discourse must be flattened out into a two-dimensional plane subject to exact definition and demarcation so as to function as a political space, a platform for regional administrative reform.

In what follows I investigate an endeavor to territorialize and contain the geographs pertinent to provinces as lived social spatiality. I wish to explore the distance from governmental to popular geographs – the gap from territory as a lived space to the territoriality of politico-administrative discourse, which constitutes institutional power vested in knowledge, but also provides a space for resistance to that power. I end my discussion by outlining this occasion for popular resistance to institutional discourses.

A genealogy of politico-administrative discourses on territory

For a critical assessment of the governmental discourse on provinces it is necessary to scrutinize both the complex spatiality through which the knowledge of society is situated and becomes functional in society, and the very idea of region itself. Governmental reports dealing with the administrative reform have not functioned merely as devices of technical definition and demarcation of provinces. Such an understanding of knowledge as simply mirroring the society 'out there in the real' must be refuted, along with the conception of region as an entity with a social existence independent of its representations.

Instead, attention must be directed to the social and historical processes through which 'regions' have come to have an existence for governmental action, that is, to the processes in which regions have been made visible for government in that particular form that institutional activities require. As my case is from the Finnish society, it is necessary to look at the history of visualization of regional space as it has unfolded in the history of the Finnish state apparatus. It should be noted, however, that the applicability of this approach, or the findings that result, are not restricted to the Finnish context. Very similar lines of development can be found in all modernizing European states, albeit in different configurations depending on the particular political and societal contexts in which the developments have taken place (e.g. Giddens 1985; Dandeker 1990).

When thinking of the modernization of states' governmental practice, one persistent trend has perhaps not caught enough attention: the growing dependency of government on the visualization of society. Yet, it can be argued that it is precisely the capability to visualize a society as both an object of administration and a subject of politics that has made the efficient and rational functioning of the modern state apparatus possible in the first place (Dodds 1993: 364–365; Edney 1993b: 63; Häkli 1994a: 51–52). There are two aspects of the governmental visualization of society that deserve attention and that also are closely connected to the territorializing discourses on provinces produced over the history of the Finnish state.

First, the production of empirical knowledge of society has not been motivated by mere scientific curiosity but rather visual devices like statistics and maps have emerged as part of the modern technologies of governmental power (Hacking 1991: 181). The connection between empirical knowledge of society and its use in the centralized organization of governmental activities has set particular requirements to the formal qualities of knowledge. Bruno Latour (1986: 15) has described these requirements with the term 'optical consistency,' meaning those aspects of knowledge that standardize it and regularize its relation to objects, thus making the combining, scaling, and analytical utilization of different units of knowledge possible (see also Edney 1993b: 63). Not surprisingly, maps and statistics meet these requirements easily. Maps are standardized projections of the social and geographical world, whereas statistics are collected by centrally directed organizations and by means of fixed routines and regulations.

Second, from a governmental point of view an optically consistent knowledge holds a particularly important dual quality of being simultaneously *fixed* as a representational form and *movable* across territory as inscribed on paper (Latour 1986). This makes possible a centralized accumulation of society and space into the archives of government, resulting in a technically mediated yet 'realistic' enough image of the territory and society. The importance of this 'abstract space' to the scientific rationalization of government can hardly be exaggerated. Through systematic mappings of territory and population governments have been able to catch a synoptic view of the society within their territorial bounds, thus making it subject to scrutiny by 'a single pair of eyes' (Widmalm 1990: 36).

The fact that all European states began to build bases of empirical knowledge had several consequences for the territorialization of governmental practice. On the one hand, the enhanced visibility of all corners of the kingdom directly contributed to states' tightening their governmental grip over their entire territories. On the other hand, these territories became objects of knowing and analysis like never before. The cartographic and statistical data produced and compiled could be utilized as raw material in analytic operations and theoretical syntheses, through which governments sought to make visible social phenomena previously unseen, unknown, and uncontrollable. The statistical surveillance of economic activities is a perfect example of how new, territorially defined, and visible society was constructed by and for policy making. A concern for the nation's wealth and the state's tax revenues brought about 'national economy' as an increasingly important field of governmental action and discourse (Latour 1986; Procacci 1991).

The proliferation of empirical knowledge of society in the state's use also marks a crucial point in the territorialization of the provincial geographs in Finland, a latecomer state on the European political map. The history of the Finnish polity really begins only in 1809, shortly after Sweden's defeat by Russia in the Napoleonic wars. Ceded to Russia, a collection of 'eastern provinces' of the Swedish kingdom became recognized as a political and economic whole, the autonomous Grand Duchy of Finland in the Russian Empire (see Figures 6.1 and 6.2) (Jutikkala 1962; Jussila 1987: 55–56).

The early Finnish state was a consolidating governmental apparatus in which the Swedish influences were clearly visible. Russia had allowed the Finns to retain their old Swedish constitution and administrative systems, but also population statistics and land survey, which had been instituted by the state of Sweden (Hjelt 1900; Gustafsson 1933). The latter were the result of large mapping projects launched by the rapidly modernizing Swedish state in the eighteenth century in order to expand the government of its sparsely populated large territory, covering also the areas that later became Finland.

The first large mapping project was the gathering of population statistics. After a few years probationary period an institution called *Tabellwärket* was established in 1749 with the task of making statistical surveys of the kingdom and its population (Hjelt 1900; Kovero 1940). Later renamed the Statistical Central Bureau, the institution was among the most advanced of its time, a fact that reflected Sweden's position as one of the European great powers. Interests behind the project were unambiguously politico-administrative, and filled with enthusiasm for the novel ways of knowing the kingdom. The increasing visibility of the society stimulated the minds of politicians and administrators, and encouraged the government to proceed with the effort despite some drawbacks in the early days of statistical survey (Johannisson 1988: 176; see also Porter 1986: 25–27).

The same excitement accompanied another large mapping project that served the state's land reform politics. *Storskifte* ('great partition') was one of the first

Figure 6.1 A late seventeenth-century map of Sweden including the eastern provinces, e.g. Finland (original in the Finnish National Archive).

state-wide programs where geographical imagination furnished with maps played a central role. By means of maps it was possible first to codify the available land and then to redistribute it according the adopted economic, and notably, visual rationale (Pred 1986). The goal was to minimize the number of individual land parcels in favor of larger clusters, enabling more labor-effective farming (Kuusi 1933; Kain and Baigent 1992). The project quickly showed the government the utility of maps as tools of administration – as devices with which the kingdom could be subjected to the observing gaze and guiding hands of its governors.

Through these and many other ambitious projects, the mapping of society and territory grew into an integral part of rational state administration in Sweden, as it did in all modernizing states in Europe (e.g. Giddens 1985: 179–180; Wittrock 1989; Edney 1993a). It also gave a tangible guise to 'abstract space,' a development where space was systematically inscribed on paper and transferred to the central archives of the government (Häkli 1994b). Mapping gave social space the qualities of being both fixed as a representation

136

and movable across time and space. As David Harvey (1989: 240) has pointed out, the 'compression' of time and space in the nineteenth century had a vast impact on how the world came to be conceived of and lived in. Likewise, the centralized accumulation of time and space revolutionized the ways in which societies were governed (see also Harley 1988; Revel 1991; Ó Tuathail 1994).

The exciting new power of mapping was based on its potential to visualize – literally to make visible – phenomena that otherwise could not be observed at all. Furthermore, the new kind of empirical data made possible, and in fact often invited more advanced analysis of the intricacies of the society. In discussing the rise of the art of government in the seventeenth and eighteenth centuries, Michel Foucault (1991: 93) points at new ways of understanding society and territory as a relational totality: 'The things with which . . . government is to be concerned are in fact men, but men in their relations, their links, their imbrication with those other things that are wealth, resources, means of subsistence, the territory with its specific qualities, climate, irrigation, fertility etc.' An analysis using maps and statistics could associate the prevailing social relations between populations with a certain spatial context and geographical area. In this respect, the machineries of mapping instituted by the Swedish state also played an important role in the emergence of provinces as governmental geographs in nineteenth-century Finland (Häkli 1998a).

Provinces territorialized

What did the Finnish provinces come to look like when represented by governmental bodies for governmental purposes? One way to answer this question is to recall the distinction between the abstract space of maps, statistics and archives, and the contextuality of social life. While easily apprehensible when pointing at different modalities of social action, this distinction is by no means a simple one when it comes to popular geographs, because they also have their origin in the abstract space – they reflect a geographical imagination akin to that which informed the new governmental conceptions of society and space. If the governmental approach to provincial space is thoroughly discursive, so is the provincial symbolism as lived social spatiality (Berdoulay 1989). It is, therefore, the ways of using provincial geographs that accentuate the distinction between governmental discourse and the contexts of deep space, and make explicit the gap between governmental and popular geographs.

As mentioned above, a new regional imagination had grown into the governmental practice in consequence of the novel ways of representing and knowing the society (see also Rabinow 1989: 197–202; Häkli 1998a). However, it was not only within the realm of politics and administration that the will to measure and map the territory changed conceptions of space. Although clearly lagging behind its institutional forerunners, changes were also brought about in the popular imagination of space, which was growing in geographical extent, but was also becoming increasingly diversified and contested. A consciousness of provinces was burgeoning among the larger population in the economically, politically, and culturally increasingly dynamic society of the late nineteenth century. Here the rise of regional newspapers and school geography played an important role (Salonen 1974; Paasi 1986). These new mediators of knowledge could more rapidly than before circulate among the 'great reading masses' provincial geographs, some of which resulted directly or indirectly from the governmental curiosity and mappings, while others were produced alongside various cultural activities involved in the 'discovery' of the Finnish territory and nation.

Among these cultural activists were the students at the University of Helsinki who, in a manner typical of European universities, were organized into students' unions according to their provincial background, usually the place of birth (Strömberg 1987: 291–293). The students' unions collected ethnographic materials and founded local museums, exhibitions, and archives, which contributed to the idea of provinces as historical entities dating back to the period of Swedish rule (Figure 6.2). Thus, the unions effectively socialized the educated elites into the imagination of Finland as consisting of distinct provinces. Among the unions' self-proclaimed tasks was to foster patriotic sentiments, which by the turn of the twentieth century were clad not only in romantic ethnology, but increasingly also in overtly nationalistic concerns (Kolbe 1996).

The educated elites of a young polity aspiring for national self-determination realized that 'Finland' had to be produced as a symbolic landscape in the popular realm to form and support consciousness of a coherent ethnic nation. As the provinces could conveniently be used in representing the stereotypical character and history of different parts of the Finnish territory, the provincial geographs played an important part in the Finnish nation-building process. A sense of national belonging together surpassing, but not suppressing, regional, linguistic, and ethnic identities was strengthened by the geographical imagination of a diverse yet unified nation inhabiting a shared fatherland (Wilson 1976; Paasi 1986; Alapuro 1988). The idea of a larger territorial unity meant that the Swedish-speaking Finns, as well as the Sámi peoples of Finnish Lapland, were just as important 'elements' of Finnishness as were the Finns themselves (Häkli 1998b).

However, it should be noted that the symbolism of provinces that gradually became embedded in various social activities did not come to imply clearly demarcated or defined regions. On the contrary, a considerable confusion has prevailed for instance in people's regional identities (Palomäki 1968; Paasi 1996: 212). The historical provinces have stood out as providing and reflecting the most established identity, but even there the particular meanings of provinciality have been highly contested, ranging from economic appropriation of cultural images to enlightened disapproval of regional parochialism. As the discussion below shows, this is still the case today. Regional identities have been profoundly affected by the processes of spatial restructuration, industrialization, and urbanization of Finnish society during the twentieth century. New 'functional' provinces have emerged, reflecting the mobility within and sense of belonging to areas defined by urban centers (Heikkinen 1986; Rasila 1993).

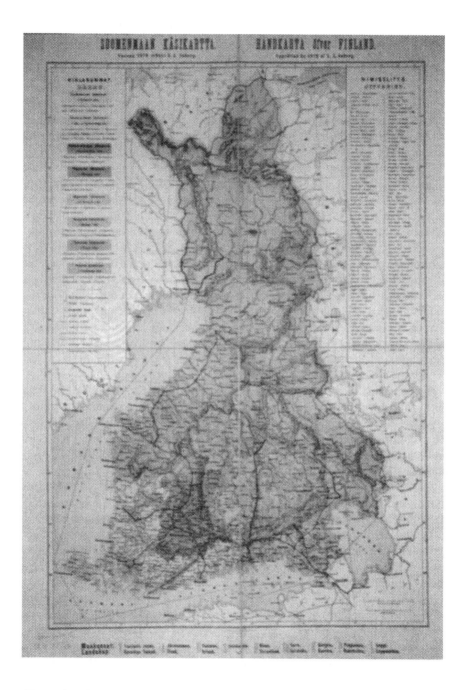

Figure 6.2 A nineteenth-century map of Finland with the historical provinces listed at the bottom (original in the Finnish National Archive).

Tensions in provincial imagery have intensified, particularly in times of economic distress, and in the peripheral areas of eastern and northern Finland, as the evident dependency of regional economies on the larger national and international market has upset the idea of local self-sufficiency. While in some this has provoked regionalistic ideas and actions, for others the economic downturn and restructuring has meant a forced migration to southwestern Finland, or even abroad. The largest mass migrations took place in the early twentieth century and in the 1960s and 1970s, stirring up the old regional identities and creating new mixes of people from different parts of the country (Neuvonen 1990: 22–23).

Understandably, amidst such economic and cultural turmoil, the provincial geographs have occupied a consistent but multifaceted place. Sometimes serving as the only lasting connection to 'roots,' the provincial identities have become important signposts in the new situations into which urbanized and industrialized ways of life have brought people. However, provinces as lived social spatiality are contextually appropriated and (re)produced, which explains why ephemerality and vagueness have prevailed in the popular symbolisms of provinces. For most 'ordinary' Finns, provinces are merely convenient labels with which to categorize experiences of people and places and to structure the social and geographical world. Furthermore, sometimes these provincial stereotypes and identities are resorted to, but sometimes not. There are considerable differences between people in how significant these regional identities are, and their significance varies depending on the particular social situation in which 'us' or 'others' are encountered.

Characteristic of an open-ended appropriation, this volatility has not been an option for governmental action. On the contrary, the government has sought to reduce the tensions and contradictions of lived provinciality by territorializing these cultural spaces, that is, by determining exactly what is the area, function, and extent of each province. A concrete endeavor to institutionalize, fix, and contain the contested and tension-laden provincial symbolisms, a series of regional administrative reforms has produced images of clearly demarcated, exactly located, objectively and unambiguously representable regions neatly parcelling out the continuous and homogeneous space of Finnish territory.

In the self-understanding of the intellectuals of statecraft – politicians, high officials of the state, reform committees, political scientists, regional activists, etc. – the territorialization of provinces has often been viewed as a step toward more humane and responsive government. In reality, while technically this may well result from a province-based administration, it also marks the confinement of provincial geographs by the governmental rationality. What presents itself as a seamless bringing together of 'the people' and 'political empowerment' is actually a conflation of two very different relationships to provinces: a contested and appropriative social spatiality on the one hand, and a detached and territorializing discourse on the other.

The *longue durée* development of governmental power/knowledge structures, the machineries of visualization and territorial imagination, must not be thought of as a historical curiosity distant from events at the present day. On the contrary, a territorialization of provincial geographs has explicitly and concretely taken place as a series of regional administrative reforms first launched in Finland in the latter half of the nineteenth century. Since 1867, altogether 34 initiatives have been taken up in the highest governing bodies of the state (Pystynen 1993). The most authoritative attempts to accomplish a reform have been made by state-appointed committees. Over the last 125 years, as many as eight state committees have delivered their reports proposing reform, until finally in 1994 more political power and administrative tasks were transferred from the state's district administration to 19 provinces (Figure 6.3).

The committee reports exemplify well the institutional requirements of official discourse. Inherently of and for political power, official discourse lacks the contextual flexibility and contradictory qualities of the 'language games' embedded in other social activities. The strategies of knowing and representation, by which the committee reports have contained and reduced the diversity of provincial symbolisms, have been dictated by the need to territorialize space in a universal and universally applicable guise. The challenge has been to arrive at a legitimate spatial division amidst conflicting economic, symbolic, and political interests vested in provincial symbolisms.

The difficulty of meeting this challenge has been evident in the processes of instituting new administrative regions in the state's established county divisions. Setting up new administrative units has been arduous, and sometimes the conflicting interests have prohibited governmental decisions from being put into effect (Ylönen 1994). Municipalities have at times not been able to decide what county they belong to, while the government has not always been able to convince itself of the necessity for new administrative regions in the first place. As in European states at large, the reform of territorial divisions has been a conflict-prone area of policy demanding the most sensitive practice of statecraft to succeed.

So as to territorialize provinces in an acceptable manner, governmental discourse has resorted to an evolutionary conception of space. The idea of natural development of provinces has elevated the essence of these regions from the mundane world of human interests up to the transcendental realm of historical telos. As objectively knowable regions of the 'external reality,' provinces exist and appear the same to everyone. It is no surprise, then, that in the textual strategies of committee reports the provinces have been portrayed organically, as naturalistic outgrowths of history, culture, and human behavior ready to be territorialized and taken in political use (e.g. committee reports 1881, 1923, 1953). The Finnish case well illustrates what elsewhere has been described as the lasting value of organicist ideas to the powers that be (Hepple 1992: 141–146).

The proponents of reform have tended to view the process in terms of the victory of people's provinces over the state's administrative rule (e.g. Mennola 1990; Kirkinen 1991; Pystynen 1993). After all, when in 1994 the provinces were finally established as governmental bodies in charge of, for instance, regional policy, the state's control over local matters loosened somewhat. Yet,

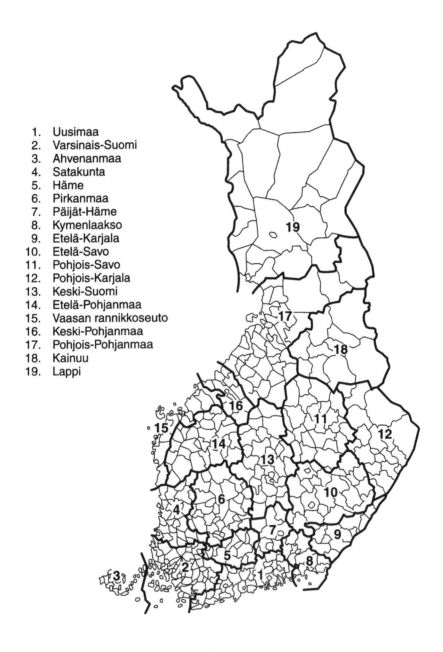

1. Uusimaa
2. Varsinais-Suomi
3. Ahvenanmaa
4. Satakunta
5. Häme
6. Pirkanmaa
7. Päijät-Häme
8. Kymenlaakso
9. Etelä-Karjala
10. Etelä-Savo
11. Pohjois-Savo
12. Pohjois-Karjala
13. Keski-Suomi
14. Etelä-Pohjanmaa
15. Vaasan rannikkoseuto
16. Keski-Pohjanmaa
17. Pohjois-Pohjanmaa
18. Kainuu
19. Lappi

Figure 6.3 The division of Finland into 19 provinces (source: committee report 1992).

contrary to this it can be argued that the governmentalization of provinces has not marked the fulfillment of their cultural potential, but rather sealed their territorialization and confinement in two interconnected ways.

First, the governmental discourse on provinces has by necessity been embedded in the abstract space of maps, statistics, and spatial analysis, which have enabled the objectivist representation of territory. Only in terms of this 'second reality', allegedly more real than the contestable social space, have the committees been able to define, demarcate, and objectify provinces and arrive at an official, universally valid regional division (Häkli 1994b: 193). In this sense, the state committees have *manufactured* provinces as much as they have re-presented them, and furthermore, have done this in their own exact, definite, and controllable image. Rather than false accounts of a true reality, governmental geographs have functioned as ideas that make things become real, that is, as geographical imaginations with tangible consequences (Painter 1995: 146).

Second, governmental discourses have followed their own logic grounded in institutional strategies unable to grasp the contexts of lived social spatiality, even when governmental geographs have been conflated with popular ones. This is evident already in the language of the reports produced by provincial authorities. Replete with politico-administrative jargon dealing with 'LFM-frames' and 'regional gateway-projects,' provincial issues have largely failed to address people other than the politicians, officials, and scientists involved. In other words, instead of bringing about a politico-administrative system of a 'more humane scale and quality,' the conflation of governmental and vernacular provinces has handed the provinces to a numerically small group of experts.

Yet, it should be noted that sometimes this conflation and confinement has taken place, but often it has not. The contradictory and contested ways of using provincial geographs constitute numerous blind spots, which evade governmental gaze and its suggestive power to conflate perspectives. These 'operational models of popular culture' (de Certeau 1984: 25), taking place in the deep space, may also be the occasion for popular resistance to the encroaching governmental rationality.

The revenge of the people?

The relations of power/knowledge embedded in geopolitical discourses can be very complicated, particularly when analyzed within a relatively homogeneous cultural environment, such as a particular state. Overt differences in the frames of meaning between institutional actors and various social groups may be subtle, and the forms of resistance almost invisible. Accordingly, it may be difficult to pin down the power relations inherent in discourse, or to account for forms of popular dissidence towards the official projections of space. In attempting to tackle this problematic, I have found Michel de Certeau's (*ibid.*: 24–31) discussion of the 'tactics' of resistance particularly helpful: 'the actual order of things is precisely what "popular" tactics turn to their own ends, without any illusion that it will change any time soon. Though elsewhere it is exploited by a dominant power or simply denied by an ideological discourse, here order is tricked by an art.'

Distinguishing between 'tactics' as ways of using, manipulating, and diverting spaces imposed by the 'strategies' of the powerful institutions, de Certeau sheds light on an important dimension in the relationship between the intellectuals (of statecraft) and (ordinary) people, one in which the supposedly repressed are in fact able to resist, albeit in a covert manner, the postulation of power and the imposition of knowledge 'proper' by official thought and action. Unlike institutional strategies engaged in the constitution of readable, controllable, fixed spaces, a tactic does not 'have the options of planning general strategy and viewing the adversary as a whole within a district, visible, and objectifiable space. It operates in isolated action, blow by blow . . . in short, a tactic is an art of the weak' (*ibid.*: 37).

Importantly, the spaces of tactical resistance are not restricted or reducible to the spaces of discourse. This is where the notion of deep space becomes significant. Even if we acknowledge the fact that everyday life is discursive just like the practices of statecraft, and that popular geographs can assume a geopolitical meaning like their institutional counterparts, the notion of deep spaces points at the different contexts and ways in which geographs are used in society at large. In discussing how the practice of everyday life is filled with propitious moments for 'the weak' to 'turn to their own ends forces alien to them,' de Certeau (*ibid.*: xix) points out that such an endeavor 'takes the form, however, not of a discourse, but of the decision itself, the act and manner in which the opportunity [to manipulate events] is "seized".'

Paying attention to the operational models of popular culture helps in understanding that the spaces discursively produced by political and administrative institutions are not necessarily resisted through an engagement in an overt political discourse (centered around, for instance, economic development, regionalism, or cultural values). The resistance may equally dwell in the countless instances of social, cultural, and economic life, that is, in deep spaces that evade the strategies of geopolitical discourses built upon the machineries of visualization. Still, it should be noted that in the institutionalization of provinces there is no balance of power between tactics and strategies: 'strategies are able to produce, tabulate, and impose these spaces . . . whereas tactics can only use, manipulate, and divert [them]' (*ibid.*: 30).

Tactics seldom leave traces that could easily be followed, such as committee reports, official documents, memoranda, and so forth. Hence, while the spatial patterns of contemporary provincial identities can be mapped (e.g. Palomäki 1968), the different ways of using provincial geographs among the larger population cannot be reached by means of standard questionnaires or Gallup interviews. These little nuances, which easily escape the analyst, are nevertheless very important when it comes to the ways in which people make use of spatial categories in constructing their identities and rationalizing their actions.

An example of how the contested meanings of provinces have been utilized in popular resistance to the state's institutional discourses is the most recent turn in the Finnish regional administrative reform. At the end of 1995, the Finnish government introduced a plan to reduce the number of state's administrative districts (*lääni*) from twelve into a mere five counties. Although no changes were proposed to the newly instituted provincial government as such, many in North-Karelia were provoked by the ensuing loss of the region's administrative status. The local political and economic elites, who mobilized against the reform, raised popular support by resorting to the image of province under threat by the central government. By conflating the concepts of *maakunta* (province) and *lääni* (county), which both refer to the North-Karelian region, yet to totally different governmental systems, the local elites were able to take advantage of a contradictory and contestable spatial imagery. In early 1996, a manifesto signed by 70,000 North-Karelians, roughly half of the adult population, was collected. The signatures were delivered to President Martti Ahtisaari in May 1996 by a group of local notables, including the president of the university, the city mayor, and the chief editor of a local newspaper (Figure 6.4) (*Karjalainen*, 8.5.1996).

Although politically not successful, the movement demonstrates the hidden powers of contradictory and contested popular geographs. In this case, popular identities and sentiments were brought into the discourses of the local elites who were fiercely arguing against the changes. The movement itself, of course, was tension-laden internally and unevenly structured in terms of power. In criticizing the uses of 'everyman' of 'anyone' in elite discourses, de Certeau (1984: 2) writes as follows: 'but when elitist writing uses the "vulgar" speaker as a disguise for a metalanguage about itself, it also allows us to see what dislodges it from its privilege and draws it outside of itself: an Other who is no longer God or Muse, but the anonymous.'

There are several ways of interpreting these enigmatic words, but in alignment with the arguments put forward in this chapter, one could read them as a satirical description of a society fraught with expert discourses. No matter what amount of expertise there is to support an argument, the everyman still has an important task as an 'oracle' whose authorization is recurrently resorted to when claims are made about society, life, politics, or the future. Hence, administrative divisions can also be argued for and against in the name of 'the people', that multipurpose faceless mass who need not be known, but whose feelings and wishes nonetheless are transparent to the elites. In the movement against the reform, there were governors who suddenly wished to be 'near the people,' scientists who expressed fear that 'the new larger districts lure people under the bureaucratic power of the state by offering public services,' and politicians who wished to 'protect the people's regional identities' (*Helsingin Sanomat*, 8.11.1996; *Karjalainen*, 11.11.1996; Kirkinen 1996).

In all, an important occasion for popular resistance to institutional discourses lies in the contested and contradictory voices of the 'ordinary people,' the endless

Figure 6.4 The delivery of signatures as reported in a local newspaper. The headline reads: 'No direct support for the struggle from Ahtisaari' (source: *Karjalainen* 8.5.1996, reproduced by permission).

murmur that in fact produces and reproduces regional identities and geographs. Even though lacking means by which to resist the territorialization of the Finnish provinces, they nevertheless dwell in deep spaces replete with blind spots, contradictory definitions, and heterotopic places empowering the everyday tactics of resistance: ways of using old, stereotypical, and 'defunct' provincial geographs in a manner that necessarily does not present itself as resistance. However, these very contradictions and tensions also open popular geographs for exploitation and political calculation. Therefore, it can be asserted that popular resistance is itself a contradictory process: while people mobilize against the state's governmental discourses, the movements themselves are structured by expert discourses that are involved in the reproduction of social power relations.

Conclusions

The provincial concept has provided the governmental discourse with a formula representing 'in the flesh' the ideal conditions for governmental rationality: the alliance of spatial demarcability, organic community and common good. Yet, paradoxically enough, as inscribed into governmental reports the official discourse has projected the Finnish provinces on a geometric, two-dimensional plane, and thus petrified an indeterminate, multidimensional symbolism into a visible and definite space of the politico-administrative system. In Pierre Bourdieu's terms, this event bears all the characteristics of legitimate symbolic violence: 'the power to constitute and to impose as universal and universally applicable within a given "nation", that is, within the boundaries of a given territory, a common set of coercive norms' (Bourdieu and Wacquant 1992: 112).

Importantly, this objectification and universalization of provinces has had consequences reaching far beyond the realm of symbolism, language, or imagery. The conflation of institutional and popular geographs discloses a persistent search for renewed governmental legitimacy through popular appeal: a spatial platform from which new political action could spring, new institutions be established, new political structures forged, new economic relations built (see also Lefebvre 1991: 275). While a strategic maneuver performed on paper, the reform discourse's seams have always been sewn in the corporeal world of social life, people, events, interests, and material projects.

The Finnish administrative reform is a showcase of the ways in which governmental and popular geographs are conflated in modern political practice. Needs and meanings that apply in politics and administration are discursively equated with 'everyman's' benefit and spatial imagination. While here approached through the Finnish context and in local circumstances, the case represents a routine procedure in modern geopolitical practice based on government by and through knowledge.

However, as my intention has been to show, popular geographs are not all vulnerable in the face of dominant discourses. In the myriad contexts of social life there are silent and scattered occasions for resistance to the official projections of territory. In also pointing out the relations of power within popular movements I have not wished to belittle their significance. Rather my aim has been to show that popular resistance to dominant discourses may just as well reside in 'noisy' and incoherent street parlance as in an organized resistance movement.

References

Alapuro, Risto (1988) *State and Revolution in Finland.* Berkeley: University of California Press.

Berdoulay, Vincent (1989) Place, meaning, and discourse in French language geography, in Agnew, J. and Duncan, J. (eds) *The Power of Place: Bringing Together Geographical and Sociological Imaginations.* Boston: Unwin Hyman, 124–139.

Bourdieu, Pierre and Wacquant, Loïc J. D. (1992) *An Invitation to Reflexive Sociology.* Cambridge: Polity Press.

Committee report (1881) Komitealta ehdoituksen tekemistä varten läänin-edustuksen aikaansaamiseksi (From a committee appointed for submitting a proposal for creating county representation) (no. 3).

Committee report (1923) Komitealta, joka on asetettu valmistamaan ehdotuksia läänien itsehallinnosta, alempiasteisista hallinnollisista tuomioistuimista sekä nykyisten läänien ja niiden hallinnon uudelleenjärjestämisestä (From a committee appointed for making a proposal for self-government in counties, courts of administrative justice, and existing counties and the reformation of their administration) (no. 13).

Committee report (1953) Aluejakokomitean mietintö (Report from the committee for regional division) (no. 4).

Committee report (1992) Maakuntaitsehallinnosta maakuntayhtymiin. Selvitysmies Kauko Sipposen ehdotus maakuntahallinnon kehittämisestä (From provincial government to provincial associations: Governmental delegate Kauko Sipponen's proposal for the development of provincial self-government) (no. 34).

Dalby, Simon (1991) Critical geopolitics: discourse, difference, and dissent. *Environment and Planning D: Society and Space* 9, 261–283.

Dalby, Simon (1993) The 'Kiwi disease': geopolitical discourse in Aotearoa/New Zealand and the South Pacific. *Political Geography* 12(5), 437–456.

Dalby, Simon and Ó Tuathail, Gearóid (1996) Editorial introduction. The critical geopolitics constellation: problematizing fusions of geographical knowledge and power. *Political Geography* 15(6/7), 451–456.

Dandeker, Christopher (1990) *Surveillance, Power and Modernity.* Cambridge: Polity Press.

de Certeau, Michel (1984) *The Practice of Everyday Life.* Los Angeles: University of California Press.

Dodds, Klaus-John (1993) Geopolitics, cartography and the state in South America. *Political Geography* 12(4), 361–381.

Edney, Matthew H. (1993a) The patronage of science and the creation of imperial space: the British mapping of India, 1799–1843. *Cartographica* 30(1), 61–67.

Edney, Matthew H. (1993b) Cartography without 'progress': reinterpreting the nature and historical development of mapmaking. *Cartographica* 30(2/3), 54–68.

Entrikin, J. Nicholas (1991) *The Betweenness of Place: Towards a Geography of Modernity.* Baltimore, Md.: The Johns Hopkins University Press.

Foucault, Michel (1991) Governmentality, in Burchell, G., Gordon, C. and Miller, P. (eds) *The Foucault Effect: Studies in Governmentality: With Two Lectures by and an Interview with Michel Foucault.* Chicago: The University of Chicago Press, 87–104.

Giddens, Anthony (1985) *The Nation State and Violence: Volume Two of A Contemporary Critique of Historical Materialism.* Cambridge: Polity Press.

Gustafsson, Alfred A. (1933) Maanmittarikunta ja mittaustyöt Ruotsinvallan aikana (Land surveyors and surveys during the period of Swedish rule). In *Suomen maanmittauksen historia. I osa, Ruotsinvallan aika* (The History of Land Survey in Finland. Volume I: The Period of Swedish Rule). Porvoo: Wsoy, 1–176.

Hacking, Ian (1991) How Should We Do the History of Statistics? in Burchell, G., Gordon, C. and Miller, P. (eds) *The Foucault Effect: Studies in Governmentality: With Two Lectures by and an Interview with Michel Foucault.* Chicago: The University of Chicago Press, 181–196.

Häkli, Jouni (1994a) Territoriality and the rise of modern state. *Fennia* 172(1), 1–82.

Häkli, Jouni (1994b) Maakunta tieto ja valta: tutkimus poliittis-hallinnollisen maakun-tadiskurssin ja sen historiallisten edellytysten muotoutumisesta Suomessa (Region, knowledge and power: the emergence of and historical preconditions for the politico-administrative discourse on provinces in Finland). Tampere: *Acta Universitatis Tamperensis, ser A* (415).

Häkli, Jouni (1998a) Discourse in the production of political space: decolonizing the symbolism of provinces in Finland. *Political Geography* 17(3), 331–363.

Häkli, Jouni (1998b) Cultures of demarcation: territory and national identity in Finland, in Herb, G. H. and Kaplan, D. (eds) *Nested Identities: Identity, Territory and Scale.* Rowman & Littlefield (forthcoming).

Harley, J. Brian (1988) Maps, knowledge, and power, in Cosgrove, D. and Daniels, S. (eds) *The Iconography of Landscape: Essays on the Symbolic Representation, Design and Use of Past Environments.* Cambridge: Cambridge University Press, 277–312.

Harvey, David (1989) *The Condition of Postmodernity.* Oxford: Basil Blackwell.

Heikkinen, Antero (1986) *Kainuun historia, osa III: Kulttuuri ja hallinto 1720–1980* (The History of Kainuu, Vol. III: Culture and Government 1720–1980). Kajaani: Kainuun Sanomain Kirjapaino Oy.

Helsingin Sanomat 8.11.1996 'Maaherrat tyrmäsivät lääniuudistuksen eduskunnassa' ('The governors knocked out the proposition for larger counties in the parliament').

Hepple, Leslie W. (1992) Metaphor, geopolitical discourse and the military in South America, in Barnes, T. J. and Duncan, J. S. (eds) *Writing Worlds: Discourse, Text and Metaphor in the Representation of Landscape.* London: Routledge, 136–154.

Hjelt, August (1900) Det svenska tabellvärkets uppkomst, organisation och tidigare verk-samhet. Några minnesblad ur den svensk-finska befolkningsstatistikens historia (The beginning and early activity of the Swedish statistical institution). *Fennia* 16(2), 1–109.

Johannisson, Karin (1988) *Det mätbara samhället. Statistik och samhällsdröm i 1700-talets Europa* (The Measurable Society: Statistics and the Utopian Society in the 18th Century Europe). Arlöv: Norstedts Förlag.

Jussila, Osmo (1987) *Maakunnasta valtioksi: Suomen valtion synty* (From a Province into a State: The Birth of the Finnish State). Porvoo: Wsoy.

Jutikkala, Eino (1962) *A History of Finland.* New York: Frederick A. Praeger Publisher.

Jutikkala, Eino (1989) Colonisation and the roots of the Finnish people, in Engman, M. and Kirby, D. (eds) *Finland: People, Nation, State.* London: Hurst & Company, 16–37.

Kain, Roger J. P. and Baigent, Elisabeth (1992) *The Cadastral Map in the Service of the State: A History of Property Mapping.* Chicago: The University of Chicago Press.

Kaplan, David (1994). Two nations in search of a state: Canada's ambivalent spatial iden-tities. *Annals of the Association of American Geographers* 84(4), 587–608.

Karjalainen 8.5.1996 'Ahtisaarelta ei suoraa tukea läänitaisteluun' ('No direct support from Ahtisaari to the struggle for county').

Karjalainen 11.11.1996 'Tuomioja arvostelee suurlääniesitystä' ('Tuomioja criticizes the proposed large counties').

Kirkinen, Heikki (1991) *Maakuntien Eurooppa ja Suomi* (Finland and the Europe of Provinces). Helsinki: Otava.

Kirkinen, Heikki (1996) Suurläänit riistävät ihmisiltä itsehallinnon (The larger counties deprive self-government of the people). *Helsingin Sanomat* 25.10.1996.

Kolbe, Laura (1996) *Eliitti, traditio, murros: Helsingin yliopiston ylioppilaskunta 1960–1990* (The Elite, Tradition and Change: The Student Union of the University of Helsinki 1960–1990) Helsinki: Otava.

Kovero, Martti (1940) Tilastollisen päätoimiston perustamisen alkuvaiheet (The foundation and early days of the statistical central bureau). *Kansantaloustieteellinen aikakauskirja* (12), 219–248.

Kuusi, Sakari (1933) Maataloudelliset uudistusvirtaukset ja maanmittauslaitos Suomessa 1725–56 (Agricultural reforms and land survey in Finland 1725–56), in *Suomen maanmittauksen historia. I osa, Ruotsinvallan aika* (The History of Land Survey in Finland. Volume I: The Period of Swedish Rule). Porvoo: Wsoy, 1–55.

Latour, Bruno (1986) Visualization and cognition: thinking with eyes and hands. *Knowledge and Society* 6, 1–40.

Lefebvre, Henri (1991) *The Production of Space*. Oxford: Basil Blackwell.

Mennola, Erkki (1990). Maakunnan vapaus (Province's freedom). *Acta Universitatis Tamperensis, ser A* (301).

Neuvonen, Lasse (1990) Väestönkasvun kehityspiirteitä (Trends in population growth), in Heikkinen, A. (ed.) *Maakuntien nousu* (The rise of provinces). Kuopio: Kustannuskiila, 13–24.

Ó Tuathail, Gearóid (1989) *Critical geopolitics: the social construction of space and place in the practice of statecraft*. Unpublished PhD dissertation, Syracuse University.

Ó Tuathail, Gearóid (1994) (Dis)placing geopolitics: writing on the maps of global politics. *Environment and Planning D: Society and Space* 12, 525–546.

Ó Tuathail, Gearóid (1996) *Critical Geopolitics: the Politics of Writing Global Space*. Minneapolis: University of Minnesota Press.

Paasi, Anssi (1986) Neljä maakuntaa: maantieteellinen tutkimus aluetietoisuuden kehittymisestä (Four provinces: a geographical study on the development of regional consciousness). Joensuu: University of Joensuu. *Publications in Social Sciences* (8).

Paasi, Anssi (1996) Alueellinen identiteetti ja alueellinen liikkuvuus: suomalaisten syntymäpaikat ja nykyiset asuinalueet (Regional identity and regional migration: birth places and present dwelling places of the Finns). *Terra* 108(4), 210–223.

Painter, Joe (1995) *Politics, Geography and 'Political Geography': A Critical Perspective*. London: Edward Arnold.

Palomäki, Mauri (1968) On the concept and delimitation of the present-day provinces of Finland. *Acta Geographica* 20, 279–295.

Porter, Theodore M. (1986). *The Rise of Statistical Thinking 1820–1900*. New Jersey: Princeton University Press.

Pred, Allan (1986) *Place, Practice and Structure: Social and Spatial Transformation in Southern Sweden: 1750–1850*. New Jersey: Barnes & Noble Books.

Procacci, Giovanna (1991) Social economy and the government of poverty, in Burchell, G., Gordon, C. and Miller, P. (eds) *The Foucault Effect: Studies in Governmentality: With Two Lectures by and an Interview with Michel Foucault*. Chicago: The University of Chicago Press, 151–168.

Pystynen, Erkki (1993) Maakuntaitsehallinnon ongelma Suomessa (The problem of provincial self-government in Finland). *Kunnallistieteellinen aikakauskirja* (3), 211–223.

Rabinow, Paul (1989) *French modern. Norms and forms of the social environment*. Cambridge, Mass.: MIT Press.

Rasila, Viljo (1993) Pirkanmaan historia (The history of Pirkanmaa). Tampere: *Pirkanmaa Association Publications* B (12).

Revel, Jacques (1991) Knowledge of the territory. *Science in Context* 4(1), 133–161.

Routledge, Paul (1996) Critical geopolitics and terrains of resistance. *Political Geography* 15(6/7), 509–531.

Salonen, Erkki (1974) Maakunnallinen tiedonvälitys (Provincial media). *Kotiseutu* (6), 199–206.

Smith, Neil (1990) *Uneven Development: Nature, Capital and the Reproduction of Space.* Oxford: Blackwell.

Strömberg, John (1987) Ylioppilaat (The students), in *Helsingin yliopisto 1640–1990* (The University of Helsinki 1640–1990). Helsinki: Otava, 291–354.

Widmalm, Sven (1990) *Mellan kartan och värkligheten. Geodesi och kartläggning 1695–1860* (Between map and reality: Geodetic survey and mapping 1695–1860). Uppsala: Centraltryckeriet.

Wilson, William A. (1976) *Folklore and Nationalism in Modern Finland.* Bloomington: Indiana University Press.

Wittrock, Björn (1989) Social science and state development: transformations of the discourse of modernity. *International Social Science Journal* (122), 497–508.

Ylönen, Ari (ed.) (1994). Keskusteluja Pirkanmaan tulevaisuudesta (Discussions of the Pirkanmaa's future). University of Tampere, *Working Papers of the Social Sciences Research Institute* (10).

7

REEL GEOGRAPHIES OF THE NEW WORLD ORDER

Patriotism, masculinity, and geopolitics in post-Cold War American movies

Joanne P. Sharp

Introduction

The end of Cold War and the loss of its communist alter ego has apparently heralded for hegemonic American culture a period of crisis that has raised profound questions about both national identity and purpose (Engelhardt 1995). American politicians (e.g. Christine Todd Whitman (see Carlin 1995: 9)), radical intellectual commentators (e.g. Ó Tuathail and Luke 1994) and popular culture gurus such as movie director Paul Verhoeven (*RoboCop, Total Recall*) alike have suggested that the USA is desperately searching for a new enemy against which to define itself.

That the Cold War was constitutive of American self-identity rather than a threat to it has now become quite clear in the clamor to find an alternative source of danger against which to define the boundaries of the USA. America's Cold War goal of containing the USSR can be understood in terms of a moving frontier between the USA and its alter ego, not unlike the frontier that characterized narratives of the country's initial western expansion. If there is a story about the nation's origins at the origin of every nation (Bennington 1990: 121), then for America this originary story is of the Frontier; the apparent closure of this narrative at the end of the Cold War must have profound effects on US identity.

Global geopolitics has dissolved from its apparent Cold War coherence into a series of political geographies at all scales seemingly rendered as fragmentary sound bites: American economic decline, ethnic cleansing in the Balkans, multinational exploitation, rising crime rates, nationalist revival in the former Soviet Union, and so on, stand side by side as unconnected events in nightly news reports. Mackenzie Wark (1994: 21) has suggested that the 'media vector renders equivalent a tiny gesture or a major battle.' Popular geopolitics would appear to be entirely fragmentary, reduced to media vectors flowing in sound bites between globalized media outlets.

Commentators have attempted to 'triangulate' (Ó Tuathail 1996b) new territorial geopolitics from the coordinates of Cold War reasoning: as Dolan (1994: chapter 3) has argued, although the Cold War may be over, Cold War metaphysics is not.

However, even the more informed spectators of the new world order often seem to despair of its complexity and fragmentation, some even apparently longing for the old stability of the Cold War. Fukuyama (1989), for instance, seems to lament the lack of challenge offered to America's historic sense of mission and destiny in this new period. John Mearsheimer (1990) threatened that soon we would all start to miss the stability offered by Cold War binary geopolitics now that the world had dissolved into a fragmentary array of competing regional powers and interests.

The mass media produce geo-graphs of world politics and international relations for public consumption alongside the more erudite highbrow texts mentioned above. Media provide global cognitive maps that allow their audiences to slot the sound bites into a larger picture and to draw out connections and causalities between them. In many cases, the media have provided more imaginative reconceptualizations of international relations, and demonstrate a greater willingness to cast off the older ways of understanding global politics, than formal theorists of international politics.

There are a number of narratives rescripting America's map of the world order, and I cannot hope to do justice to this range in an essay of this length. The narrative I will consider here is what Susan Jeffords (1989) has termed the 'remasculinization of America.' Jeffords argues that reworking of national identity is achieved through renegotiations of gender identities and relations. Although her work discusses film narratives of Vietnam, I think the insights it raises can be brought to wider film production in the USA, which is the aim of this chapter. Masculinity is not incidental to film narratives; as Richard Dyer (1993: 111) has noted, even when it is not an overt topic, 'we are looking at the world within its terms of reference.'

After a discussion of the role of the mass media in the construction of popular images of identity and geopolitical relations, I will consider the ways in which post-Cold War representations of relations and events are entwined with renegotiations of masculine identity, particularly patriotism.

Geopolitical maps and popular culture

Although the role of the media in deciding upon agendas for domestic political debate and action is well documented, there is less acknowledgment of the effect of popular forms of cultural production on the 'high politics' of international relations. Indeed, some commentators have gone so far as to deny that media have any impact whatsoever (for a discussion of this see Dalby 1994). But it is possible to see the influence of these productions on state practice, whether directly in the central role of CNN as a source of information during the Gulf War for American leaders, or indirectly in the role of popular culture in the construction of hegemonic cultural values that shape both the actions of politicians and the expectation of societies.

The neo-realist domination of international relations and geopolitics has focused upon the state as *the* individual actor in international events. As Cox (1986: 216) has remarked, neo-realist theory in the United States 'has treated civil society as a constraint upon *raison d'état*, which is conceived of, and defined as, independent of civil society.' On the other hand, a Gramscian approach opens up the state actor to expose its constituent elements, internal contradictions and power relations. It exposes the concealed workings of civil society in the construction of international politics.

The ever more powerful mediascapes of contemporary global capitalism mean that any reliance upon a division between 'high' cultures of statecraft and 'low' or 'middlebrow' cultures of the media and popular culture is becoming ever more tenuous. The circulation of images in the media, and the almost total mediation of any but the most local political activity, has lead to an ever-increasing importance of the media in our (Western) everyday understandings and conceptualizations of global geopolitics.

First-hand experience is not dependent upon physical presence. Few hear much politics from the politicians who supposedly make policy and set the agenda. Instead, events are reproduced in the media, and to attract attention these are made into spectacles, so that political reporting 'continuously constructs and reconstructs social problems, crises, enemies, and leaders' as spectacles of media entertainment (Edelman 1988: 1). What this has led to is the 'creation, transmission, and adoption of political fantasies as realistic views of what takes place' (Nimmo and Combs 1983: xv). More often than not, it is the media that are taking the initiative and political figures 'increasingly find themselves responding to rather than initiating public issues' (Denton 1992: xiv). Political issues and events are linked to media-friendly individuals, the major actors who most competently deliver the scripts of geopolitical reasoning to the public and to other players. To a greater and greater extent, the public are being asked to select between actors rather than policies or ideologies. Reagan, 'the Movie President,' typified this in that he spent one-third of the working hours of his time in office answering 'fan mail' (Gibson 1994: 266).

Increasingly, questions of national identity are becoming hybridized: in the media, notions of American national identity are always already entwined with the demands of Americanized global capitalism colonizing and commodifying all aspects of global culture. This obviously complicates questions regarding intended audience. Although the narration of much American popular culture is undoubtedly centered on an understanding of national identity, it is written in such a way as to reproduce a more generic sense of identity for consumption within the globalized media, 'an image so generic, so affecting, so ubiquitous,

and so empty that it will no longer be recognized as American, it will just be'
(Barber 1995: 94). For example, films such as *Braveheart* and *Rob Roy* are
ostensibly about Scottish national identity (and indeed it is considered that their
impact may have been to boost this cultural identity), but the narrative structure
and themes produced in them (freedom from colonization, power of heroic figures over feminized elites, and foundation of heroic citizenship in the protection
of women, in this case) are central to hegemonic visions of American identity
written into the narratives comprising that nation's origins. The structure of
American national narration is projected onto the stories of other nations. The
narrative structures and themes described above are rendered natural and in-
disputable, as non-politicized myths. American aspirations and culture are
projected as inherently human characteristics. This is not to argue, however,
that in all places the films' narration is unproblematically internalized: violent
spectacle and special effects translate well into other cultures, better than the
narrative that connects them.

Wark suggests that audiences are becoming wise to the distortions and outright lies of the 'instant history' presented by sensationalist media; of such
'media spectacles' as the Gulf War, for instance. However, it is the media itself
that provide the possibility of critical analysis. Audiences can be critical of media
distortions only 'through other media. Slower and more considered media, like
articles in the highbrow monthlies, or earnest, truthful hour-length documentaries, but media all the same' (Wark 1994: 6).

Ó Tuathail claims that geopolitics 'has itself become a televised and entertainment phenomenon' (1996b: 250). Certainly the power of the media was made all
too evident during the Gulf War, when viewers were continually reminded that
the conflict was running to 'schedule,' a term more reminiscent of TV programming that the requirements of war tactics. In other words, now that the media
have attained such importance, for an event to be considered important or
significant it must be registered and acknowledged as such in the media. The distinction between the truly factual and fictional blurs in this society of the media
spectacle. Certainly, it is the sensational(ized) elements of any event that achieve
reportage; consider the issues debated in global environmental debates that get
the greatest media coverage: it is always those forecasts that project the most damage and destruction that are covered. The question of climate change becomes
one of the spectacle of disaster rather than a more rational debate on the range of
possible effects and occurrences that might arise in different future scenarios.

It is important to point out that although global politics is mediated, this does
not mean that all media present a seamless or coherent image. The media are a
diverse group of knowledge producers, influenced by differential material support, dependence on the market, political backing, type of product and intended
audience, amongst other things.

This chapter will consider the geopolitics presented in post-Cold War films.
There is obviously a question of factuality when considering such a genre. None
of the films that I will consider claims to be factual, none claims to offer a true

155

story. We are asked to suspend disbelief wilfully from the opening scenes to the final credits, yet these films are powerful sources in understanding the current American reconstruction of maps of world politics.

Film sites are of significance to the construction of geopolitical narratives and maps because of the permeability of boundaries between producers of knowledge in the USA. There is a flow of information and knowledge between works of fact and fiction, between high, middle and low brows, and between all of these and the official geopolitics in statecraft, in that each attempts to identify a narrative from the chaos of the international realm. As Der Derian suggests:

> we can see how the American popular literature of international intrigue shares and privileges a narrative of the truthsayers of the security state: beyond our borders the world is alien, complex, practically incoherent – an enigma but one which can be unraveled by the expert story-teller.
>
> (1992: 41)

A national culture represents a common source of narratives and understandings that provide a sense of belonging. These narratives and beliefs are drawn upon to define and explain new situations and their importance to individuals in the community. Both politicians and the media are storytellers – and in order for their stories to be accepted by their audience they have to resonate with meta-level hegemonic cultural values. Those values that flow between sectors of hegemonic culture are those that facilitate the narration of events and processes in an acceptable or meaningful way in the context of national self-identification. At the end of the Cold War, it is possible to see a common struggle to redefine a cartography based on traditional Cold War narratives – but also recognition of the limitations of these stories. The solutions found in fictitious and factual media will not be entirely separate.

Fictional productions, especially high-budget Hollywood thrillers, rely upon a 'reality effect' to make their film narratives credible and exciting. The end of the Cold War has required new geographies of danger in order to make believable the conflict and tension in the stories. Such films have drawn upon narratives of increased danger from fragmentation of superpowers and danger. However, film texts are of particular interest because this process of intertextuality works both ways, as Der Derian explains:

> We are witnessing changes in out international, intertextual, inter*human* relations, in which objective reality is displaced by textuality (Dan Quayle cites Tom Clancy to defend anti-satellite weapons) . . . representation blurs into simulation . . . imperialism gives way to the Empire of Signs (the spectacle of Grenada, the fantasy of Star Wars serve to deny imperial decline).
>
> (1992: n1)

156

There is clear evidence of the efficacy of the realism of such movies. It is accepted that the film *The Manchurian Candidate* provoked the CIA to pursue the possibility of controlling secret agents without their knowledge, and to envisage methods of combating the prospect of the Soviets having already developed such technology. The film starts from the plausible premises of both the Cold War exchange of spies and the 'brainwashing' of soldiers in the Korean War. It then develops the possibility of the programming of people who are unknowing of their condition and therefore undetectable, the perfect spies.[1] Ultimately, the film influenced the political situation that it initially sought to reflect. More recently, Reagan suggested enthusiastically, after seeing *Rambo* that the next time American hostages were taken in the Middle East he would know what to do about it.

Obviously, the media are always rewriting the world order – if only to allow it to remain constant – but at times of great change, their scripting and siting of it is most evident. In this paper I will take the end of the Cold War in America as a theme around which to explore the role of various media in the reproduction of hegemonic values that relate to the international but, in doing so, also construct an image of American national identity. For although on one level the Cold War presented the immediate political threat of territorial expansion and the potential of missile destruction, on another level it is possible to read it as something not threatening to 'America' but in fact constitutive of it. Through the repetitive drawing of Cold War geopolitics, the imaginary essence of 'America' has been produced. The containment of the USSR acted simultaneously to contain 'America': it acted to discipline the myriad possible characterizations of 'America' into a coherent moral agent. The Cold War offered a set of scripts in a quite literal sense: it wrote parts for the 'bad guys' and for the 'good guys' (or 'black hats' and 'white hats' in Reagan's western-influenced terminology) with a set of stories that demonstrated the nature of each: it meant that viewers knew where various actors fitted in. Viewers could understand the role of each character(-type) in the unfolding drama. They would anticipate the conclusion: there was an expectation of the triumph of good over evil. The end of the Cold War has disrupted the repetition of this geopolitics and the identity it reflected into the space of the USA.

John McClure has suggested that the end of the Cold War has presented authors of fictional works with a dilemma. The ideological geography of the Cold War dualism had facilitated the construction of spaces of conflict and terror, and safe spaces of home. The destruction of this geopolitics has forced authors to reinscribe the world with a structure that audiences will accept as credible within which they can locate characters and plots. McClure thus sees the end of the Cold War as offering novelists a fundamental challenge – and I think that his argument can be extended to the scripters of global geopolitics more generally. They are presented with a scenario devoid of commonly accepted dangerous areas within which heroic (masculine) subjectivity is constructed through narratives of struggle and triumph. As McClure suggests:

Without the unordered spaces, or spaces distorted by war, it is impossible to stage the wanderings and disorientations, the quests and conquests and conversions, the ordeals and sacrifices and triumphs that are the stuff of romance. The ultimate enemies of romance, then, are not the foreign foes confronted on the field of battle in the test itself, but the foes held at bay by these essential antagonists: the banal, quotidian world of calculation and compromise from which the heroes of romance are always in flight.

(1994: 3)

McClure discussed relatively highbrow works of contemporary literature, but the end of Cold War romance can also be seen to run though a range of popular culture from supposedly educational and definitely moralistic middlebrow productions such as the magazine *Reader's Digest* (Sharp 1993, 1996) to the more clearly entertaining popular medium of Hollywood movies.

Further, the media are creators of 'public-ity'[2] in that they can be considered as institutions through which individuals are interpolated as political subjects into 'publics,' in which individuals are incorporated into the political process. Media present the importance of events to the operation of international politics, but also to the individual observer to explain why he or she should care what is going on across the globe: now that capitalism has triumphed, why should Americans care what is happening in Central America or in Africa? The media offer explanations: they offer geographies linking the audience in their living rooms, at work or in the cinema with those on the battlefields, in the prison camps, on the firing lines. The geographic narratives produced in these media act to tie places together through causal links of economy, politics, espionage, or more simply in a cinematic cut between scenes: '. . . meanwhile in Iran . . .'

Perhaps more obviously than other media, films position the viewer in relation to the characters and narratives being displayed. Laura Mulvey's (1989) celebrated paper on voyeurism in the cinematic tradition explains the role of cinematic techniques in maintaining a particular relationship between audience and film. Mulvey argues that film narrative constructs a particular viewing space for the audience that directs the viewer towards one rather than other interpretations of the screen action. This position is characteristically a variation on the theme of heterosexual masculinity, for example in thrillers offering the viewer superiority towards female characters, 'because we either know more than her (we know that the psychopath is there but she hasn't spotted him yet), or because we can see what any sensible person would do but she, foolishly and pathetically, doesn't' (Dyer 1993: 118).

Of course, there is always room for resistance, room for alternative readings – but often movies have the effect of reinforcing what is already known: the fictitious stories pick up on other media, either resonating with or amplifying the already known. This is particularly true for political thrillers, which require a degree of realism in order to convince the audience of the plausibility of their tale.

Film scripts

In this part of the chapter, I want to look at the geopolitical scripts of/for the new world order offered by popular cinema. I will concentrate on post-Cold War thrillers, those films that most directly engage questions of geopolitics and national identity and purpose. I want to use the concept of 'scripts' in two connected senses. On the most obvious level, I mean scripts in their literal sense of film narrativization. On a more metaphorical level I want to look at these scripts as a set of directions for the performance of global politics. Such directions participate in the construction of 'the real' in that subsequent events are interpreted in terms of them. At any time, one script dominates over another; one way of explaining or narrating a situation becomes primary, it is chosen rather than alternative forms of explanation. In this sense, the dominant script becomes accepted as a value-neutral description: 'One can understand scripts, in a Gramscian fashion, as particular productions of "common sense" upon which a consensual political mythology can be constructed' (Ó Tuathail 1992: 158). However, as Gramsci theorized for hegemonic culture more generally, hegemony is never complete: alternative scriptings and interpretations of situations are always available. I will return to this in the conclusion.

I have attempted to choose some of the most successful – and so probably the most influential – of the range of films. Of the most obviously geopolitical scripts, I have focused upon the series of films made from Tom Clancy's novels. Clancy's work has been among the most popular of this genre in recent years, drawing large audiences and being commended by American leaders, including Dan Quayle and Ronald Reagan. With the erosion of Cold War dualist geopolitics, Clancy places danger within the previously safe space of America. He fractures the traditional espionage division of global space so that the central location of danger moves from Russia/USSR to allied governments (in *Patriot Games* (1992)) to the heart of America, the White House, in *Clear and Present Danger* (1994) and the made-for-TV version of his book *Op Center* (1995). The movie of John Le Carré's book, *The Russia House* (1990), and the submarine thriller, *Crimson Tide* (1996), can also be regarded as broadly within this genre. I have also decided to look at less obviously 'geopolitical' films for their encodings of American patriotic identity at the end of the Cold War, and again I have engaged with a handful of the most influential, in this case *Independence Day* (1996), *The Rock* (1996) and, to a lesser extent, *Mission: Impossible* (1996).

What I want to pull out from this apparently diverse list of films is the project of redefining a sense of national identity in the apparent chaos of international relations in the 1990s. Following Susan Jefford's work, I have chosen to organize my analysis of these themes through an engagement with a project of 'remasculinizing America' (Jeffords 1989), in other words to engage with the re-establishment of a hegemonic masculinist sense of patriotism.[3] This is achieved in two main ways in the films listed above. First masculinity is rewritten through the exclusion of women from the film narratives. The heroics of individual triumph are a predominantly male preserve. Similarly, many scripts are centered around a male bonding, where male characters overcome their differences in their attempts to overcome danger. Women are simply prizes to be protected. In *The Rock*, the two central characters generate a bond as a result of their adventures. At one point, one, Mason, chastises the other, Goodspeed, for his lack of confidence. 'You'll do your best?' he questions, 'Losers always say that. Winners go home and fuck the prom queen.' Goodspeed replies: 'Carla [his fiancee] was the prom queen,' confirming his credentials to the other. Gender relations are engaged with more generally in films' attempts to rewrite hegemonic male values of patriarchal power and protection onto a society perceived to have been compromised both by the dissolution of the moral evil of the USSR and an America seen to have embraced both multiculturalism and women's equality to the detriment of 'normal' values.

The second invocation of remasculinization is more metaphorical. Here narratives are structured around heroic individuals' attempts to beat a feminized bureaucratic state structure.

Masculinity and moral relativism

This narrative takes on two themes: first, the dangers of moral equality with the previous alter ego of the USSR that emerges from the apparent chaos of a multi-polar world order, and, second, attempts to reinvigorate a strong sense of patriotism from what is perceived on the right of center as the moral relativism of contemporary American society.

Chaos/disorder in the international

The concept of a script of chaos suggests that in the realm of international relations, any neat division between the 'good guys' and the 'bad guys' has broken down. The reliable structure of the Cold War has deserted global politics, and in the absence of this geopolitics, international relations are often presented in films as being in a state of disarray.

Much writing on the end of the Cold War suggests that the USA required a new enemy to take the place of the Soviet Union. A number of others have been highlighted as possible alternative threats to the American state. Terrorists, especially 'narco-terrorists' and Middle Eastern fundamentalists, have been of particular interest. Terrorism has the potential to present the USA with the same scenario of total war as did the Cold War: perpetual vigilance and pre-emptive action are required to combat what is often described as an incessant threat. Anyone could be a terrorist just as anyone in the past might have been a communist. As in America's fight against communist subversion during the Cold War, for the USA in the late 1980s anti-terrorism could not take the form of a traditional military battle. There was no front line, in that terrorist intervention might be enacted anywhere throughout society.

However, it seems that for big budget Hollywood films, terrorists have presented too weak an opponent in comparison with the might of the Evil Empire of the USSR. In the film *Patriot Games*, international links had to be forged between terrorist groups in order to present a sufficiently powerful threat:

> Apparently sensing that the IRA isn't good enough on its own (pains are taken to point out that this group has broken off from the IRA proper; one wonders if the sensibilities of Irish-Americans were a consideration), the Irish terrorists are teamed for a while with Middle Eastern terrorists, who on their own don't really seem to scare Americans much.
>
> (Baseline's Movie Picture Guide Review 1993)

Evidence of the disarray of the world system is perhaps clearer in the narrative structure of some films. In *Clear and Present Danger, The Rock, Mission: Impossible* and *Op Center* in particular, the lines of good and bad are unclear. Political relations – legitimate and covert – are represented to range across territories and involve an array of secret and/or private espionage operations. In *Clear and Present Danger*, the division between Us and Them is unclear: a drug lord tells a member of Congress that 'You and I want the same thing.' The intrigue is tied to the President, who is clearly manipulated by his aides, and will lie to Congress and deal with drug lords if it gets him a second term in office. Danger is not something clearly located at the borders of the USA. In *The Rock*, the audience is presented with an American state that has secretly undertaken illegal military interventions across the globe. This internalizing of danger is also evident in the seduction of the US President by TV anchorwoman Kate Michaels in *Op Center*. Here the ultimate danger to America is manifest: located in the figure of the President of the USA himself. Michaels was using the President to attain secret information for a conservative Israeli political faction. The viewer is shown Michaels reading files marked 'Top Secret' in a bedroom while the president showers, presumably after sex. This device presents a powerful cinematic narrative in its linkage of the long-held cinematic fear of female sexuality and woman as seducer with the power of the media to seduce and manipulate the political sphere.

Noting this political confusion as a symptom of postmodern politics, Frederic Jameson (1992: 16) suggests that it

> . . . is at the point where we give up and are no longer able to remember which side the characters are on, and how they have been revealed to be hooked up with other ones, that we have presumably grasped the deeper truth of the world system (certainly no one will have been astonished or enlightened to discover that the head of the CIA, the Vice President, the Secretary of State, or even the President himself, was secretly behind everything in the first place).

The sense of disorientation and confusion mentioned by a number of critics is reproduced in film narratives that twist and turn, producing enemies and danger in the most unlikely of places.

Another effect of this chaos is represented in narratives that suggest that the transition from one geopolitical system to another had not been complete. The dramatic tension in such films emerges from the danger that might arise from the use of inappropriate maps of world politics: someone following the wrong script, acting as though the Cold War was still with us rather than relying upon the new script structures around *détente, perestroika, glasnost* and the new world order.

The film *The Hunt for Red October* offers a clear initial picture of some of this confusion. Despite in some respects being a last remnant of the Cold War, it also indicates the fissures where Cold War narratives were beginning to lose coherence. Released in 1990, it told of a top Soviet Navy skipper, Ramius, leaving dock in a new, and virtually undetectable, Soviet submarine, *Red October*. The American Navy charted the sub leaving dock and then it disappeared. The Soviets too lost touch. Fearing a defection, the Soviets try to convince the Americans that Ramius has had a mental breakdown and is heading to the US coastline to start a one-man World War III. They request American help in capturing *Red October*. However, CIA agent Jack Ryan (also the protagonist of *Patriot Games* and *Clear and Present Danger*), an 'expert' on Ramius, feels that he may indeed be defecting to the West with the new technology. The remainder of the plot revolves around Ryan's attempts to convince the powers-that-be of his belief. The film neatly illustrates the anxieties of this period of transition, the anxieties of letting go of the natural reactions of the last half century.

The dramatic tension in *Crimson Tide* emerges from the disjuncture between the old-style career naval officer, Ramsey, and the family man, Hunter. The tension is between Ramsey, a man who leaps heroically into each situation, following orders without question, and Hunter, who has never seen active combat but has a list of academic accolades, the man who always asks 'why' of any order. Although it is the more feminized figure of Hunter who saves the world from renewed conflict, the film does present Ramsey as a tragic figure, as a man out of time unable to operate in the chaotic and fragile system of the new world order. But the film also produces a sense of loss: it was in Ramsey's world where right and wrong could be differentiated and ultimately where someone could make a difference. In the new world order, negotiation and compromise seem all that is possible: difference has to be tolerated.

Ramsey explains his irritation with the new Russian nationalist leader, Radchenko, early in the movie. He claims that in the days of the Cold War, 'the Russians could be depended upon to do whatever was in their own best interests, but this Radchenko's playing a whole new ball game, a new set of rules.' Similarly, in the film *Patriot Games*, the terrorists seem irrational and obsessive rather than following the agreed set of power political performances that scripted the actions of such figures in the stories that constituted the geography of the Cold War.

These and other films show the danger of a renegade undermining the fragile *détente* of the new world order. These characters play a similar role to that of the brainwashed character in *The Manchurian Candidate*: they are programmed how to react to certain situations, programmed to interpret situations and then act from the point of view of the Cold War. These films suggest that it is such Cold War figures that most threaten the fragile stability of the new world order. Characters face the chaos of the changing world, unable to control the fluidity, uncertainty and flows emerging in the new world order in the confident, clear ways of the past. McClure's heroic figures are rendered impotent in the spaces of flows that resist the territorial–moral encodings of the past.

This fragmentation is apparently manageable only in science fiction films, where the Other transcends the differences between humans. The turning point for the President in *Independence Day* was when the captured alien tells him that there can be no negotiations and no peace: humans must die. Here a geopolitical territory of good and evil is reinscribed with Cold War certainty. The President's uplifting speech to the fighter pilots before the final battle scene bore a striking resemblance to Reagan's speeches in which he spoke longingly of the arrival of a truly alien threat to bring the world together. Addressing the National Strategic Forum in 1988, he stated that:

> I've often wondered what if all of us in the world discovered that we were threatened by a power from outer space – from another planet. Wouldn't we all of a sudden find that we didn't have any differences between us at all – we were all human beings, citizens of the world – wouldn't we come together to fight that particular threat?
>
> (quoted in Der Derian 1992: 126n)

In *Independence Day*, the climax offered just that: the world united behind America (even – and symbolically – its old enemies, Russia, Vietnam, Iraq) to fight the new enemy and celebrate in real American style in a restyled Fourth of July. Again, Vietnam is refought in the media and America wins, not in the unpolitically correct violent spectacle of *Rambo*, but in a spectacle of global cooperation behind America's lead. As the movie progresses, one after another of America's old enemies lines up behind the warrior-nation, as if admitting that the USA was right all along, or if not that much, that America had finally won their admiration and support.

Patriotism in a multicultural America

Some films have sought to control and reorder the chaos of international politics. As already mentioned, the film narratives themselves offer crude maps of new world orders. But these attempts at reordering are often also represented metaphorically in the sexual relations between the protagonists in the films. Increasingly, the heroes attempt to maintain domestic order and structure (protect or reunite families) rather than establish (hetero-)sexual union at the narrative climax of the films (of which the James Bond films have surely been the epitome[4]). Such scripts promote not conquest but protection.

In both *Patriot Games* and *Clear and Present Danger*, the hero figure struggles to protect his family from harm, Hood strives to hold his marriage together in *Op Center*, and Hunter is established as a family man in the early scenes of *Crimson Tide*. Film analyst Steven Prince has suggested that images of the family reconstituted serve as a symbol for a nation healed (1992: 66). This rescripting of the family (a center for American national identity and the cause of great concern for hegemonic projects at present) also clearly flags a desire to return to a nuclear-style family protected and ordered under the patriarchal sign of the father.

But it is *Independence Day* that offers the clearest example of the dangers of a fragmented society and the power of its reunification. At the start of the film, when America is welcoming the aliens (with the naiveté of so many peaceful liberals in the détente period with the USSR?), the cast is peopled with individuals representing the apparent decay of American society: the alcoholic Vietnam veteran, the single mother working as a stripper, the fighter pilot who cannot commit to serious relationships, the brilliant scientist who lacks ambition, and the First Lady who neglects her daughter for the advancement of her own career. By the end of the film, the alcoholic has redeemed his pride and self-respect, the single mother and the pilot are married, and the scientist saves the day, abandoning his ecological beliefs to nuke the aliens and save the world. It is this last figure, the ecologist, David, who perhaps illustrates this most clearly. At the beginning of the film, he is a typical '90s New Age guy:

> He's sensitive and caring, he weaves his bicycle through city streets clogged with cars, he recycles his refuse, he objects to his father's cigar smoking, he's aware that a nuclear explosion might block out the sun, he rejects monetary measures of success, and he is unconcerned about getting ahead in the competitive world of business and politics. These last character traits have cost him his marriage.
>
> (Parkin 1997: 3)

Over the course of the film, David's masculinity is reconstructed and loses its feminized traits. In the end he does get to save the planet:

> but only by leaving behind his quirky environmentalism, and embracing the power of military technology. He sets off the nuclear explosion that brings down the aliens, and returns to Earth a swaggering, cigar-smoking war hero. And as a result of his warrior masculinity, he Gets the Girl.
>
> (*ibid.*)

Even David's estranged wife, who 'by now [had] exchanged her business suit for softer casual-wear' (*ibid.*) is convinced of this transformation. And so at the end of the film, all is well once more in American society: men are men and women rush to them in congratulatory zeal.

Tyranny of a feminized state

Jeffords (1989) and Gibson (1994) have pointed out the consistent presence of the state or bureaucrats as an opposition to the heroic protagonist in American films in the aftermath of Vietnam. Rambo's question 'do we get to win this time?' raises the issue of whether the state will this time commit to the men she [sic] sends to fight, to defend her borders and protect her 'women and children' (Enloe 1989). Jeffords demonstrates the feminization of the government in many films: that it is the weak and indecisive state that causes the problem by not allowing protagonists (as metonyms for patriotic interest) to perform their tasks without hindrance. This theme continues, perhaps more so given paramilitary and right-wing antipathy towards government in recent years.

The film version of John Le Carré's *The Russia House*, released in 1990, was one of the first espionage thrillers to take place in a world where *glasnost* had begun to erode the old certainties about the Cold War. The film starts with the delivery of a manuscript to a cynical and weary, drunken book publisher, Barley, which is intercepted by British intelligence. The manuscript is highly technical and calls into question the quality of Soviet defense. The persistence of 'Cold War metaphysics' is evident. The British intelligence officers need to know who wrote the manuscript in order to judge whether it is true or false. As a result of decades of distrust they are suspicious of whether the author knows what 'he' is talking about. The film shows well the problems faced by Cold War institutions adjusting to the new world order, the anxiety of 'men who have been spies too long, and cannot unlearn their old habits' (Ebert 1993). The film suggests that change is difficult because of 'gray men' on each side of the old Cold War division whose interests are best served by perpetuating the arms race. The CIA men who attempt to take over from British intelligence believe that the Russians have only changed because they are on their knees. Yet, others are keen to welcome Russia into the fold and so pose a challenge to the CIA's role and purpose. One CIA agent reflects on the problems of this new asymmetry: 'How the hell do you peddle an arms race when the only asshole you've got to race against is yourself?'

But the film has homed in on some of the more subtle potential problems facing the West's reconciliation with the USSR:

> *The Russia House* is a profound study in miscommunication and misinterpretation, contrasting a jaded West, made complacent by its long history of freedom, and the East, passionate and sincere as it tests the bounds of that same freedom it is only beginning to test.
>
> (Baseline's Movie Picture Guide Review 1993)

165

The new Russia is emerging as a passionate young power not so dissimilar to the way that the Founding Fathers understood a youthful USA in comparison to tired Europe. The scenario that the film hints at is what future might await America if there is a new young power taking over its mission and destiny. America's role as the patriarchal hegemon faces challenge.

It is degeneration in the state that caused problems for the heroes in *Clear and Present Danger* and *Mission: Impossible*. But it is most clearly expressed in *The Rock*, where Hummel, hero of Vietnam and the Gulf War, becomes so angered by the government's refusal to honor the soldiers sent on covert missions that he decides the government must at last make a commitment to its warriors. He steals chemical warheads, takes them to Alcatraz Island and threatens to fire them at San Francisco if the families of the dead soldiers are not compensated for their deaths. He claims that he has been forced to make the ultimate sacrifice, to be labeled a traitor in order to do what is best for the country. But in doing so he is forced to break the masculine bond of the armed forces: the bloodbath of Navy Seals as they enter the building in their attempt to defeat Hummel is emphasized as mythological through the use of slow-motion photography, as evidence of the seriousness – the historic importance – of the issue. 'Look what you have made me do,' Hummel spits through a dead soldier's video camera at the men in suits sitting around TV screens. The difference between the figures is emphatic: the heroic, battle-weary soldier at the front line and the men in suits in their comfortable chairs, watching the action on TV from a safe distance.

Postmodern America? identities at the end of the Cold War

David Campbell (1992: 195) claims that the set of practices comprising the Cold War represented a series of boundaries between civilization and barbarity, and as a result, rendered a contingent identity of 'America' secure. Thus, he argues (*ibid.*) that containment is not just a historically significant foreign policy strategy, rather 'containment is a strategy associated with the logic of identity whereby the ethical powers of segregation that make up foreign policy constitute the identity of an agent in whose name they operate, and give rise to a geography of evil.' However, the territoriality of today's 'geography of evil' is not so easy to define as its Cold War counterpart was: terrorists, drug dealers and corruption in the US government power as often compound the break-up of Cold War identities as repair them.

But this inability to set boundaries and to confine identity and contain meaning might also be regarded, at another level, as a failure of masculinity in the face of fluid, messy and uncontrollable (feminized) forces (Theweleit 1987). As it has been argued that masculinity and nationalism and patriotism are intimately entwined (Mosse 1985; Jeffords 1989; Enloe 1989) then this offers similarly severe challenges to both gender relations organized around patriarchal authority and coherent, exclusionary national identity. The films mentioned above both exploit this sense of unease and attempt a reconciliation of it.

This does not mean that the films are singularly patriarchal and nationalist in their narration and, more importantly, in receptions of them. Clearly, the media are not a seamless unity; alternative and progressive readings of events are presented in some sources, such as Maggie O'Kane's reporting of the Bosnian War, subverting the distanced geopolitics of most news coverage (Ó Tuathail 1996a), or 'alternative' media sources such as *Mother Jones* and the myriad information now available on the Internet. Big budget films rarely adopt a more critical stance. However, these films do not present coherent wholes. It has been suggested that media audiences are becoming more conscious of media genres and styles, so leading to the current predilection for 'postmodern' ironic films, which wear their sources of inspiration and self-doubts on their sleeves. The construction of films such as *Independence Day* is obvious to anyone cognizant of the development of science fiction genres in American cinema. The film makers seem to play with an identikit of aspects of American culture and manly patriotism in the face of adversity. The irony would seem to suggest that the producers of the film acknowledge that for America, 'there will indeed be external danger, but it is the US *national identity*, not the United States as a nation, that is truly at risk' (Der Derian 1992: 94).

Acknowledgements

Thanks to Gearóid Ó Tuathail and James Sidaway for valuable comments on earlier versions of this chapter.

Notes

1 The concept of an unknown influence on people was a popular theme in horror and science fiction genres of the early Cold War period, where it was difficult to tell whether or not people had been affected, and as a result, difficult to know who was good or bad (see Dolan 1994: 70–71)

2 I have adopted Hartley's term 'public-ity' rather than 'knowledge production' because 'public-ity' evokes not only the creation of knowledge but also the audience or readership's self-identification within this knowledge (see Hartley 1992).

3 This is not to suggest that at some time masculinity was absent from American patriotism and nationalism but that the *idea* that this had happened has been used to produce a belief in the need to reinforce a certain form of masculinity. Jefford's explanation is that a sense of emasculation in the failure of America to win in Vietnam was used in hegemonic culture to reassert a form of patriarchal masculinity in a US society increasingly characterized by a tolerance for multiculturalism and women's rights.

4 Although even here there is change. In an era of 'political correctness' and the threat of global AIDS, 007 is allowed to seduce only one 'Bond girl' per film, and in the latest release, *Golden Eye*, the female 'M' tells the agent that she thinks he is 'a sexist, misogynist dinosaur . . . a relic of the Cold War.'

References

Barber, B. (1995) *Jihad vs. McWorld*. New York: Times Books.

Baseline's Movie Picture Guide Review (1993) 'Patriot Games.' *Cinemania CD-ROM*.

Baseline's Movie Picture Guide Review (1993) 'The Russia House.' *Cinemania CD-ROM*.

Bennington, G. (1990) 'Postal politics and the institution of the nation,' in Homi K. Bhabha (ed.) *Nation and narration*. New York and London: Routledge.

Campbell, D. (1992) *Writing Security: United States Foreign Policy and the Politics of Identity*. Minneapolis: University of Minnesota Press.

Carlin, J. (1995) 'A gentlewoman's challenge to Newt's Republican revolution.' *The Independent on Sunday* 9 April, p. 9.

Cox, R. (1986) 'Social forces, states and world orders: beyond international relations theory,' in Robert O. Keohane (ed.), *Neorealism and its Critics*. New York: Columbia University Press.

Dalby, S. (1994) 'Gender and geopolitics: reading security discourse in the new world order.' *Environment and Planning D: Society and Space* 12(5), 525–546.

Denton, R. (1992) 'Series foreword,' in Prince (1992).

Der Derian, J. (1992) *Anti-Diplomacy: Spies, Terror, Speed and War*. Oxford: Blackwell.

Dolan, F. (1994) *Allegories of America: Narratives–Metaphysics–Politics*. Ithaca, NY, and London: Cornell University Press.

Dyer, R. (1993) *The Matter of Images*. London: Routledge.

Ebert, R. (1993) 'Review of *The Russia House*.' *Cinemania CD-ROM*.

Edelman, M. (1988) *Constructing the Political Spectacle*. Chicago: University of Chicago Press.

Engelhardt, T. (1995) *The End of Victory Culture: Cold War America and the Disillusioning of a Generation*. Basic Books.

Enloe, C. (1989) *Bananas, Bases and Beaches*. Berkeley: University of California Press.

Fukuyama, F. (1989) 'The end of history.' *National Interest* supplement, summer, 16pp.

Gibson, J. (1994) *Warrior Dreams: Violence and Manhood in Post-Vietnam America*. New York: Hill & Wang.

Hartley, J. (1992) *The Politics of Pictures*. London: Routledge.

Jameson, F. (1992) *The Geopolitical Aesthetic: Cinema and Space in the World System*. Bloomington and Indianapolis: Indiana University Press.

Jeffords, S. (1994) *The Remasculinization of America: Gender and the Vietnam War*. Bloomington and Indianapolis: Indiana University Press.

McClure, J. (1994) *Late Imperial Romance*. London and New York: Verso.

Mearsheimer, J. (1990) 'Why we will soon miss the Cold War.' *The Atlantic* 266(2): 35–50.

Mosse, G. (1985) *Nationalism and Sexuality*. Wisconsin University Press.

Mulvey, L. (1989) 'Visual pleasure and narrative cinema.' *Visual and Other Pleasures*. London: Macmillan, pp. 14–26.

Nimmo, D. and Combs, J. (1983) *Mediated Political Realities*. New York: Longman.

Ó Tuathail, G. (1992) 'Foreign policy and the hyperreal,' in T. Barnes and J. Duncan (eds) *Writing worlds*. Routledge, pp. 155–175.

Ó Tuathail, G. (1996a) 'An anti-geopolitical eye: Maggie O'Kane in Bosnia, 1992–1993.' *Gender, Place and Culture* 3(2): 171–185.

Ó Tuathail, G. (1996b) *Critical Geopolitics*. University of Minnesota Press.

Ó Tuathail, G. and T. Luke (1994) 'Present at the (dis)integration: deterritorialization and reterritorialization in the new wor(l)d order.' *Annals, Association of American Geographers* 84(3): 381–398.

Parkin, A. (1997) 'Moral majority takes on aliens . . . and wins!' *Peace Studies* 1(2): 3–5.

Prince, S. (1992) *Visions of Empire: Political Imagery in Contemporary American Film.* New York: Praeger.

Sharp, J. (1993) 'Publishing American identity: popular geopolitics, myth and the *Reader's Digest*.' *Political Geography* 12(6), 491–503.

Sharp, J. (1996) 'Hegemony, popular culture and geopolitics: the *Reader's Digest* and the construction of danger.' *Political Geography* 15(6/7): 557–570.

Theweleit, K. (1987) *Male Fantasies, Volume One: Women, Floods, Bodies, History.* Cambridge: Polity.

Wark, M. (1994) *Virtual Geography.* Bloomington and Indianapolis: Indiana University Press.

8

ENFRAMING BOSNIA

The geopolitical iconography of Steve Bell

Klaus Dodds

Jebes Zemlju Koja Bosnu Nema! [Fuck the country that does not have a Bosnia].

(cited in Thompson 1992: 90)

Introduction

Steve Bell is the 1996 British Cartoonist of the Year and a regular contributor to popular magazines and newspapers. Since 1981, *The Guardian* newspaper (a centre-left broadsheet) has published his singled-framed images and cartoon strips, which have represented polemically a range of international events from the 1982 Falklands War, the Ethiopian famine and the ending of the Cold War, to Operation Desert Storm and the 1992–1995 Bosnian crisis. His widespread popularity appeared to be cemented when he represented John Major as a figure who wore his underpants over his trousers. His vision of the prime minister as an inadequate superman became the defining image of the Major administration (1990–1997) and has become part of the iconography of modern British politics. In doing so, Bell's images have been widely interpreted as continuing a long-standing tradition of political cartoonists contributing to the evolution of particular national iconographies (Aulich 1992; Farren 1993).

The images produced by Steve Bell during the Bosnian crisis are subjected to critical evaluation for the purpose of investigating how this political cartoonist employed icons and symbols to represent ethnic nationalism, violent wars and the subsequent reaction of the international community. The imagery employed to symbolize how Bosnia was or should have been represented can be investigated through a consideration of his selection, siting and arrangement of visual themes. In the process, it will be argued that these images effectively problematized many of the common assumptions not only about Bosnia as a place but also the reactions of the European Union and the United States to the unfolding horrors in the former Yugoslavia. Cartoonists, along with the endeavours of novelists and poets, contributed to a critical vein of opinion in both the United Kingdom and the United States over the handling of events in Bosnia. Many of

170

Bell's images about Western failures to intervene in a socially and geographically proximate place echo the sentiments of writers such as Salman Rushdie, who noted that:

> There is a Sarajevo of the mind, an imagined Sarajevo of whose ruination and torment exiles us all. That Sarajevo represents something like an ideal, a city in which the values of pluralism, tolerance and co-existence have created a unique and resilient culture.
>
> (1994: 17–18)

This chapter is organized into four main parts, and the initial section explores the geopolitical iconography of Steve Bell. Drawing upon the recent literature in critical geopolitics and cultural geography, the cartoonist's images are situated in an emerging body of literature on visual material that seeks to investigate the historical context and the ideas implemented in the imagery (see Daniels 1993; Rose 1993). Subsequently, discussion turns to the five images that have been selected for the purpose of exploring their symbolic importance *vis-à-vis* the unfolding events in Bosnia. These interpretations are supplemented with contextual material relating to the political and geographical disintegration of Yugoslavia. The final section offers some conclusions on the enduring importance of visual images such as political cartoons within the geopolitical imagination.

Geopolitical iconography of Steve Bell

Within cultural geography and related disciplines such as art history, there has been a sustained interest in the critical investigation of visual imagery (Cosgrove and Daniels 1988; Daniels 1993; Ryan 1994). Moreover, expanding interest in the visualization of war in newspaper photographs, films, video games and paintings has attached great importance to the systematic reading of texts in order to elucidate broader relationships such as the production and consumption of war images (Aulich 1992; Brothers 1997). The term 'geopolitical iconography' has been adopted in order to highlight interest in the icons and symbols employed by Steve Bell to represent international political events such as the Bosnian crisis. This approach is complementary to the recent work in critical geopolitics that has been concerned with critical evaluation of popular geopolitical sources such as magazines, newspapers, television and cartoons (Ó Tuathail and Luke 1994; Myers *et al.* 1996; Sharp 1993, 1996). In contrast to the existing literature on iconography within cultural geography, critical geopolitics has not engaged in close and detailed readings of visual material. Images have either been employed to illustrate a general analysis or used occasionally to illuminate specific issues such as media war reporting.

James Aulich has argued: 'Bell's work belongs to a distinctive tradition, originally divorced from the mainstream, and characterised by the "underground" work of Robert Crumb on the one hand, and the mischievous anarchy of *The*

Beano's "Desperate Dan" [*sic*] and the "Bash Street Kids" on the other. In both instances, his style is divorced from the "heroics" of many orthodox strips and oppositional stance of liberal scepticism' (1992: 93–94). His images, therefore, signal various strands of visual culture from the comic tradition of *The Beano* to works of high art. Bell's work is also close to the English tradition of political satire of cartoonists such as Gilray, Strube, Low and Rowlandson. The nineteenth-century cartoonist Gilray, renowned for his images of bottoms and bodily functions, was emblematic of a bawdily satirical tradition that was very popular within the coffee houses of English society. Alternatively, the *Daily Express* cartoonist, Sidney Strube, used figures such as the Little Man in the 1930s to be satirical in a polite rather than bombastic manner. Strube's images enjoyed considerable influence because they were used to promote Lord Beaverbrook's editorial policies based on promoting newspaper readership for middle-class Britain. His cartoons were given unprecedented coverage on the editorial page and were intended to make readers laugh while at the same time conveying a specific political message (Brooks 1990). In contrast, Bell's depictions of political and social life in the 1980s and 1990s have appeared in a newspaper, *The Guardian*, which was subjected to relentless legal pressure and political criticism from the Thatcher government over its reporting of the 1982 Falklands campaign, the handling of the 1984–1985 miner's strike the 1990–1991 Gulf War and the Spy Catcher affair.

Bell's editorial page images and/or his cartoon strip characters such as the Falklands War anti-hero Kipling or President Clinton's special advisor Socks the Cat were designed to convey a political message through the deployment of farce and satire. The image of the suburban garden surrounded by the barbed wire fence (see Figure 8.1) is a case in point. The issue was the massacres of Bosnian Muslims in central and eastern Bosnia. Western European leaders such as Helmut Kohl and John Major appear unconcerned with what is happening in a neighbouring garden. The bundle of wire in Kohl's garden seems to indicate a concern that the fence might not be of sufficient strength to prevent bedlam breaking out onto his property. In contrast with Sidney Strube's famous cartoon of October 1934 (Well, everything in the garden is lovely), the purpose of the cartoon is to highlight the narrow-minded and parochial nature of these countries and their leaders rather than to warn against the dangers of meddling in foreign affairs motivated by the irrational forces of nationalism and ethnic cleansing (see Brooks 1990: 31). Steve Bell later confirmed to this author that he had not been aware of Sidney Strube's 1934 garden image (Bell 1996). The motivation for Bell's image had been derived from the on-the-spot reporting by *Guardian* journalists such as Maggie O'Kane and Ed Vulliamy, who had already provided chilling accounts of the Bosnian Serb concentration camps and ethnic cleansing strategies. The cartoon of the garden employs bitter satire to convey a sense of Western European indifference to the plight of the Bosnian Muslims.

The subversive imagination of the dissident cartoonist, therefore, extends the current portfolio of critical geopolitics through consideration of the geopolitics

Figure 8.1 The Bosnian peace process is all but dead (reproduced with permission of Steve Bell).

of satirical representation of places and the boundaries of supposedly sovereign territory. As 'outsiders' within the journalistic community, cartoonists are often able to represent satirically events that may not be possible for journalists (with the attendant pressures of proprietorial influence and the political complexion of the newspaper) working within large newspaper corporations such as News International. In conjunction with other cartoonists, the self-employed Steve Bell has a direct working relationship with the editor of *The Guardian* and is able to submit material without any form of editorial interference. In a recent interview, Bell reflected on the production and circulation of his political images and argued that 'Cartoons are very good at reflecting on contemporary prejudices . . . they draw upon a whole web of meaning . . . you use double edged irony in the op-ed [opposite the editorial] cartoons' (*ibid.* 1996). In the case of the Bosnian crisis, Bell noted that his position was equivocal. His intention, therefore, was to produce images of ethnic cleansing and war that were ironic and double-edged. The single-framed representations of Bosnia were intent on exploring European complicity in the destruction of the former Yugoslavia. The failure of the EU to act against ethnic cleansing and the death camps was a key theme throughout his representations of Bosnia (see Bell 1994). As he noted to the author: 'You have to take sides in order to establish some basis for moral action . . . The way it was allowed to happen . . . nothing was done, nothing was ever said' (Bell 1996). His personal opposition to the 1993 Vance–Owen peace plan was powerfully illustrated through various images of the Bosnian corpus being cut into various pieces by the European peace negotiators.

I have chosen five images produced by Steve Bell over a three-year period of the Bosnian crisis. The reasons for selecting these particular images are in the timing of their production and their political significance. This production period ranges from September 1991 to February 1994. The periodization is important, because it effectively encapsulates the outbreak of ethnic cleansing and the Vance–Owen and Owen–Stoltenberg peace plan process. These plans for the partition of Bosnia were a crucial feature of European endeavours to secure peace on the basis of dividing up the former Yugoslav Republic into various segments. It was argued by some commentators that the failure to endorse either of these proposals had dire consequences for the inhabitants of Bosnia. However, Bosnian Muslim parties rejected both plans on the basis that the 1994 Owen–Stoltenberg peace plan rewarded ethnic cleansing by apportioning 52 per cent of the territory to the Serb parties and 17 per cent to the Bosnian Croats. Importantly, however, the Clinton administration also refused to endorse either of those plans. As a consequence of this failure to gain any form of agreement, the European Union and the United Nations appeared helpless to prevent further outbreaks of conflict and to limit the extent of ethnic cleansing by the Serb and Croat forces.

For some commentators, the political cartoon may appear either an irrelevance or silly distraction from the serious business of international politics. This is a restricted view of the importance of cartoons to the enframing of international politics, because these images are part of a network of mass-mediated culture composed of film, music, comic books, advertising, radio and television (see Kellner 1988). Whilst this chapter is concerned with Bell's cartoons, these varied artefacts played their part in constructing public knowledge of the Bosnian wars and subsequent reaction to these war stories. More specifically, Bell's cartoons appeared next to the leader page of *The Guardian*. With a circulation of over 300,000, *The Guardian* was a major British contributor to critical reporting on the impact of war on civilians and the failure of the international community to respond effectively to ethnic cleansing and allegations of mass war crimes. Bell's cartoons can thus be represented as a powerful iconographic critique of ethnic cleansing, Western inaction and the lack of recognition of the social proximity of Bosnia. Cartoon and graphic representations of war dramatize events, and the stage-like compositions can force readers to confront these events differently to the often detached and objective terms of journalistic reporting.

Warring factions?

The geographical enframing of the Bosnian crisis was one element in the subsequent news reporting of events in the former Yugoslavia. Within media studies, the concept of enframing refers to the ways in which news frames 'encourage those perceiving and thinking about events to develop particular understandings of them' (Entman 1991: 7). In alliance with the recent work within critical geopolitics, 'frames' have been analysed for the purpose of detecting how information about places and peoples contributes to the structuring and ordering of particular events. The geographical frames underlying news stories have been held to exert considerable influence on the opinion of readers and the modes of justification adopted by policy makers. For a place that played a minor part in the maps and stories of the Cold War, the geographical enframing of Bosnia became a contested and controversial undertaking in the 1990s (Glenny 1992; Denitch 1993; Bowman 1994; Ignatieff 1994; Salecl 1994). The varied frames used to represent the Bosnian crisis had implications for possible policy responses. For the newspaper editors and media managers charged with coverage of the disintegration of Yugoslavia, various terms such as 'warring factions' or 'genocidal violence' constructed different understandings of unfolding events and processes such as ethnic cleansing.

Within the British media, the dominant enframing of the collapse of the former Yugoslavia was to represent the wars as a product of competing ethnic factions. As a consequence, events such as the 'ethnic cleansing' of eastern Bosnia and the siege of Dubrovnik were frequently labelled as the actions of different factions rather than overwhelmingly the product of a particular element such as the Bosnian Serbs and the former national army (Shaw 1996). In the midst of apparent confusion for the responsibility of events, the British media frequently followed the British government in cautioning against active intervention in the former Yugoslavia. In an alternative vein, the American print media were over-whelmingly inclined towards an approach to the Bosnian crisis that stressed the genocidal aggression of the Bosnian Serbs against the Bosnian Muslims (Guttman 1993). Implicit within such a stance was the belief that the Americans and their European partners had a moral obligation to respond to the plight of the Bosnian Muslims. In the geopolitical eye of power, Bosnia had to be restored to its existing territorial boundaries, given this presumption of obliga-tion on the part of Western governments.

Steve Bell's depiction of the Serbian and Croatian political leaders, Milosevic and Tudjman, respectively (produced on 17 September 1991, see Figure 8.2), looking on as two faceless soldiers shoot each other with the aid of a machine gun (the Russian-made AK-47) and bazooka is both dramatic and shocking. The stern-faced depiction of both leaders conveys little obvious emotional reaction to the intensity of the unfolding conflict, and their bodily postures further indi-cate a certain resignation to its inevitability. At the onset of the disintegration of Yugoslavia, a number of Croat and Serb commentators had warned that the conflict between the two largest republics of the former Yugoslavia would be horrendous and that contrary to liberal commentators this was a tragic inevitability given the ethnic and racial hatreds in the Balkans. In particular, the Serbian leader, Slobodan Milosevic, frequently referred to the geographical and historical composition of the Balkans in order to convince much of the world that his aggression was merely part of an ancient, obscure and intractable ethnic conflict. Misha Glenny has argued, however, that such an argument effaces the responsibilities of modern state leaders for their own violent actions:

> Many commentators, some from the Balkans themselves, have encour-aged the notion that the peninsula's inhabitants are incorrigibly violent, mired in the blood of five centuries . . . [However] Mass killing in the Balkans has always taken place in times of political and constitutional crisis whose origins are thoroughly modern.
>
> (1996: 35)

The brutality of the Croat–Serb confrontation is powerfully illustrated through the depiction of these soldiers' bodies being blasted to pieces by the bullets of the machine gun and the bomb of the bazooka. Although the soldiers appear recognizable largely on the basis of their lapels, which identify them as either Serbian or Croatian, they wear identical uniforms and are faceless. The mere fact that the reader is unable to identify the faces of the soldiers may well accurately reflect the lack of conviction by outside political bodies to intervene in order to end the violence. The evident inability of the European Union and the United States to decide whether they should intervene at this stage to prevent escalation of the violence further compounded the misery in Yugoslavia. Whilst the daily news coverage of raging wars in Croatia had played a part in persuading the EU to act as peace brokers between the 'factions' on the basis that they would never

Figure 8.2 The disintegration of Yugoslavia continues, supervised by Presidents Franjo Tudjman of Croatia and Slobadan Milosevic of Serbia (reproduced with permission of Steve Bell).

reward those nations that had participated in ethnic cleansing or spatial expansionism, the EU and the international community implicitly consolidated new patterns of place domination and ethnic irredentism. The strategy of the EU was essentially two-fold: to negotiate a settlement with the protagonists whilst providing humanitarian aid to the victims of ethnic cleansing. However, when violence broke out in Bosnia in April 1992 the EU (and other international bodies) refused to commit itself to the reversal of illegal territorial aggrandizement and ethnic purification.

Since the summer of 1991, therefore, the Yugoslav National Army and the Serb militias had waged a fierce war against the Croatian army on the pretext that Serbs living within Croatia would face ethnic cleansing and possibly genocide. At the time of the publication of Bell's image of Croat–Serb aggression in September 1991, some *Guardian* readers actually complained to the editor and the cartoonist that this image was unduly critical of Croatian actions in the face of unparalleled Serbian aggressiveness. One of the apparent purposes of Bell's image was to highlight the fact that Tudjman controlled a fascistic government intent not only on protecting Croatia's territorial boundaries but also on seeking to incorporate 'Croat territories' in Bosnia and parts of Slovenia. Whilst the veteran American reporter Peter Brock noted that the Western media did not report (in the same detail) the atrocities committed by Tudjman's regime, this underestimates the fundamental inequalities between those who perpetuated violence (Bosnian Serbs and the Yugoslav National Army) and those who were on the receiving end of ethnic cleansing and widespread massacres (*cf.* Brock 1996).

Timothy Garton Ash's observations about the subsequent Bosnian crisis (1992–1995) not only reflect the personal position of this author but also highlight a failure of the British and American governments to distinguish between the extent to which one particular faction was overwhelmingly responsible for ethnic violence. As Ash noted: 'There is an important difference in the *degree* of responsibility between the Serbian and Croatian regimes, but there is a difference in *kind* between the responsibility of the Serbian and Croatian regimes, on the one hand, and that of the Bosnian regime on the other. Bosnia was the victim of aggression, first from the Serbs and then from the Croats. So also with the results. Bosnia and the Bosnians have suffered the most, lost most, and are still most likely to lose more' (Ash 1995: 30). From a critical geopolitical perspective, the most striking aspect of the destruction of the former Yugoslavia compared with the 1991 Gulf War was that Western political leaders and the mainstream media implicitly treated these events in a quasi-anthropological manner, insisting that one could not take sides (because of the historical and geographical complexity of the Balkans) and instead would have to wait for those adversaries to finish their barbaric business. In contrast, Saddam Hussein's aggression in Kuwait was quickly transformed into a psychological battle between the forces of evil (as personified by Saddam himself) ranged against the civilized Western world. It was considered imperative, therefore, to act swiftly against Iraq's aggression and restore Kuwait to its 'rightful' owners. In contrast, relatively little information was presented to Western audiences about the religious and territorial complexities of the Middle East (Kellner 1995).

Restating Bosnia?

The disintegration of Bosnia was precipitated by a meeting between Tudjman and Milosevic held in March 1991, in which the decision was taken to carve up Bosnia between the Croatian and Serbian republics (Ali and Lifschultz 1993; Irvine 1993). In contrast, it has been argued by some commentators that the subsequent recognition of Bosnia's independence by Germany in 1991 not only created considerable uncertainties amongst the European Union and the United States but also effectively precipitated the onset of conflict in Bosnia in April 1992. This argument was demolished by a number of reporting journalists and Balkan experts, who have stated that if European peace negotiators were exposed to an awkward position it was largely one of their own making. In December 1991, the EU decided at a meeting of foreign ministers that the Yugoslav republics could gain recognition as independent states if they fulfilled certain obligations regarding human and ethnic minority rights and settled territorial claims (Thompson 1992: 167). Whilst this mechanism sufficed for Croatia and Slovenia, it failed Bosnia because rather than accepting its claim for independence it proposed an internal referendum to decide the matter. The referendum of March 1992 confirmed an overwhelming desire of voting adults (63 per cent voted and of those 99 per cent were in favour) to opt for independence, the Bosnian Serb population vetoed the procedure and their political leader Radovan Karadzic condemned the Muslims and Croats for attempting to exclude the Serbs from Bosnia.

The subsequent attempts by European negotiators to devise a partition plan for Bosnia that reflected the interests of the three major parties (the Muslims, the Croats and the Serbs) has been interpreted as a catalyst for the ethnic and territorial partition of Bosnia. The evident confusion amongst the EU negotiators as to the ultimate objective of their peace package was reflected in their simultaneous recognition of Bosnian independence in 1992 (and territorial integrity) whilst appearing to encourage territorial partition through the Vance–Owen peace plan process. In February 1993, David Owen presented a partition plan for Bosnia that effectively meant that ten new provinces would be created and distributed amongst the three major ethnic parties (see Figure 8.3). The Vance–Owen plan for Bosnia was also based on the principle that the United Nations would be responsible for the demilitarization of the capital city, Sarajevo. Bell's cartoon of the implementation of the Vance–Owen peace plan (see Figure 8.4) was accompanied by a critical leader in *The Guardian* that focused on the inability of the international community to secure peace in the Bosnian region (see Figure 8.5). Published in February 1993, the geographical restitching of Bosnia is represented against the backdrop of UN peace-keeping

Figure 8.3 Revised Vance–Owen peace plan, 8 February 1993.

forces uncertain of their political and military mandate. David Owen, perched on the head of the Bosnian cow, is depicted as stitching together the hide, which had already been divided into ten territorial sections. The biblical analogy with the sacrificial cow seemed appropriate given the widespread condemnation of the Vance–Owen partition plan by Bosnian observers, who objected to the disintegration of the unitary Bosnian nation-state. The transparent contradictions of the peace plan were exposed by Branka Magas in his letter dated 14 February 1993 to *The Guardian*:

> The map proposed by Vance and Owen openly ratifies the results of ethnic cleansing, which international public opinion has condemned. In reducing the country's multi-national government and army to its Muslim component alone, the plan betrays all Serbs and Croats who,

Figure 8.4 Lord Owen's mission to Bosnia to implement the Vance–Owen peace plan continues (reproduced with permission of Steve Bell).

Figure 8.5 Bosnia-Herzegovina, 1996–.

by remaining faithful to the ideal of a single Bosnian state, have given the lie to the claims of Karadzic and Boban to represent all Serbs and Croats in Bosnia-Herzegovina. Besides being unprincipled, the plan is also unworkable because it contains no provisions for disarming the aggressor, while denying the central state authority that 'monopoly of legal violence' without which no normal state can function ... It is obvious that a Bosnia-Herzegovina constructed in accordance with the

Vance–Owen plan could not pretend to sovereignty, integrity or independence.

(Magas 1993, cited in Ali and Lifschultz 1993: xxxvii)

Moreover, Bell's depiction of the United Nations troops standing around the
façade of the Bosnian cow can be interpreted as a powerful indictment of the
role of UNPROFOR. As a protection force, these troops were supposed to
guard aid deliveries, clear mines and prevent looting. With their limited mandate
and lightly armed, UNPROFOR not only failed to protect aid conveys but also
ensured that aggressive action by the international community against Serbian
warmongering would be postponed in the face of possible reprisals against the
UN force in Bosnia. The question marks placed over the plane and the submarine located off the coast of the former Yugoslavia also convey that sense of
confusion as to the purpose of the protection force, which had already witnessed
the *de facto* ethnic partition of Bosnia. The dismemberment of the Bosnian cow
into distinct cuts of meat is a distinctive feature of Bell's satirical images of the
Vance–Owen peace plan, which consistently emphasised Western failure to prevent either the partition of Bosnia or the cessation of violence. By perpetuating
the mythology that UNPROFOR's reflected the EU's neutral engagement,
European governments ensured not only that Serbian aggression in Bosnia was
not directly confronted but also that Croatian opportunism in eastern Bosnia
was not stopped.

In his book *Balkan Odyssey*, David Owen recorded how the Bosnian Muslims
would not accept the Vance–Owen peace plan (VOPP) because President Alia
Izetbegovic believed that the United States would offer a superior territorial deal
(Owen 1995: 196–7). In contrast, a number of observers pointed to other reasons for the VOPP's collapse and the subsequent Owen–Stoltenberg plan of
1993–1994 (which had envisaged a series of mini ethnic states and some form of
international control of the cities of Sarajevo and Mostar): first, the Bosnian government would not accept any form of partition to their nation-state; second,
the Croats were never confronted by the West over their military activities in
eastern Bosnia; and, third, the Bosnian Serbs were unlikely to desist from further
territorial aggrandizement, because the West had failed to commit itself to military intervention. With considerable confusion in Europe and the United States
as to whether to launch air strikes against the Bosnian Serb forces between 1993
and 1994, President Clinton's refusal either to pledge US troops to a UN peace-
keeping force in Bosnia or to countenance a European-led military consortium
did not assist the process of conflict resolution or the pursuit of territorial justice
for the Bosnian government.

The depiction of Bosnia's ethnic and territorial boundaries was not just a
representational crisis but also a very bloody and material struggle for human
security. The fate of the town of Srebrenica in eastern Bosnia indicated the failure
of the international community to protect particular territorial and imaginative
boundaries. In June 1993, UN Resolution 836 declared Srebrenica a safe area

for Bosnian Muslims in a predominantly Serbian-held area of Bosnia. Within the boundaries of these areas, the Bosnian Muslims were allowed to keep arms, and the UN peace-keeping force was given the task of protection. Two years after this declaration, however, Srebrenica witnessed a massacre of over 6,000 Bosnian Muslims by the Serbian army, which had decided that the town was being used covertly by Bosnian Muslim troops in order to make attacks on their forces (see Honig and Both 1996). The failure of the Dutch UN peace keepers, stationed in the town, to protect the fleeing citizens of Srebrenica was widely seen as a damning indictment of the safe area policy and of the operational effectiveness of the UN (see *ibid.*; Vulliamy 1997).

Social proximity of Bosnia

The question of responsibility for the fate of Bosnia between 1992 and 1995 had been enframed by most Western political leaders as a question of possible military intervention. In doing so, the problem of moral responsibility became reduced to an issue of how powerful UN member states might intervene on the basis of their own strategic interests rather than a critical consideration of the potential for other political actors to contribute to the peace process (see Figure 8.1). Steve Bell's portrayal of Bosnia in May 1993 was indicative of his critical evaluation of European endeavours to secure peace in the former Yugoslavia. Based on a painting (entitled 'The Triumph of Death') by the sixteenth-century Flemish artist Pieter Bruegel the Elder, Bell's cartoon links two radically different representations of place: Western European suburban tranquillity, and the hellish horrors of modern-day Bosnia. The cartoon is striking because, in spite of the barbed wire fences, it depicts Bosnia as a part of Europe that not so long ago other Europeans used to visit as a holiday destination. A returning David Owen calling out 'I'm home darling' to a relaxing John Major seems therefore out of place in the midst of our social and geographical proximity to these unfolding horrors. Overlooking Major's garden fence are President Mitterand of France and Chancellor Kohl of Germany, presumably discussing events in the former Yugoslavia. The timing of the cartoon was significant, given that there had been well-placed fears that Europe was politically unwilling to prevent the mass rapes of Bosnian and Croatian women, the construction of concentration camps in western Bosnia and the televisual reports of atrocities such as the Croatian destruction of Mostar and mass murders of Bosnian Muslims in Ahmici (Gowing 1996). The discovery of the Ahmici massacre by Colonel Bob Stewart, accompanied by the journalists Martin Bell and Paul Davies, in April 1993 revealed to wider audiences the 'viciousness of the Croat–Muslim war in central Bosnia which whilst well known to Western governments was not publicly acknowledged' (*ibid.*: 87; Gow *et al.* 1994).

The barbed wire fence separating the Bosnian genocide from the gardens of suburban homes of Western political leaders can highlight three additional dimensions. First, in European culture the garden has long been held to represent a peaceful and secure space at the expense of the wild landscapes out-side the boundaries of the garden fence. The geographical enframing of Bosnia as a 'wild' space had implications for the willingness of Western governments to send troops to defend the Bosnian Muslims against genocide. The inability of the British government, for instance, to promote a firm policy against Serbian aggression was in part related to a fear that the Balkans would become a morass that would entrap British troops in a long and costly engagement. The physical landscape of Bosnia was frequently described as unsuitable for rapid military action: mountains, thick woodland and a harsh winter climate. Second, the barbed wire reminded this author of the televisual reporting of a British ITN (Independent Television News) team led by Penny Marshall, who first exposed the Serbian-run concentration camp in Trnopolje and the presence of emaciated figures behind barbed wire. For many reporters and newspapers this videotape footage (taken on 5 August 1992), later dubbed 'Belsen 92' by the *Daily Mirror*, became indicative of an unfolding holocaust against the Bosnian Muslims. The significance of those images was widely recognized, as Ed Vulliamy of *The Guardian* noted that: 'stripped to the waist, in their thousands, against the wire in the relentless afternoon heat . . . They wait, stare at nothing, sweating and wondering what will happen next' (Vulliamy 1994: 150). These television pictures, according to journalist Nik Gowing, created a considerable stir in Whitehall and Washington: 'The camp story came out of the blue. The duty Foreign Minister Lynda Chalker in our studio [Channel 4 News] that night was not only moved; she was politically flustered by it. President Bush was in the White House briefing room within an hour to condemn the camps and to promise America "will not rest until the international community has gained access to all detention camps." This was policy panic' (Gowing 1996: 89–90).

By the summer of 1993, the Vance–Owen peace plan process had transferred from Geneva to New York. In the following months, the EU and the United States effectively perpetuated further ethnic violence through their refusal either to lift an arms embargo on the Bosnian government or to commit themselves to halting the Serbian and Croatian armed forces. The attempted demarcation of Bosnia had tragic consequences as arbitrary boundaries were imposed upon eth-nically mixed towns and villages. The European proposal for the construction of ten provinces within Bosnia actually encouraged further ethnic cleansing, as Serbs cleansed their provinces and Croats expelled non-Croats from Gornji Vakuf, Vitez and Busovaca. While the massacres continued in western Bosnia, it was reported by the *New York Times* in July 1993 that Secretary of State Warren Christopher had 'made no secret of his desire to move attention away from Bosnia' because the former Yugoslavia was no longer considered a foreign policy priority (Guttman 1993). Within Western Europe, Bell's image of European political leaders relaxing in their respective gardens conveys the sense of compla-cency that many Bosnian commentators detected throughout 1993–1994. While ethnic cleansing was condemned by political leaders such as Major and

Clinton, NATO or European military resources were not going to be deployed in order to prevent further genocide.

To intervene or not to intervene?

Steve Bell's cartoon of Prime Minister Major and President Clinton considering the possibility of intervention in February 1994 effectively captured the unwillingness of the international community to take responsibility for the Bosnian genocide (see Figure 8.6). The two Western political leaders are depicted watching the submerging 'hand of Bosnia' slipping into the dark and uninviting ice-filled lake. Perched precariously on the ice, John Major is encouraging Bill Clinton to intervene first in order to rescue the sinking figure. Their incapacity to take military action on behalf of Bosnia in the light of further evidence of Serbian and Croatian atrocities is acknowledged by the five bear-like judges to have been pitiful. Ten years earlier, Jane Torvill and Christopher Dean had triumphed in the 1984 Sarajevo Olympics with their performance of 'Bolero'. In contrast to their perfect scores of '6', Clinton and Major have been adjudged to have scored '0' points for their performance. As a satirical representation of ice-skating judging or perhaps even the Eurovision Song Contest, Bell's cartoon is a powerful illustration of Western inaction in the face of the obvious distress of the Bosnian population, compounded further by the harsh Bosnian winter climate.

The timing of this cartoon in the aftermath of television reports of a mortar bombing of a crowded market square in Sarajevo that had killed 69 people was pertinent. The demands for NATO air strikes against the Serbian forces in Bosnia returned to the political agenda for the first time since the US government had discussed the possibility of taking military action in May 1993. However, American commentators such as the former Department of State official George Kenny have alleged that Christopher's endorsement of air strikes and the lifting of the arms embargo against the Bosnian government was half-hearted. Unsurprisingly, the British and French governments continued with their arguments that the lifting of arms embargoes would heighten fighting in the Balkans. While expressing its unwillingness to either protect the territorial integrity of Bosnia or unilaterally lift the arms embargo, the United States formally recognized the former Yugoslav republic of Macedonia after a two-year delay during which it had failed to persuade the political leaders of Macedonia to change the name of the state. The Clinton administration, between 1992 and 1994, was sensitive to the fact that Greece had objected to the choice of that name (Greece also has a province of Macedonia) and was conscious of the Greek-American voting constituency in the USA, which could have been alienated (see Fowler and Bunck 1996: 402). Tragically, US and Western European political leaders waxed and waned over intervention and refused to allow the Bosnian government even to defend itself against Croat and Serb forces.

While the mainstream British media largely accepted the British government's

Figure 8.6 Intervention in Bosnia is discussed (reproduced with permission of Steve Bell).

assurances that air strikes would only prolong the Bosnian war rather than terminate hostilities, the British Muslim press argued that Major's reluctance to intervene was driven by racial prejudice against the Bosnian Muslims. As an editorial in the English language magazine *Impact* reflected in March 1996:

> For months and then years, as the Christian world in the midst of which the Bosnians had the misfortune to find themselves watched intently, the politicians staged their theatre of negotiations. A war of genocide, of course, a holocaust, is not to be dealt with by negotiation; but of course Bosnia was 'not a war of genocide'. It was a civil war involving mysterious tribal people whom even the BBC could call 'warring factions'. And the good guys, who to the horror of the press appeared to be the Muslims, were, after all, Muslims, and hence 'we should hesitate before giving them weapons'. Muslims are a single, Oriental, aberrant human type with no real place in white Christian Europe. Noel Malcolm reported in the Telegraph that Sir Michael Rose, the UN Commander in Sarajevo, privately referred to the Muslims as 'the wogs'.

While the orientalist prejudices of the mainstream Western media and government officials are beyond the scope of this chapter, the fact remains that the international community had failed to accept responsibility for ending Serbian atrocities in spring 1994. Subsequently, the apparent reluctance of the Americans to endorse the peace proposals of the Contact Group allegedly encouraged the Bosnian Muslims to launch a counter-offensive in central Bosnia during the autumn of 1994 (see Petras and Vieux 1996). The United States defended this offensive action and again called for the lifting of the international arms embargo against Bosnian Muslim forces. Petras and Vieux have argued, perhaps unreasonably given the evident confusion in Washington, that tacit American support for the Bosnian Muslim government was based on cynical evaluation of US–European relations and US domestic mood rather than a principled response against ethnic cleansing: 'more dead Bosnians meant more favourable propaganda for Washington, more moral discredit for Europe, and more demand for US intervention ... With leadership passing to the US and NATO, the policy of peacekeeping was replaced by that of answering Bosnian Serb provocation with massive air-strikes' (*ibid.*: 21–22). With the assistance of media stories of Serbian war crimes, this pro-interventionist script became more entrenched within the dominant US narratives of intelligibility of the crisis between 1994 and 1995. Croatian atrocities, such as the expulsion of 150,000 Serbs from the Krajina region, remained relatively unreported in the face of stories about mass rapes of Bosnian Muslim women and the destruction of mosques in central Bosnia.

The eventual troop-based involvement of the USA in 1995 confirmed that the political debate over Bosnia had shifted towards armed intervention and the reinforcement of the principle of state sovereignty. The US Secretary of State, Warren Christopher, reminded the Senate Foreign Relations Committee of the principles at stake: 'There will be no peace agreement in Bosnia unless NATO and the United States, the United States in particular, take the lead in the implementation of a peace agreement' (Christopher 1995, cited in Petras and Vieux 1996: 4). President Clinton's decision to implement the Dayton Agreement in 1995 effectively codified a form of ethnic cleansing that had not been anticipated by European negotiators (see Figure 8.5). The Bosnian Serbs were allocated 49 per cent of Bosnian territory and given land corridors in the Posavina region. The Croats were joined with the Muslims in a federation that also included eastern Slavonia. The crucial dimension to the Dayton Agreement was the local agreement between Milosevic and Tudjman over land corridors and the ringing endorsement of the deal by the American administration in the face of Muslim protests. As Petras and Vieux concluded:

> What was crucial in this chain of events was the replacement of the European led security arrangement, co-ordinated under UN auspices, by the US–NATO command ... Once European leadership in the region was broken, the Muslims were given short shift ... An integral Bosnia was no longer the centrepiece of US policy. NATO and the reestablishment of US global leadership was the principal victory in the Dayton Agreements ... The Dayton Agreement reconfirmed the division of Bosnia according to an ethnic distribution of power, but not until more years of slaughter.
>
> (1996: 23)

Bosnia had been effectively divided into three parts, and the prospects for any form of unification must appear bleak given the presence of three national armies and the impediment of refugee movement in the region.

The post-Dayton occupation force of the UN was composed mainly of American troops intent on policing the ethnic division of Bosnia in 1995–1996. The mediating endeavours of the EU and other organizations such as the Organisation of the Islamic Conference had been banished from the political scene. Political leaders in the United States rapidly celebrated in late 1995 the reimposition of US leadership in Europe and the political reemergence of NATO. From the Clinton perspective, the Bosnian crisis revealed a growing European capacity to negotiate with a range of actors without the military umbrella of the Cold War apparatus of NATO and the United States, while other commentators such as the British Muslim magazine *Impact* (1996) have interpreted the Dayton Agreement as just another ethnic partition plan:

> Absurd long borders now weave across the Balkan map, promising years and generations of tensions and future wars to come. Like the border between Pakistan and the occupied Kashmir, or that between

Israel and the Palestinian bantustans, it is a wound cutting into living flesh that is unlikely to be healed.

The recent elections in Bosnia's different regions have been widely perceived as failing to enhance the credibility of the Dayton partition. The Croats' refusal to accept the election results in Mostar, which witnessed a Muslim majority in the muncipal city council, was indicative of the fragility of the democratic process in Bosnia (Glenny 1996). The long-term prospects for Bosnia must appear bleak if the nationalist parties cannot implement power-sharing and territorial agreements cemented in Dayton, Ohio.

Something must be done!

Recent studies have highlighted the importance of the print and televisual media in representing global politics to a range of audiences (see Dalby 1996; Ò Tuathail 1996b; Sharp 1996). Media studies scholars such as Steve Livingston have referred to the development of a worldwide array of real-time print and electronic media coverage of breaking events as the CNN factor (Livingston and Eachus 1995; Neuman 1996; Hopkinson 1993). As Ed Turner, vice-president of CNN, noted in the aftermath of Operation Desert Storm:

> For the first time because of technology we had the ability to be live in many locations around the globe and because of the format of these networks we can spend whatever time is necessary to bring the viewer the complete context of that day's portion of the story.
>
> (cited in Channel 4 programme *Noam Chomsky and the Media*, 1995)

However, there has been some disagreement over the role and importance of media reporting in shaping government decision making. On the one hand, it has been argued that televisual images have played an important role in shaping government responses to particular places undergoing war or suffering in places scattered around the globe; South Africa, Bosnia, Somalia and the Gaza Strip. As former American National Security Advisor Anthony Lake noted in 1993:

> Public pressure for our humanitarian engagement increasingly may be driven by television images, which can depend, in turn, on considerations as where CNN sends its television crews. But we must bring other considerations to bear as well.
>
> (cited in Luke and Ó Tuathail 1997: 710)

On the other hand, British journalist Nik Gowing cautioned against assuming a simple cause-and-effect relationship between media images of war and suffering and instant policy responses by government (1994, 1996).

Moreover, academic commentators such as Kevin Robins have proposed that media representations of crisis regions such as Bosnia, Rwanda and Zaire often have paradoxical consequences:

> Yet through the distancing force of images, frozen registrations of remote calamities, we have learned to manage our relationship with suffering. The photographic image at once exposes us to, and insulates us from, actual suffering; it does not, and cannot, in and of itself implicate us in the real and reciprocal relations necessary to sustain moral and compassionate existence . . . It may no longer be a question of whether this strengthens conscience and compassion, but of whether it is actually undermining and eroding them [senses of compassion].
>
> (1996: 77)

Working journalists such as Kate Adie (whose reporting of the Dunblane massacre in 1996 later coined the Adie factor: things that cannot be reported) have argued that forms of compassion fatigue may become more entrenched: 'Have we grown more wary of instant response to disaster, more indifferent to the stream of seemingly baffling conflicts which flit past on the screen? Do the pictures of the displaced, the homeless and injured mean less when they are so regularly available? Have we in short begun to care less?' (Adie 1996: 14). This seems a far contrast to the acknowledged significance of the BBC's reporting of the Ethiopian famine in 1984 and the subsequent Bandaid appeal by Bob Geldof, when televisual reporting played a crucial role in mobilizing public outrage in Britain and a determination to contribute to the famine-relief operations (Glasgow University Media Group 1985).

Television's ability either to inform or to insulate the viewer from the actual horrors of war and crisis is influenced by the actual coverage of place and its inhabitants (Gow *et al.* 1994; Kellner 1995; Morley and Robins 1995). Television can play an important role in extending human awareness of suffering but it can also promote forms of non-responsibility. The television pictures of Trnopolje camp in August 1992, for instance, undoubtedly had a profound impact on the British and American governments. Former Department of State official George Kenny noted that television pictures of the concentration camps in eastern Bosnia forced the Department of State to reconsider its Bosnian policy. Within a week of the television pictures, President Clinton had approved National Security Council resolution 770/771, which called for the use of military support for the provision of humanitarian aid. At the same time, the British government also promised troop support for Bosnia in order to support humanitarian activities. Whilst the inspiration for UNPROFOR may have been derived from the exposure of Serb concentration camps, subsequent television coverage of mass graves and ethnic cleansing did not encourage or stimulate governments to take military action against the Serb forces.

Steve Bell's cartoon of a family watching pictures of atrocities in Bosnia on 11

August 1993 (see Figure 8.7) goes to the heart of a particular dilemma: does television coverage of horrors promote moral responsibility for distressed others and/or influence the government decision making of watching states? The family watching the televisual pictures of Bosnia appears numbed with incomprehension and disbelief as the British media carried stories of mass rape, detention camps and events such as the destruction of Mostar. The over-sized television screen alongside the video dominates this particular cartoon image. The watching family including their pets appears mesmerized by the television pictures of the former Yugoslavia depicting further evidence of atrocities in Bosnia. In the same month, the British print and television media concentrated on injured children located in Sarajevo hospital. After hearing of the children's suffering and in particular the plight of one Bosnian girl, Irma Hadzimuratovic, the Prime Minister John Major apparently ordered that the injured children should be airlifted from Sarajevo to Britain for emergency treatment. While acknowledging the plight of the children, the British government refused to airlift wounded adults. The decision to launch Operation Irma was widely regarded as a cynical exercise rather than a principled intervention against the continued suffering of Bosnian Muslims. As a UN representative, Sylvana Foa, noted in August 1993, 'Does this mean Britain only wants to help children? Maybe it only wants children under six, or blond children, or blue-eyed children?' (cited in Morley and Robins 1995: 145).

Former Foreign Secretary Douglas Hurd was an outspoken critic of the emotional blackmail of the television screen. During the Bosnian crisis, he frequently condemned reporters such as *The Guardian*'s Maggie O'Kane for promoting the 'something must be done' approach to foreign policy (Ó Tuathail 1996a; Luke and Ó Tuathail 1997). Bell's image of the father declaring 'Why don't they do something' would probably fall into the same Hurd category as the reporting of field journalists such as O'Kane, Ron Guttman and Ed Vulliamy. Underlying this hostility to television reporting of war and crisis is the fear that journalists might determine the foreign policy making of nation-states. Political leaders fear losing control over decision making and hence Warren Christopher's comment that television pictures should not become the north star of American foreign policy. This image produced by Steve Bell in August 1993 acts as a form of visual agitation because it refuses to judge statesman on their own policy terrain. The ability of the visual image to disturb and disrupt foreign policy managers is disturbing to politicians such as Hurd and Christopher not through their influence on policy options but rather because media exposure might induce a loss of control over the depiction of global political space.

Cartoon images of television pictures or reporting can alert readers to the importance of media reporting of war and crisis. As a form of videocameralistics, the television screen could emerge as a new register for judgement and analysis of global political space (Luke and Ó Tuathail 1997). However, the capacity of the media to report on the destruction of Bosnia was constrained in part by the geographical inaccessibility of parts of Bosnia and the lack of suitable technology

Figure 8.7 Atrocities continue in the war in Bosnia (reproduced with permission of Steve Bell)

such as the satellite phone and video link-up. During 1993, for instance, media coverage of the violent Croat siege of Mostar received minimal coverage, and Croatian atrocities in places such as Ahmici were not supported by filming because news channels such as the BBC lacked a battlefield satellite. It has been alleged by Nick Gowing that the failure to report such incidents in 1993 may have had implications for the capacity of some UN member states to pressure the NATO coalition to intervene in Bosnia (Gowing 1996).

The dominant enframing of Bosnia as a place filled with ancient hatreds and warring factions appeared to undermine 'the something must be done' sections of British public opinion because it legitimated Western non-intervention between 1992 and 1994 (Ó Tuathail 1996b: 192–193). As a result of this particular geopolitical enframing, televisual pictures of ethnic cleansing and war were often used by political leaders to argue that nothing could be done. In the aftermath of the Srebrenica massacre, Douglas Hogg of the Foreign Office told Radio 4's *Today* programme that 'If you are asking me if we have a policy that will certainly save Srebrenica in a few hours, the answer I regret to say is no' and then argued that the safe haven policy was ill-conceived (cited in Gowing 1996: 86). Moral responsibility for Bosnia was often submerged in an unforgiving quagmire, and televisual coverage of massacres such as Srebrenica in 1995 may have motivated political rhetoric but it did not help to implement effective action. The eventual intervention of the United States and NATO in 1994–1995 were empowered in part by the televisual reinscription of Bosnia as a morally proximate place via renewed coverage of shocking atrocities. Ironically, the 1995 Dayton agreement for Bosnia reinscribed a form of ethnic cleansing onto Bosnia rather than undermining the televised Serb and Croat fantasies of ethnic and spatial purification.

Conclusions

Steve Bell's cartoons have been used to investigate the unfolding contours of the Bosnian crisis and geopolitical issues associated with Western inaction, social proximity and moral responsibility. However, the employment of interview material with the author of these images was also designed to enframe geopolitics as more than representational practices. Geopolitical iconography, as a form of analysis, should be sensitive to authorial context and audience impact rather than just another privileged act of academic interepretation. The persistent moral indifference, political ineptitude and humanitarian confusion of Western political leaders was exposed by many artists, journalists and writers in the face of repeated televisual and print media reports of atrocities, religious desecration and genocide (see Ali and Lifschultz 1993; Rushdie 1994). Bell later refused to contribute some cartoon images on 'Europe' for a film of the former Yugoslavia, because as he noted: 'This was a place which produced the Holocaust on an industrial and modern scale, which sells bombs to nasty dictators and did nothing over Bosnia ... It made me want to be sick' (Bell 1996). Stark televisual images of suffering and brutalities may not stimulate effective political and moral action by watching states and citizens. However, a considerable degree of research remains to be carried out on the impact that these images actually make on audiences in Britain and elsewhere in the world (Shaw 1996).

Whilst conscious of the enduring perils of advocating another universalizing position, critical geopolitics should contribute to the necessary and urgent task of criticizing *inter alia* the violent activities of political leaders, evaluating televisual coverage of wars and suggesting alternatives to these injustices and inequalities. This call to critique needs to be situated within an academic and political context that is sensitive to the fact that geopolitics should not be reduced to interpretation. In the case of Bosnia, the televisual and print media enframing of that crisis had consequences not only for moral responsibility but also for forms of action (see Thrift 1995). Understanding of the Bosnian crisis was enframed through the circulation of two geopolitical scripts. These scripts in turn influenced policy options and public support for particular strategies of intervention (see Pugh 1996). On the one hand, Bosnia was frequently compared to a potential quagmire, so that intervention appeared hopeless and inappropriate. The consequences of this position was that 'the place is outside the universe of obligation of US foreign policy. It is beyond the limits of duty, beyond the boundary within which open-ended, conditional moral responsibility is exercised' (Ó Tuathail 1996b: 220). On the other hand, Bosnia was televised in the grip of ethnic genocide and in doing so this offered: 'the possibility . . . of producing proximity, of reclaiming Bosnia within the universe of obligation' (*ibid.*: 222). The tragedy of the Bosnian crisis is that the 1995 Dayton Agreement reinforced not only the ethnic partition of that republic but also confirmed that Western humanitarian involvement was unable to resist ethnic cleansing and genocide (Campbell 1995).

Acknowledgements

I owe thanks to Steve Bell for granting me an interview in November 1996 and for giving permission to reproduce some of his images in this chapter. Denis Cosgrove, Steve Daniels and Gearóid Ó Tuathail made a number of very helpful comments on an earlier version. The usual disclaimers apply.

References

Ali, R. and L. Lifschultz (eds) (1993) *Why Bosnia?* Stoney Creek: Pamphleteers Press.
Adie, K. (1996) 'C is for compassion fatigue,' *The Observer*, 17 November 1996.
Arnett, P. (1996) 'The clash of arms in exotic locales,' *Media Studies Journal*, 10: 21–26.
Ash, T. (1995) 'Bosnia in our future,' *New York Review of Books* 21 December 1995: 27–32.
Aulich, J. (1992) 'Wildlife in the South Atlantic: graphic satire, patriotism and the fourth estate,' in J Aulich (ed.) *Framing the Falklands War: Nationhood, Culture and Identity*, Milton Keynes: Open University Press: 84–116.

Bell, S. (1994) *For Whom the Bell Tolls*, London: Mandarin.

Bell, S. (1996) Interview with Klaus Dodds, 25 November 1996.

Bowman, G. (1994) 'Xenophobia, fantasy and the nation: the logic of ethnic violence in former Yugoslavia,' in V. Goddard, J. Llobera and C. Shore (eds) *The Anthropology of Europe*, Oxford: Berg 143–172.

Brock, P. (1996) 'Greater Serbia versus the greater western media,' *Mediterranean Quarterly*, 7(1): 55–74.

Brooks, R. (1990) 'Everything in the garden is lovely: the representation of national identity in Sidney Strube's Daily Express cartoons in the 1930s,' *Oxford Art Journal*, 13: 31–43.

Brothers, C. (1997) *War and Photography*, London: Routledge.

Campbell, D. (1994) 'The deterritorialization of responsibility: Levinas, Derrida and ethics at the end of philosophy,' *Alternatives*, 19: 455–484.

Campbell, D. (1995) 'Violent performances: identity and the state in the Bosnian conflict,' in Y. Lapid and F. Kratochwil (eds) *The Return of Culture and Identity in International Relations Theory*, Boulder, Colo.: Lynne Rienner: 163–180.

Channel 4 (1995) *Noam Chomsky and the Media*, London: Channel 4 Productions.

Cosgrove, D. and S. Daniels (eds) (1988) *The Iconography of Landscape*, Cambridge: Cambridge University Press.

Dalby, S. (1996) 'Reading Rio, writing the world: the New York Times and the Earth Summit,' *Political Geography*, 15(6/7): 593–614.

Daniels, S. (1993) *Fields of Vision*, Cambridge: Polity.

Denitch, B. (1993) 'Learning from the death of Yugoslavia,' *Social Text*, 34(1): 3–16.

Dodds, K. (1996) 'The 1982 Falklands War and a critical geopolitical eye: Steve Bell and the If . . . Cartoons,' *Political Geography*, 15(6/7): 571–592.

Dragnich, A. (1995) 'Greater Serbia and the West's miscalculation,' *Mediterranean Quarterly*, 6: 49–60.

Entman, R. (1991) 'Framing US coverage of international news: contrasts in the narratives of the KAL and Iran Air incidents,' *Journal of Communication*, 41: 6–27.

Farren, M. (1993) 'Underground comix,' in R. Sabin (ed.) *Adult Comics*, London: Routledge: 36–51.

Fowler, M. and J. Bunck, (1996) 'What constitutes the sovereign state?,' *International Studies Quarterly*, 22(4): 381–404.

Glasgow University Media Group (1985) *War and Peace News*, Milton Keynes: Open University Press.

Glenny, M. (1992) *The Collapse of Yugoslavia*, Harmondsworth: Penguin.

Glenny, M. (1996) 'Why the Balkans are so violent,' *New York Review of Books*, 19 September 1996: 34–39.

Gow, J., R. Paterson and A. Preston (eds) (1996) *Bosnia by Television*, London: British Film Institute.

Gowing, N. (1994) *Real-time television coverage of armed conflicts and diplomatic crises: does it pressure or distort foreign policy decisions?* The Joan Shorestein Barone Center on the Press, Politics and Public Policy, John F Kennedy School of Government, Harvard University.

Gowing, N. (1996) 'Real-time TV coverage from war: does it make or break government policy,' in J. Gow *et al.* (eds) *Bosnia by Televisison*, London: British Film Institute: 81–91.

Guttman, R. (1993) *A Witness to Genocide*, New York: Macmillan.

Honig, J. and N. Both (1996) *Srebrenica: Record of a War Crime*, Harmondsworth: Penguin.

Hopkinson, N. (1993) *The Media and International Affairs after the Cold War*, Norwich: HMSO.

Ignatieff, M. (1994) *Blood and Belonging*, London: Verso.

Impact (1996) 'Getting the Agreement back on track,' 26 March, 13–14.

Irvine, J. (1993) *The Croat Question*, Boulder, Colo.: Westview.

Kellner, D. (1988) 'Reading images critically: towards a postmodern pedagogy,' *Journal of Education*, 170: 31–52.

Kellner, D. (1995) *Media and Culture*, London: Routledge.

Livingston, S. and T. Eachus (1995) 'A quantitative content analysis of the 1992 Somalian Crisis,' *Political Communication*, 12: 413–429.

Luke, T. and G. Ó Tuathail (1997) 'On videocameralistics: the geopolitics of failed states, the CNN International and (UN)governmentality,' *Review of Political Economy*, 4: 709–733.

Morley, D. and K. Robins (1995) *Spaces of Identity*, London: Routledge.

Myers, G., T. Klak and T. Koehl (1996) 'The inscription of difference: news coverage of the conflicts in Rwanda and Bosnia,' *Political Geography*, 15(1): 21–46.

Neuman, J. (1996) *Lights, Camera, War: Is Media Technology Driving International Politics?* New York: St Martins Press.

Owen, D. (1995) *Balkan Odyssey*, New York: Alfred Knopf.

Ó Tuathail, G. (1996a) 'An anti-geopolitical eye: Maggie O'Kane in Bosnia 1992–3,' *Gender, Place and Culture*, 3(2): 171–185.

Ó Tuathail, G. (1996b) *Critical Geopolitics: the Politics of Writing Global Space*, Minneapolis: University of Minnesota Press.

Ó Tuathail, G. and T. Luke (1994) 'Present at the disintegration: deterritorialization and reterritorialization in the new world order,' *Annals of the Association of American Geographers*, 84: 381–394.

Petras, J. and S. Vieux (1996) 'Bosnia and the revival of US hegemony,' *New Left Review*, 218: 3–25.

Pugh, M. (1996) 'Humanitarianism and peacekeeping,' *Global Society*, 10(3): 205–224.

Robins, K. (1996) *Into the Image*, London: Routledge.

Rose, G. (1993) *Geography and Feminism*, Cambridge: Polity.

Rushdie, S. (1994) 'Bosnia on my mind,' *Index of Censorship* 23: 16–20.

Ryan, J. (1994) 'Visualising imperial geography: Halford Mackinder and the COVIC, 1902–1911,' *Ecumene*, 1: 157–176

Salecl, R. (1994) 'The crisis of identity and the struggle for new hegemony in the former Yugoslavia,' in E. Laclau (ed.) *The Making of Political Identities*, London: Verso: 205–232.

Sharp, J. (1993) ' Publishing American identity: popular geopolitics and the Readers Digest,' *Political Geography* 12: 491–503.

Sharp, J. (1996) 'Hegemony, popular culture, and geopolitics: the Readers Digest and the construction of danger,' *Political Geography*, 15(6/7): 557–570.

Shaw, M. (1996) *Civil Society and Media in Global Crises*, London: Pinter.

Thompson, M. (1992) *A Paper House*, London: Verso.

Thrift, N. (1995) *Spatial Formations*, London: Sage.

Vulliamy, E. (1994) *Seasons in Hell*, London: Simon & Schuster.

Vulliamy, E. (1997) 'Dutch ordered retreat,' *The Observer*, 7 February 1997.

9

OUTSIDES INSIDE PATRIOTISM

The Oklahoma bombing and the displacement of heartland geopolitics

Matthew Sparke

> Outside and inside form a dialectic of division, the obvious geom-
> etry of which blinds us as soon as we bring it into play in
> metaphorical domains. It has the sharpness of the dialectics of *yes*
> and *no*, which decides everything. Unless one is careful, it is made
> into a basis of images that govern all thoughts of positive and neg-
> ative. . . .The dialectics of *here* and *there* has been promoted to the
> rank of an absolutism according to which these unfortunate
> adverbs of place are endowed with unsupervised powers of onto-
> logical determination.
>
> (Gaston Bachelard 1969: 17)

> The tragedy in Oklahoma City must remind Americans of the
> obvious but insufficiently stressed reality that the end of the cold
> war did not end the dangers that Americans face from their collec-
> tive involvement in the world.

> It is clear, I think, that there must almost certainly have been a for-
> eign origin to this, and probably one in the Middle East, although,
> of course, I have no facts to confirm that yet.

> I would have no objection if we picked out a country that is a
> likely suspect and bombed some oil fields, refineries, bridges, high-
> ways, [and] industrial complexes.
>
> Three US commentaries after the Oklahoma bombing
> (Naureckas 1995a: 7)

As the words of Gaston Bachelard remind us, and as work by international relations critics such as the political philosopher Robert Walker (1993) repeatedly
underlines, the spatial dualisms of 'outside' and 'inside' are a blindingly sharp
force of division in modern Western thought, ultimately governing even moral
considerations of good and bad. Both Bachelard and Walker are interested primarily in the philosophical and disciplinary force of such divisions but, in the

hope of displacing an epistemology/empirics dualism in this chapter, my concerns move between such philosophical themes and their reactivation in the domains of cultural commentary and criticism. The particular domains in question are those discourses that were so evident in the commentary that followed the bomb blast that killed 168 adults and children in the US federal office building in downtown Oklahoma City on 19 April 1995. From the very first headlines about terror reaching into the heart of the heartland to the later responses concerning the need for renewed government guarantees of heartland security, these discourses were characterized by a common popular geopolitical construction of a 'heartland' space. While nobody may have been able to draw it on a map and while it proved nationally expandable from the rural mid-west outwards (with politicians in 1996 starting stump speeches addressing so-called heartland audiences the length and breadth of the country) it was always clear – or so it seemed – what the heartland was not. It was not the outside. It was the heart of the inside. Thus, as the three quotations collected in the epigraph above make manifest, immediate post-bombing responses cast the attack as originating beyond the heartland and beyond the borders of the USA. In so doing, they divided the 'national' from the 'international' in terms of 'inside' and 'outside', placing the heartland at the very centre of a now seemingly vulnerable inside (see Figure 9.1). This inside/outside discursive pattern clearly paralleled and, in some obvious cases, directly borrowed themes from other discourses that had characterized earlier commentary on the Gulf War in the USA (see Sparke, 1999). However, as a result of the hegemonic geopolitics dividing the outside from the inside in American popular culture, the two events were generally viewed as unrelated. It is this rigidly dualistic, moralistic and, as such, powerful spatialization of the political that I have entitled here heartland geopolitics. As such, I argue that it is a geopolitical graphing of the geo to which critical geopolitics can bring the same kind of critical deconstructive analysis that Gearóid Ó Tuathail (1996) has brought to analysing the formulation of the paradigmatic geopolitical heartland by Halford Mackinder.

Just as Mackinder's massive simplification of global politics into a tripartite spatialized scene of struggle can be recontextualized and thereby reunderstood as a faltering discursive attempt to manage the crises of late imperialism, so too can American popular and political discourse surrounding the Oklahoma bombing be reinterpreted as a form of geopolitical crisis management. In both cases, it is possible to chart the deployment of a geopolitical gaze. Generally speaking, this gaze can take many diverse forms. It may be the basis of either individual geopolitical insights or more socially generalized constructions of popular geopolitical perspectives. Nevertheless, as Ó Tuathail makes clear, the gaze remains chronically, albeit ironically, characterized by the wholesale '*suppression* of geography and politics' (1996: 53). With Mackinder's modern gazing, Ó Tuathail shows how such suppression was linked to the ex-soldier's preoccupation with practical, 'hands-on' knowledge and his parallel disavowal of written and mediated knowledge production. By contrast, with the post-modern

Figure 9.1   If it can happen here . . . (courtesy of *Las Vegas Sun*, reprinted by permission of Mike Smith).

hyper-mediatized production of heartland geopolitics in the wake of the Oklahoma bombing it is possible to perceive another kind of suppression of geography and politics by the media gaze. In this case, it took the more socially forgetful form of disavowing precisely the practical, 'hands-on' connections that *linked* outside and inside, and showed them thus – contrary to the hard and fast dualisms of heartland geopolitics – to be intimately interrelated.

Specifically, the hands I am referring to here are those of Timothy McVeigh, the man found guilty on 2 June 1997 of using the bomb as a weapon of mass destruction (Thomas 1997a: A1). Yet as well as being the hands of the killer of 168 people in Oklahoma City, these were also the hands that operated a machine gun atop a Bradley fighting vehicle as it charged across Iraqi lines – presumably killing Iraqi conscripts – during the major land attack phase of Operation Desert Storm. These then are hands that directly connect the violence of the Gulf War with the violence of the Oklahoma bombing. And yet while McVeigh has been found guilty and sentenced to die for the heartland bombing, little has been made of his past service as a heartland hero, a decorated soldier-patriot of the Gulf War. It is this fundamental form of cultural forgetfulness that the following pages question as a way of unpacking the geopolitical constructions and contradictions of contemporary American patriotism more generally.

The chapter is divided into three main sections. To begin with, I chart in greater detail the involuted contours of heartland geopolitics by documenting how the supposedly *outside* threat to the heartland was ultimately found to be a Gulf War gunner, a heartland hero whose very vision of patriotism was nurtured in the *inside* of what Arjun Appadurai has so aptly termed 'The Heart of Whiteness' (1993: 421). The crisis in patriotism created by the disconcerting and geopolitically destabilizing fact of an outside turning out to be an inside was subsequently successfully covered over by the policy makers and spin doctors of state, and it is this form of geopolitical crisis management that I proceed to examine in the second section. I outline how heartland geopolitics was effectively resuscitated by a form of hegemonic remapping that recentered the US government. Beginning by scripting the professional bureaucracy of government as the victim of the bombing, and moving to a scripting of the same professional system as the ultimate protection against future acts of terror, the remapping enabled a territorialized and patriotic picture of defensive government to eclipse any attention that might have otherwise been paid to the fact that McVeigh himself was a product of government training in patriotism. It thereby allowed the old inside/outside analytic of heartland geopolitics to be redeployed in the shape of the 1996 Anti-Terrorism and Effective Death Penalty Act. Having examined this Act as an act of crisis management, I turn thirdly and finally to re-emphasize how, despite the seemingly obdurate character of heartland geopolitics, it is a fundamentally fragile and contradictory construction. To do so, I explore the way in which McVeigh's own act of terrorism was itself ideologically underpinned by the very same inside/outside analytics. By describing some of his favourite movies, books and journeyings into a violent subculture organized and known in the US as the militias, I argue that McVeigh simply left the broader templates of inside and outside in place while simultaneously scripting the US government as the ultimately alien, outside force. In short, he himself identified the outside inside, and thus, to the extent that this provoked him to become the heartland bomber, the bombing represents a doubly displacing performance of heartland geopolitics and the American patriotism it subtends.

Structured environments

> They call it the Heartland, a cliché that claimed its truth at 9:02 one warm Wednesday morning when a nine-story concrete and marble government office building in a quiet midwestern city was ripped apart by a powerful explosion. To the millions of Americans who watched body after body borne from the wreckage – some of them so small that weeping fire-fighters couldn't bear to look down at what they so gently cradled – it seemed as if the heart of the nation itself had been sundered.
>
> *Life* magazine (quoted in *Oklahoma Today* 1995: 100)

The murderous impact of the bombing and the terrible pain, misery and suffering forced on both the victims' loved ones and the survivors should serve before all else as a saddening and brutally obvious reminder of the life-and-death issues at stake in the discourses surrounding the bombing. No critical geopolitical examination of these issues can afford to ignore or speak glibly about the horrendous impact of the explosion. The bloodied dead bodies of the infants carried by fire-fighters were indeed part of the mediatized spectacle of an innocent and vulnerable heartland under attack, but they were no less painful scenes and experiences to witness. Thus, while I proceed to examine the cultural construction and contradictions of the discourses surrounding the bombing, I want to emphasize that the suffering caused by the violence needs to be remembered through the process of cultural contextualization. Indeed, it is precisely because of the need to come to *embodied* as opposed to *abstract* terms with such violence and its causes that a critical cultural analysis is necessitated. Such analysis enables us to go beyond the geopolitical abstractions implicit in such standardized summaries as the following: 'The bombing was the worst terrorist act on American soil and shook the nation's sense of security within its borders' (Thomas 1997b: A1). Instead, an investigation of McVeigh's soldierly background and its cultural context makes it possible to ask how such quotidian cultural abstractions of security may themselves have underwritten the terrible act of violence.

Contrary to the construction of McVeigh as an evil outsider bringing terror to the heartland, a consideration of his own schooling in heartland geopolitics highlights the obvious but ignored parallel between turning the people working for the federal government into minions of an evil state apparatus and turning the people of Iraq into minions of an evil state apparatus. Both discourses essentially involve the same form of geopolitical abstraction, and both in turn can be understood as informed by a larger cultural–political discourse about America and its enemies. Thus, just as the Gulf War was frequently scripted as a high-tech battle against a new, post-Cold War evil empire, so too, it seems, did McVeigh justify his killing of the people in the federal building using the same sort of Reaganite rhetoric. He told his friend Michael Fortier, who testified at the trial, that he 'considered all those people to be as if they were storm troopers in the movie "Star Wars." They may have been individually innocent, but because they are part of the evil empire they were guilty by association' (Thomas 1997c: A1). Ultimately, my goal in underlining the national disavowal of McVeigh's military involvement in the Gulf War and his embeddedness in this more general cultural discourse of heartland geopolitics is to suggest that the very sense of heartrending loss forced on Americans by the killing of 168 people in Oklahoma City could have been usefully turned from the inside out to bring about a critical reevaluation of the estimated 5,000 to 13,000 killings inflicted by the American military in Operation Desert Storm (Andrews 1992: Y5; Tyler 1991: A6). However, in fact very few of the same expressions of sympathy, solidarity and care were ever actually extended to the Iraqi dead, including amongst them many more children and infants. Instead, their pain and suffering became mere 'collateral damage' in an abstract geopolitical struggle against the forces of chaos and evil, outside forces that became metonymically symbolized by the ultimate Arab enemy outsider, Saddam Hussein. Such geopolitical scripting is of a piece with the wider cultural–political discursive formation of American patriotism that the sociologist James Gibson (1994) calls the 'New War'. Following feminist critics such as Susan Jeffords, Gibson argues that this so-called New War comprises the whole hegemonic masculinist cultural crisis management of post-Vietnam patriotism (see Jeffords 1989, 1994). As such it includes everything from action movies like *First Blood*, *Dirty Harry* and *Red Dawn* to *Soldier of Fortune* conventions, to the organization of militias, to the actual deployment of the American military in events like the Gulf War. My own point here is that the Oklahoma bombing can itself be seen as one more deployment in this same New War. Rather than make this connection, however, most mainstream analysis of the bombing foreclosed any critical re-evaluation of New War culture by actually deploying it yet again to script the terror as emanating instead from an outside threat. It is therefore to this scripting and its short-lived coherence that I turn first.

The rush to blame outsiders began from the assumption that the morally abhorrent attack on the heartland could never have come from inside: it almost certainly must have come, in the words of the former CIA official and terrorism 'expert' Donald Jameson, from a foreign origin. The other quotations I have cited in the epigraph at the start of the chapter argue from the same New War script, aligning good and evil with inside and outside, respectively. The bombing must thus have been related, in journalist Jim Hoagland's terms, to Americans' collective international involvements. And, in Mike Rokyo's chillingly glib radio reporting, it therefore demanded prompt international retaliation. Clearly, then, in the terms of the hegemonic geopolitics of American patriotism, the earliest reporting on the Oklahoma bombing shared much with the New War discourses surrounding the Gulf War. With the bombing scripted as another case of Middle Eastern terror originating in an outside space of crazed turmoil, its coding had directly racist implications similar to those that had led from anxieties about Arabs to anti-Arab violence during the Gulf War (Abraham 1992). Feeding racist anxieties again, CNN turned to generic orientalism in reports about authorities tracing several 'Middle Eastern-looking' men, while local media outlets like the *El Paso Times* were more specific with reports about a search for 'two men of Middle Eastern appearance,' who might have been wearing bloody clothes and who were heading toward the Mexican border in a Chevy Cavalier or Blazer (Stern 1996: 182). Such generic orientalism led to targeted racism half a world away when airport officials in London sent another 'Middle Eastern looking' man back to the USA, where he was described by the Justice Department as 'a possible witness' (*ibid.*: 183). Domestically, such assumptions sanctioned a new round of anti-Arab attacks, with several mosques receiving telephone threats, while a former Oklahoma congressman, Dave McCury, pointed to an Islamic conference held in Oklahoma City in 1992 as the reason why he knew terrorism 'could happen here' (Lacayo 1995). Meanwhile, the foreign policy 'experts' on Middle Eastern terror were in their element. Steven Emerson, for example, a supposed expert on Islamic Jihad, said: 'There is no smoking gun. But the modus operandi and circumstantial evidence leads in the direction of Islamic terrorism' (Stern 1996: 183). As Edward Said angrily notes, according to this standard US foreign policy script 'Arabs only understand force; brutality and violence are part of Arab civilization; Islam is an intolerant, segregationist, "medieval," fanatic, cruel, anti-woman religion' (1993: 299). Thus it followed that the appropriate response was the imperial one that the American media had presented as a huge, clean and virtuous success four years earlier.

Before too long, however, the other, far more complex connection with the Gulf War became apparent when the chief suspect in the bombing became McVeigh. In this moment, the outside Arab threat turned out not only to be an ultimate insider, a white American Army man, but also a 27-year-old veteran of Operation Desert Storm itself. McVeigh (along with his alleged accomplice Terry Nichols) had been in the First Division, otherwise known as the Big Red One, a division that had buried thousands of Iraqi soldiers alive by ploughing sand over their trenches (Kellner 1992: 103–109). For his services manning the machine gun on the Bradley fighting vehicle, McVeigh was awarded a Combat Infantry Badge and a Bronze Star (Kifner 1995a: A1 and A12). Upon his return to the USA, however, his hopes of becoming a 'green beret,' a member of the US Army's famed Special Forces, were ended when he dropped out on the second day of a 21-day selection course (*ibid.*: A12). From this point on he drifted from job to job, going to gun shows and reading and distributing white supremacist literature about resisting both federal and 'world' government (Kifner 1995b: A1 and A10). During the penalty phase of his trial, it became clear that McVeigh's anti-government resentment was heightened during this time by the passage of gun control legislation in Congress and by the 19 April 1993 raid by the Federal Bureau of Alcohol, Tobacco and Firearms on the Branch Davidian compound in Waco, Texas, the anniversary date of which he evidently chose for the bombing (Bernstein 1997: A9). In his anti-government anger, McVeigh is understood to have been sympathetic to the militia movement more generally, worried about the threat of 'world government' at large, distrustful of the US federal government in particular, and violently and particularly opposed to gun control (Kifner 1995b: A10). The two bumper stickers on his car said it all: 'A man with a gun is a citizen. A man without a gun is a subject;' and 'Ban guns. Make the streets safe for a government takeover' (Thomas 1997d: A14).

Given McVeigh's soldierly passion for guns, it is not without a certain irony that, 90 minutes after the explosion of the Murrah building, the Gulf War veteran's destructive odyssey finally came to an end when he was arrested on a weapons possession charge after a routine traffic stop about 60 miles away from Oklahoma City. When Assistant District Attorney Mark Gibson later approached McVeigh in court to tell him that he would be held on charges of blowing up the Murrah Building, the Gulf War gunner's reaction, according to Gibson, was 'like the dutiful soldier.' 'Emotions don't come into play, right and wrong don't come into play. My feeling was that, in his mind, that was the end of that

portion of his life. . . . He exuded nothing. . . . His mission was accomplished'
(Stern 1996: 184). Indeed, McVeigh is said to have referred to himself as a prisoner of war and, following the Michigan Militia's guide to members 'captured
by the federal government,' would give only his name, rank and date of birth
(*ibid.*: 184). According to an FBI expert on psychological profiles, his post-war
peregrinations from the army to the bomb plot to prison had been totally to
psychological type. 'These people are comfortable in a structured environment,
they do very well. But outside of a structured environment, without that rigidity, they just can't survive' (Kifner 1995a: A12).

I will return in more detail to the question of McVeigh's own character in the
third section of this chapter but my immediate point here concerns the disruption his actions represented in that more general 'structured environment'
constituting heartland geopolitics. When the FBI released sketches of suspects
who looked more like stereotypes of mid-western fraternity boys, and when
shortly thereafter they identified the already arrested McVeigh as the suspect, the
whole outside/inside logic of national identification collapsed. McVeigh was not
an Iraqi. He was a veteran of the war against Iraq. The outside threat thus
turned out to have been a native son, an *insider* who knew the Heartland well,
a soldier-patriot who had fought for his country *against* the outside. This reversal displaced the heartland geopolitics of inside and outside. The ambivalence of
national love and hate, normally so well set apart and controlled by what Homi
Bhabha calls a dialectic of division – the 'obvious geometry . . . which blinds us
as soon as we bring it into play' (1994: 149) – collapsed in crisis.

According to Bhabha, such crises are a symptom of how 'the ambivalent identifications of love and hate occupy the same psychic space' (*ibid.*). This co-occu-
pation means, he contends, that 'paranoid projections "outwards" return to
haunt and split the space from which they are made' (*ibid.*). If that space is seen
here as the imagined geography of America as nation, McVeigh's return from
the Gulf War would seem to represent a particularly murderous incarnation of
the return of the paranoid. Normally, of course, it does not happen in this form,
and as Bhabha argues, 'as long as the boundary is retained between the territories, and the narcissistic wound is contained, the aggressivity will be projected on
to the Other or the Outside' (*ibid.*). It is not necessary to embrace the whole
Freudian apparatus and its attendant assumptions about sexuality and family to
suggest that in general terms this is what happened during the Gulf War, a war
in which America as nation was cathected with love and festooned with yellow
ribbons while Iraq and Saddam Hussein became the outside objects of hate. But
the bomb blast and McVeigh's arrest exploded this easy inside/outside geography of national crisis management. The wounds were therefore delivered not
only to the children and workers who were killed and injured by the explosion
in Oklahoma. There were also ideological wounds to the structured cultural
geopolitics underwriting much of American patriotism.

The fact that the response of US commentators immediately after the bombing was to blame Middle Eastern terrorists, and to even call for a show of

strength like that shown in the bombing of Iraq, clearly illustrated the normal inside/outside geopolitics at work. In turn, the personal itinerary of McVeigh revealed a doubly displacing reversal of this geopolitics, where the outside force turned out to be an insider involved in a far-right subculture obsessed with the most paranoid American projections about the outside. These ironies are important, but it should be remembered that, even after the FBI had released its sketches of the suspects, many commentators could not abandon the secure rigidity of the inside/outside vision of foreign terrorist involvement. John McLaughlin, for example, was reluctant to change the script for his TV political entertainment show 'the McLaughlin Group,' asking questions such as: 'Even if Oklahoma turns out to be wholly the work of domestic terrorists, will it become a wake-up call for the need to focus on international terrorism, conventional and possibly nuclear?' (Naureckas 1995b: 7). Similarly, a *New York Times* editorial, addressing the federal grand jury's release of its indictments of McVeigh and Nichols, would not let the inside/outside logic drop. It did note that Americans had wrongly 'assumed it must be the work of Islamic fundamentalists,' and that this assumption had been dropped when it was realized that 'this was a home-grown plot' *(New York Times* 1995: A14). However, the editorial concluded by 'outside'-ing the accused again, this time as a lunatic fringe: 'it is disturbing to learn just how vulnerable our wide-open but heavily armed society remains to the venomous acts of a handful of crazed outcasts' (*ibid.*). Beyond these particularistic cultural reproductions of heartland geopolitics, however, the same hegemonic discourse of inside and outside was also far more extensively re-entrenched through the official response of the Anti-Terrorism and Effective Death Penalty Act passing through Congress. It is to this far-reaching bill, which President Clinton signed into law on 24 April 1996, that I therefore turn next (Mitchell 1996: A6).

Secret truths

America cannot forget and will not forget Oklahoma City. America will not forget the victims, the courage of the rescuers, the service to our nation of the federal workers who died, or the promise of the children who left our world too soon. One of the secret truths about our human condition is that suffering binds us together. Your resoluteness has also told us something about the state of the union. In America, terror will not triumph. Let me say it again, terror will not triumph. The reason it will not is because in our nation we settle our differences with dialogue and debate. We do not steal precious human lives to express our discontent.

(Vice President Al Gore quoted in Thomas 1996: A8)

Vice President Al Gore's somber words of remembrance illustrate a non-academic form of national narrativization that is so common that it generally passes without comment. A year after the bombing, in a speech turned homily before mourners in Oklahoma City, he waxed transcendental on the 'secret truths' underpinning the 'human condition' and the 'state of the union:' on how people are bound together through suffering; how Americans in particular do not forget victims; and how, 'in our nation, we settle our differences with dialogue and debate.' Such rhetoric calls out for examination for at least two sets of reasons. On the one hand, as a heartland homily it narrativizes the essence of national identity, assuming connections between the collective identity of Americans and the personal identities of, in this case, the Oklahoma City mourners in a way that immediately occludes the complexity of these interconnected but not entirely continuous subjects of identification. On the other hand, its very call to remembrance seems simultaneously to take part in one of those rituals of amnesia that, as Ernest Renan once argued, are so important to the consolidation of the modern nation. 'Forgetting,' Renan reminded his late nineteenth-century Sorbonne audience, 'is a crucial factor in the creation of a nation' (1990: 11). In the late twentieth-century case of the remembering that marked the anniversary of the Oklahoma bombing, such constitutive forgetting was no less evident. Thus, the very day before Gore urged his audience to remember how in America 'we do not steal precious human lives to express our discontent,' Republican legislators with significant support from the White House and congressional Democrats passed a purportedly 'anti-terrorism' bill through Congress that had as one of its major purposes a form of legislated amnesia about the constitutional right to appeal, and thus the right to life, of the mainly African-American prisoners awaiting execution in US prisons. Their human lives, the legislation seemed to say, *could* be stolen in order to express discontent about the threat of terror. Their constitutional rights and their possible innocence could be forgotten. They could even be forgotten on the very next day, a day of national remembrance.

By beginning this section with these observations, my initial point is to underline the homogenizing and all-encompassing involutions of heartland geopolitics. Generalizing from individual desires, fears, hopes and dreams to collective group emotions is, to some extent, what the imagining of a national community is all about. The metonymic leap from the individual to the group simultaneously distinguishes inside from outside, organizing the geography of memory and forgetting along the way. In the case of Gore's explanation of American resolution, for instance, the words about remembrance, dialogue and debate all thus seemed constitutive of a normative nationalism that simultaneously invited his listeners to forget how the anti-terrorism legislation – approved by the Senate on Wednesday and supported by a 293 to 133 vote in the House on the eve of his Friday speech – posed a direct threat to the lives of Americans judicially deemed non-precious and thus, in a certain sense, non-national.

One of the tasks for critical geopolitics is clearly to break up such narratives of nation, examining the moments where normalizing assumptions are deployed with exclusionary effect. The debate that has taken place in the aftermath of the

Oklahoma bombing is interesting in this regard because it has increasingly turned away from McVeigh's personal journey from Gulf War gunner to heart-land bomber. Instead, a form of national crisis management has remapped the threat of terror from white male insiders like McVeigh onto a normative geogra-phy of non-white/foreign threats. This remapping accounts in turn for the amnesia about stealing life that marked Gore's remembrance speech. Exploring this discursive remapping, it is possible to chart how the crisis management turned quickly from the subject of McVeigh and the militias to questions about efficiently arresting and trying the suspects, to the altogether different subject of the sweeping anti-terrorist legislation passing through Congress one year later. After twelve months, it seemed, the reversal and displacement of inside and out-side resulting from McVeigh's actions had itself been reversed and recovered by a national New War narrative about threats from the outside and bold resolution – 'terror will not triumph' – within.

After the arrest of McVeigh, attention did turn, albeit briefly, to the subject of the alienated white American men like McVeigh in the militias. April 19 was known to be the anniversary of the Waco raid that had angered so many anti-government members of the militia movement. Thus, when news of McVeigh's own militia sympathies emerged, reporters began to ask questions about the more general threats posed by the gun-toting, largely white supremacist, mainly male clubs. Was there a connection between this particular culture of masculin-ity and the bombing? Some commentators thought there was, and generalized from this to the extent that they even found fault for the bombing in white mas-culinity itself. The terror in Oklahoma City, concluded Juan Williams in the *Washington Post*, was committed by 'white men in their natural state' (quoted in Pfeil 1996: 115). Such essentialist claims, as Fred Pfeil (*ibid.*) highlights, hardly contributed to a nuanced argument about class, gender and violence in America. Instead, they fueled more anxious cynicism about politics, a cynicism that President Clinton himself argued enabled the 'purveyors of hate and divi-sion' to 'leave the impression, by their very words, that violence is acceptable' (quoted in Cooper 1995). Reactionary critics in turn saw Clinton's remarks as an attack on right-wing talk radio. Rush Limbaugh, for example, wrote a full-page column in *Newsweek* headlined 'Why I'm not to blame.' He also com-plained that 'those who make excuses for rioters and looters in Los Angeles now seek to blame people who played no role whatsoever in this tragedy' (Limbaugh 1995).

In all the accusations and counter-accusations, interest in the militias them-selves waned to be replaced a year later by interest in the passage of an Act that might expedite the legalized killing of some of the imprisoned African-American protesters turned rioters mentioned by Limbaugh. In the interim, commentators found solace in the arrest of McVeigh. Here, at least, was comfort. Comfort, first of all, because the Gulf War veteran turned suspect had been apprehended. Comfort, second, because he was seen as an outcast loner, his actions not directly connected to a broader documentary conspiracy of terror. And comfort,

third, in the fact that the FBI and other law enforcement agencies had discovered all this with professional efficiency. They had chased down leads, and,
interviewing hundreds of people in the largest manhunt in American history,
they had effectively isolated the suspect and his small band of accomplices. Later
on, in the lead-up to the trial of McVeigh, it became clear that the FBI had been
anything but professional in its handling of evidence from the bomb scene
(Thomas 1997e). But in the meantime, the system had seemed to work. Thus,
while politicians and talk radio hosts shifted positions and traded accusations
about politicizing the tragedy, there was security at least in the efficiency with
which the law enforcement apparatus had operated. Most of all, there was the
sense of security to be found in the way that McVeigh had already been arrested
as a result of a routine traffic check by a highway patrol officer. The officer in
question, Charles D. Hangar, had noticed that McVeigh's yellow Mercury was
missing a license plate. It was only when he approached the car that Hangar saw
a bulge in McVeigh's jacket that turned out to be a 9 mm Glock semiauto-
matic pistol loaded with Black Tallon bullets (the infamous bullets known as -
'cop-killers' because of their capacity to penetrate bulletproof vests). It was this
concealed weapon that then led to the arrest (Stern 1996: 184). Subsequently,
the postponement of McVeigh's arraignment (because of an ongoing divorce
case the judge was hearing) made it possible for the young prisoner to be
matched with an artist's impression of suspect 'John Doe Number 1' being circulated by the FBI. Despite the serendipity of the judge's delay, there was still
comfort here insofar as the professional system itself, the observation of officers,
the integrity of the routine arrest procedure, and the effectiveness of the whole
managerial machinery of surveillance had been vindicated.

I summarize this comforting scenario in this way because it seems to have
been the first step in the remapping of the terror back on to the templates of the
New War discourse earlier evidenced in the Gulf War. As Susan Jeffords (1993)
argues in her examination of the Gulf War 'Patriot system,' part of what caught
the public attention, and made the war seem the neat Nintendo exercise it was
not, was the illusion of efficiency surrounding the defensive capabilities of the
Patriot missile. While this US missile may never have in fact locked on to any
incoming Iraqi Scud rockets in the Gulf War, discursively it won the media war
in the US, argues Jeffords, because of how it locked in to the mystique and
appeal of professionalism. It is this notion of security through professionalism
that Jeffords labels the 'Patriot system.' She argues that it served as such during
the Gulf War to manage the crisis created by the Iran–Contra scandals, restoring
a sense of security in bureaucratic management by repackaging professionalism
as the ultimate defense against disorder in what President Bush opined was a
new world order (Jeffords 1995: 544). The 'Patriot system' is then, Jeffords
says,

> a pervasive and complex one that permeates not only military and mil-
> itary industrial networks, but popular cultural representations as well.

> The Patriot system is more than an individual weapon, encompasses more than any single economic or foreign policy interest, and intersects with U.S. activities and interests both in the United States and abroad. It is through examining such a system as this . . . that we can begin to get a glimpse of how the New World Order that was the single greatest advertising campaign of the Persian Gulf War was being sold.
>
> (*ibid.*: 542)

My suggestion is that the Patriot system's utility outlived the new world order campaign and came to serve in new crisis-managing capacities. As such, its characteristic managerial heroism seems to have played a vital part in the recuperation of the reversals of inside and outside represented by McVeigh. Thus, while to begin with it was the law enforcement system of suspect identification and arrest that helped to restore a sense of security in the immediate post-bombing weeks, the wider Patriot system's organization as an inherently defensive, efficient, American and, thus, just system helped in turn to coordinate the remapping of the Oklahoma aftermath in terms of inside/outside oppositions. From the celebration of the efficiency of law enforcement to Gore's eulogy to the just and law-abiding 'state of the union', it was thus an easy and securely patriotic transition.

To some extent, of course, the reaffirmation of law enforcement managerial heroism represented a direct response to the forms of anti-law enforcement, anti-government proclamations made by McVeigh, the militias and their sympathizers on the right. A letter left by McVeigh on his sister's computer, for example, addressed the Alcohol, Tobacco and Firearms Bureau (ATF) as 'tyrannical agents.' The letter went on to threaten the ATF, saying '[you] will swing in the wind one day for your treasonous actions [at Waco] against the United States. . . . Remember the Nuremberg trials – "but, but, but I was only following orders." Die you spineless cowardice [*sic*] bastards' (Thomas 1997f). Yet whether or not the reaffirmation of law enforcement was targeted at such involuted patriotism, it is clear that it also served the larger purpose of reorganizing the hegemonic framework of commentary and criticism about the bombing. For one thing, it enabled those on the right to begin to counter-attack with complaints about supposedly left-wing affronts to the patriot system of managerial heroism. For example, in a post-Oklahoma discussion of talk radio on 'This Week With David Brinkley,' conservative commentator George Will was able to slot Oliver Stone and his movie *JFK* into the position of anti-patriot outsider:

> I must tell you today the clearest expression of violent hatred of the United States government was a blockbuster success of a movie called *JFK*, by Oliver Stone . . . who said, 'The federal government is a murderous conspiracy designed to kill Jack Kennedy.'
>
> (quoted in Naureckas 1995: 18)

210

More generally, as soon as post-Oklahoma discourse came to be dominated by such re-centering of the US government as the embattled but resolute heart of the nation, it became possible for the whole political debate over an adequate response to the bombing to be redirected. It was in this way, I suggest, that the national *vs* non-national coordinates of inside and outside were slowly reintroduced, leading ineluctably to the Anti-Terrorism and Effective Death Penalty Act of 1996.

As he signed the Act into law, President Clinton explained to the families of those who died in the Oklahoma bombing that: 'Your endurance and your courage is a lesson to us all. Your vigilance has sharpened our vigilance' (Mitchell 1996). Quite what vigilance on the part of the victims' families the president was referring to, it is hard to tell. But it is clear that the 'sharpened' form of vigilance he saw as instantiated in the Act bore many of the hallmarks of the Patriot system's professionalism. The president had called for the passage of the counter-terrorism legislation within days of the bombing, but it became stalled in Congress because some of its provisions had attracted the ire of gun rights and privacy rights advocates. As a result, the version he ultimately signed on 25 April was stripped of certain measures he had supported, including lowered standards for lawsuits against sellers of guns used in crimes and increased wiretap authority. The legislation also included provisions that Clinton had not directly called for, including the measure that would give death row inmates only six months from their final state court proceeding to file a *habeas corpus* petition, the process that convicts use to obtain a federal review of their convictions. Nevertheless, upon its signing into law this aspect of the Act came together with a series of anti-foreigner measures as the twin pillars of what was effectively a legislative reinscription of the professional Patriot system's inside and outside.

Under the legislation, the government, or more specifically the Patriot subsystem of the Immigration and Naturalization Service (INS), is permitted to deport suspected terrorists without presenting a judge with the evidence used against them. As well, of course, as assuming that terrorists are foreigners, that they come from outside and that they therefore have a non-domestic destination to which they can be deported, this aspect of the legislation provides a clear example of the way in which the Act places complete trust in the Patriot system to decide inside from outside. Likewise, there is another provision in the Act that allows the government to freeze the assets of foreign organizations that the Administration considers terrorist. The Act furthermore empowers the INS to exclude foreigners who belong to suspected terrorist organizations from entering the United States, even if there is no evidence that they have broken the law. For certain immigrants who manage to enter the USA under these new protocols, the Patriot system, as enshrined in the Act, has other obstacles to offer. Thus, people who say they are fleeing persecution and who arrive without valid travel documents would have their asylum claims decided by a single INS border officer (Schmitt 1996: C19). Previously, the law had required a hearing before an immigration judge, and in 1995, 3,297 asylum seekers who arrived without valid documents received such hearings. Now, instead, such asylum seekers will be given no more than a supposedly professional interview by an individual border guard. Nothing guarantees their fair treatment except the assumed professionalism of such guards of the Patriot system. As James Zogby, president of the Arab American Institute, concluded, it will therefore 'seriously erode civil liberties' (Lacayo 1995: 7).

With still wider ramifications, the Act also puts at the mercy of managerial heroism the rights of those who enter the country illegally (Schmitt 1996: C19). Previously, such immigrants were entitled to a deportation hearing that guaranteed them constitutional rights like legal representation and thereby placed the burden of proof on the INS. Now that burden of proof has been lifted. This is a change that could theoretically affect hundred of thousands of immigrants, including those who have lived in the USA for years, married American citizens or had children who are citizens. Eric Schmitt reports that in 1995 about 80 per cent of the 110,000 cases that immigration judges decided involved people who faced deportation for being in the country illegally (*ibid.*). Under the new Act, the fate of such people rests entirely on the decisions, supposedly professional ones, of individual immigration officers. Critics such as Lucas Guttentag, director of the Immigrants' Rights Project at the American Civil Liberties Union, complained that the Act's provisions were 'a radical restructuring of the immigration laws that have nothing to with terrorism' (*ibid.*). However, in terms of the Patriot system, such legislated management of inside and outside instead promised heroic management in the face of attacks on the heartland. The ugly irony of this inside/out logic was not lost on Robert Rubin, assistant director of the Lawyers' Committee for Civil Rights. 'They are not terrorists,' he said, referring to many of the immigrants affected by the Act, 'they are the victims of terror' (*ibid.*).

Similarly, the Act implemented the inside/out logic of heartland geopolitics to inscribe an 'Effective Death Penalty' as the professional response of the Patriot system to the Oklahoma bombing. According to Stephen Labaton, this aspect of the Act – the new limits set on prisoner appeals of death penalty sentences – imposed the most rigorous constraints on the constitutional right to seek federal review of convictions since Lincoln suspended the writ of *habeas corpus* in the Civil War (1996: C18). As well as restricting the *habeas corpus* right to appeal to the six-month period following the final state court proceeding, the legislation also demands that federal judges defer to the legal conclusions of state courts about when a conviction violates the constitution. Gregory Nojeim of the American Civil Liberties Union underlined the role of this part of the legislation within the broader apparatus of the Patriot system when he complained that the Act marked 'an historic expansion of Federal law-enforcement authority at the expense of civil liberties. We haven't seen anything like this since the days of McCarthyism' (*ibid.*). Rather than tackling terrorism, critics such as Nojeim have said that instead the law would 'virtually insure that a person wrongly incarcerated would not get his day in court to prove his innocence' (*ibid.*).

The cultural geography of heartland geopolitics represented and reproduced by the 1996 Anti-Terrorism and Effective Death Penalty Act can be examined on several levels. Most racistly, its restrictions on the writ of *habeas corpus* represent the start of a return to the pre-civil rights movement and even pre-reconstruction days of white supremacist definitions of citizenship and civil rights. It had been the Reconstruction Congress of 1867 that had first passed the Habeas Corpus Act in 1867 in order to protect the new rights of former slaves by giving federal judges the jurisdiction to hear any cases involving a person 'restrained of his or her liberty in violation of the Constitution' (noted in *ibid.*). By the turn of the century, the Supreme Court was giving more force to the Act, using the writ to overturn a 1915 conviction in which a prisoner's rights to due process were violated by a mob-dominated state trial. Four decades later, an increasingly activist Supreme Court, responding to the nascent civil rights movement, took the writ still further, and, in a landmark ruling in 1953, the Justices said that *habeas corpus* gave federal judges the authority to consider any constitutional claim, even if the state court had considered and rejected it (*ibid.*). Thus against this steady expansion of civil rights under the writ, the 'Effective Death Penalty' part of the 1996 Act has represented a conservative and effectively racist curtailment of what scholars of law and race like Cornel West see as the emancipatory potential in the reworking of the American liberal legal tradition (see West 1993).

Not only did it represent a reactionary rebuttal to those who, like West, seek more expansion of civil rights, the Act also deflected attention away from the background of the Oklahoma bombing in the 'Heart of Whiteness' of heartland geopolitics. The provocative rephrasing of Conrad's imagined geography is Appadurai's, but it is given substance in concrete studies of the American far right by writers such as Jessie Daniels (1997) and James Ridgeway (1990). In *White Lies*, for example, Daniels makes clear not only how disturbingly close to the mainstream far right white supremacism actually is, but also how bound up it is in masculinist notions of white male heroes struggling for order against the feminized chaos of such 'plots' as government-imposed affirmative action. McVeigh was clearly a product of this middle American milieu, and in referring to how his platoon had the worst race problem in the 16th Infantry Regiment's Second Battalion, a *Newsweek* article noted what the authors called 'the fact that McVeigh, the all-American boy next door, is almost certainly a racist' (Morganthau and Annin 1997). Nevertheless, the official response to the terrorism instantiated in the so-called anti-terrorism Act allowed Americans to forget the white supremacism of the militia subculture in which McVeigh moved by repeating that same supremacism through the curtailment of an anti-racist law. While its inside/outside analysis of terror located terror on the outside of the patriot system, it also deflected attention from the questions raised by McVeigh's position as an insider white male soldier-patriot turned self-styled patriot-bomber.

Ironically, of course, the death penalty portion of the Act will nonetheless have a bearing on McVeigh's ability to file a *habeas corpus* appeal against his death sentence. No doubt this will be seen by the Act's supporters as vindication of its relevance, but what would still be neglected in such arguments is the very peculiarity of McVeigh as a white man – 'the all-American boy next door' – on death row. Of the twelve prisoners on the *federal* death row, nine are African-American, one is Hispanic, one is Asian and one is white. McVeigh will be the second white, a rare exception to the general pattern that has seen minorities constitute 78 per cent of the defendants in federal cases against whom prosecutors have sought the death penalty (Dedman 1997: A15). Perhaps this accounts for why so many of the commentators assessing whether or not McVeigh would be sentenced to death in the penalty phase of the trial returned repeatedly to McVeigh's all-white 'boy next door' character. 'It is far more difficult to kill someone you can empathise with – someone who, despite his crime, is at least a bit like you,' explained the *Newsweek* article, also noting that this was why the strategy of his lawyers in this phase was to humanize McVeigh as precisely the 'boy next door' (Morganthau and Annin 1997: 23). To this end they called upon John and Elizabeth McDermott, who were the actual neighbors of so-called 'Tim' when he was in his teens. 'I loved him,' said Elizabeth McDermott. 'I liked him very much,' her husband testified, sobbing. 'I can't imagine him doing anything like this' (quoted in *ibid*.: 23). Clearly, though, the jury that found McVeigh guilty *could* imagine him carrying out the bomb plot. Perhaps this is why the judge hearing the case warned against a 'lynching' in the penalty phase (Thomas 1997g: A1–A15). It seemed that, as a mass murderer and therefore automatically an outcast from the heartland, McVeigh had also become an honorary non-white vulnerable to lynching. To this extent, the defense strategy of humanizing McVeigh and trying to show that he was a misdirected patriot was probably doomed from the start. It ran right up against the monolith of heartland geopolitics. However, even as it failed, it nevertheless also provided a still more detailed glimpse of the convoluted logic through which the 'boy next door' turned the outside into an inside – in this case the federal government – thereby further displacing heartland geopolitics for a second time. It is to this second major displacement that I turn in the third and final section.

McVeigh: the militia man next door

There is much violence. There is much death. There is tremendous suffering. But there is also a person at the center whom you will not be able to dismiss easily as a monster or a demon, who could be your son, who could be your brother, who could be your grandson, who loved this country, who served it , but who is just like any of us.

Richard Burr, lawyer for McVeigh (quoted in
Thomas 1997h: A15)

214

While McVeigh's lawyers argued that he was normal, that he loved America and that his anger about the federal government's handling of the Waco siege was in fact patriotic, few reports took up the question of what this actually might imply about the nature of American patriotism more generally. Like the jury, which voted unanimously against the mitigating claim that McVeigh 'believed deeply in the ideals upon which the United States is founded' (Bernstein 1997), mainstream coverage largely decided to dismiss the connection to middle America and the militias and, with it, the questions raised about American patriotism more generally. By contrast, critics of the militia movement such as Kenneth Stern have attempted to keep such questions afloat by using the Oklahoma bombing to highlight the hatreds held as a matter of so-called 'patriotism' amongst the militias. As well as noting such details as the connections between McVeigh and the militia group known as the Arizona Patriots, and the fact that McVeigh and Nichols established a paramilitary cell that they called the 'Patriots', Stern begins his book by asking critical questions about the broader system of patriotism turned paramilitary activism:

> Since April 19, 1995, and the revelation that those accused of the bombing – Timothy McVeigh and Terry Nichols – had connections to and shared the ideology of the militia movement, Americans have not needed reports with large appendices to tell them that there are people who call themselves patriots yet would commit terrorist acts. But we still need to understand why. What has drawn between ten thousand and forty thousand Americans to these private armies ready to make war with the American government? Why do hundreds of Americans – some say millions – sympathize with them?
>
> (1996: 198)

This crucial question of 'why?' was what the remapping of the bombing in the anti-terrorism Act foreclosed. But by returning to the very ordinary life depicted by the defense team in the penalty phase of his trial, and by examining McVeigh's links to the militia movement, the implicit questions raised about patriotism can be made manifest.

In a brilliant piece of post-bombing investigative reporting subtitled the 'Unravelling of One Man's Frayed Life,' John Kifner introduced readers of *The New York Times* to key aspects of McVeigh's personal life. Asking 'Why would he do it?' Kifner begins his report with a description of some 'seemingly ordinary activities' that characterized the suspect's day-to-day routine prior to the bombing.

> He rented movies, playing one about a Colorado football team over and over. He wore a favorite T-shirt with a quotation from Thomas Jefferson. He once changed cheap motels so he could watch the X-rated Spice Channel.
>
> (1995a: A1)

Kifner goes on to point out that each of these supposedly all-American male activities can be linked in turn to the peculiar world of fantasy and paranoia out of which the bomb plot appears to have been hatched. The movie, called *Red Dawn* (directed by John Milius, an ardent gun rights advocate), is about the Wolverines, a small-town football team, and their guerrilla war against invading Soviet, Nicuraguan and Cuban paratroopers. It is a cult film of the far right, and the Arizona Patriots were found by the FBI to be watching it in 1986 before embarking on a plot to rob armored cars to finance the bombing of power plants, dams, synagogues and abortion clinics. Second, and similarly symptomatic of a Rambo-like vision of violent 'patriotic' struggle, Kifner reports that McVeigh's T-shirt quotation celebrated the bloody sacrifices needed to keep the American tradition of patriotism alive: 'The tree of Liberty,' the quotation read, 'must be refreshed from time to time with the blood of patriots and tyrants.' Third, McVeigh's interest in the Spice Channel also bespeaks, albeit in a very anecdotal and unsystematic way, an alienated experience of a hypermasculinized and heterosexualized identity embodied in what Jeffords calls 'Hard Bodies.' Analysing the mainstream Hollywood movies of the Reagan era, Jeffords argues that

> the heroes of hard-body films suggest a . . . kind of social order . . . in which men who are thrust forward into heroism are not heroic in defiance of their society but in defiance of their governments and institutional bureaucracies.
>
> (1994: 19)

McVeigh, as the middle American boy next door, may well have seen his own struggle against the government as part of such Ramboesque defiance: socially patriotic and manly but also alienated, anti-bureaucratic and unappreciated. But whatever he may have imagined he was doing, our knowledge of his all-American life – his fascination with the militaristic genre of popular culture, his love of the military, and his frustrated attempts to become a hard body man – all raise questions about violence and patriotism that his dismissal as 'a monster or a demon' does little to address. In short, his defense team raised a vital issue.

Beginning with the personal life-story, Kifner notes that McVeigh was badly affected as a boy by the departure of his mother, Mildred McVeigh, from the family home in upstate New York. After she left, McVeigh almost never spoke to her again, and, even after he joined the army, he seemed to harbor a lasting resentment against her. 'I just remember him calling his momma "that no good whore, a slut," words like that,' recalled an army friend (Kifner 1995a: A12). His army buddies also reported that he never had a girlfriend and that this made him a standing joke in the unit. The driver of McVeigh's Bradley vehicle noted that 'he never had any luck with girls. He was nerdy. He used to say, "Man I've got to get me a girlfriend," but he never had any luck' (*ibid.*). Nevertheless, joining the army did provide McVeigh with another form of long-term and passion-filled relationship. The army allowed him to live out what were apparently dreams of militaristic masculine performance. These dreams had been evidenced earlier in events like his appearance at his pre-army workplace dressed like Rambo with crossed ammunition bandoliers. In the army, of course, military dress became the uniform, a uniform that McVeigh seemingly cherished. His was always pressed, his boots always shined, and, as a result, he became a sergeant when his contemporaries were still only privates. Above all, his first love was for military hardware, and, in particular, guns. He had many of his own too, keeping a private arsenal of nearly a dozen even while in the army. He spent many of his off-hours cleaning these weapons, and his fellow soldiers report that he was also an avid reader of army manuals and gun magazines like *Soldier of Fortune* (*ibid.*).

If it was thus weaponry that McVeigh cathected with a form of displaced desire, that self-defining desire was in turn directed, organized and consolidated by the wider culture of military discipline. This discipline became his way of life. He was an obsessive spit-and-polish soldier who embraced the regimented and tightly organized routines of army life with passion: hence his promotion to sergeant. Even after leaving the army he remained extraordinarily neat and tidy. When he moved into the Canyon West trailer park in Kingman, Arizona, in June 1993, the manager said he was 'an ideal tenant' who cleaned up a previous renter's mess and bought new furniture (*ibid.*). Likewise with the Fortiers – his connection with the Arizona Patriots – in whose Kingman trailer home he also sometimes crashed, McVeigh would cook, do the dishes and then berate his 'hosts' for slovenliness. Such compulsive attention to the details of cleanliness was more than just a way of living in the world. It was also a way of perceiving and understanding the world (*cf.* Theweleit 1987). For instance, the one woman found by reporters who acknowledged a passing interest in McVeigh also reported that she was put off by the way his compulsion for tidiness was linked with an anti-government, pro-Nazi vision of politics. Kifner relates Catina Lawson's account thus:

> 'He would talk about the Government a lot,' she said, remembering how his face reddened when he got mad. 'Usually the topic was brought up by him. he would shoot off his mouth and just bitch about the Government. He also talked about Hitler'; this is what made me angry. 'From what I remember, he said he didn't necessarily agree with all those Jews being killed. But he said Hitler had the right plan. I think he was talking about when Hitler tried to conquer the world how he went about it, little pieces at a time. He thought that was admirable. I didn't like him after that.'
>
> (1995a: A12)

I would argue that the Nazi visions McVeigh found admirable can be linked with the Gulf War. Fascist rhetoric about order in a new world order had a telling echo in the rhetoric of spatial control surrounding Operations Desert Shield and Desert Storm. Klaus Theweleit's description of the geography of fascist desire in an earlier version of the new world order bears repeating in this regard. He argues that the protofascist *Freikorp's* worldview turned reality into a 'lifeless monument.'

> Reality, robbed of independent life, is shaped anew, kneaded into large, englobing blocks that will serve as the building material for a larger vista, a monumental world of the future: the Third Reich. In constructions of this kind, with their massive exteriors and solid forms, everything has its proper place and determinate value: the 'army' and 'nation', 'Germanness', 'rifle-women', 'mothers', 'Scythen Castle', 'workers', 'nurses', and above all soldiers.
>
> (1987: 218)

Clearly, not all participants in the Gulf War subscribed to similarly monolithic visions of inside and outside in which different, good and bad, patriotic and tyrannical, identities were all relegated to an appropriate place. However, with McVeigh a case can be made about a personal pattern of chauvinism that may have been activated and reproduced by the staging and actual events of the war. The young man manning the gun atop the Bradley fighting vehicle might very well have seen himself in this way. Ploughing over the chaotically arranged Iraqi forces, charging through the desert and freeing Kuwait from the clutches of the tyrannical Hussein, the Jeffersonian T-shirt male fantasy of watering the tree of liberty with the blood of tyrants may have been realized momentarily. McVeigh may indeed have seen it all in terms of interiors and exteriors, with himself as the hard body pushing back the outside with the desert shield.

When he returned from the war and dropped out of the army, it seems that McVeigh's moral order of tyrants and patriots did not disappear but its inside/outside geography simply became involuted. The tyrant was no longer Saddam Hussein. It became instead the federal government: the government that was implementing gun control legislation in the form of the November 1993 Brady Bill, the government that had set the psychological tests that McVeigh failed at Fort Bragg, and the government that, McVeigh averred, had placed a microchip in his back while he was in the army. Likewise, the patriots were no longer his army buddies. Instead they were McVeigh's paramilitary mentors, arms-length literary mentors in most cases, in the militia movement. McVeigh had thus moved from the Patriot system to the system of self-styled Patriot paramilitaries criss-crossing rural America. In his mother's words, it was simply as if Timothy had 'traded one Army for another' (Stern 1996: 194)

The *Red Dawn* genre of movies that McVeigh enjoyed were not out of line with this transmogrification of identification. Rambo, after all, is a story of a dutiful Vietnam veteran turned outlawed renegade commando. Another of McVeigh's favorites, more curiously, was the paradigmatic postmodern movie

Brazil, directed by Terry Gilliam. Apparently, McVeigh identified with the commando character called Tuttle, who bursts in wearing a ski mask to fix the constantly malfunctioning air-conditioning ducts that are one of the movie's metaphors for modernity gone awry. Perhaps McVeigh saw himself as Tuttle fixing a similar problem in Oklahoma, and, in this respect it is not incidental that the last movie he rented in Kingman was *Blown Away,* a story about a mad bomber. In any event, Tuttle was the name that McVeigh used as his alias in the gun-show and militia milieu. It was in the culture of this wider and still functioning movement that we can trace – in addition to the family experiences, the army, and the popular culture he enjoyed – the fourth and final major influence overdetermining the patriotism of the man suspected of America's worst act of 'domestic' terrorism.

McVeigh was not a regular member of one of the more well-established militias. However, as Stern notes, while the militia movement is comprised of a series of particular club-like institutions, it is probably better understood as 'a mass social movement' (*ibid.*: 194). As such, it presents the complex ideological articulation of diverse interests – white supremacism, anti-semitism, anti-environmentalism, anti-feminism and, especially, anti-abortionism – all brought together by a common thread of gun-focused anti-government activism. 'The organizing principle of the militias,' says Stern, in a rather optimistic use of the past tense, 'was that the government had been taken over by evil forces and could not be reformed, that it had to be combated – with arms' (*ibid.*: 246). Insofar as he seemed to hold to such principles, McVeigh was a member of the movement. Certainly, he shared the common concerns about guns and government, and was an avid reader of the gun-lobby literature. More than this, he collected and passed on to his friends and relatives far-right literature including the anti-semitic *Spotlight,* the *Patriot Report* and a document called 'Operation Vampire Killer 2000' (Kifner 1995a: A12). Kifner reports that this latter 'strange,' 'incoherent and impossible to follow' document was written by a former Phoenix police sergeant with a view to enlisting men in the struggle against 'the ongoing, elitist covert operation which has been installed in the American system with great stealth and cunning' (*ibid.*). McVeigh, it seems, believed he had enlisted, and to enlist more patriots like himself he circulated more of the same literature.

Most notably and eerily, McVeigh distributed a book called *The Turner Diaries.* 'He carried the book all the time,' said a gun collector who encountered McVeigh on the circuit. 'He sold it at shows. He'd have a few copies in the cargo pocket of his cammies. They were supposed to be $10 but he'd sell them for $5. It was like he was looking for converts' (Kifner 1995b: A1). Written under the pseudonym of Andrew Macdonald by William L. Pierce, a physics professor turned American neo-Nazi, the novel is set in the future as the newly discovered diary of Earl Turner, a fighter in the Organization, an underground group of militant white supremacists. After the passage of a law called the Cohen Act, which makes gun ownership illegal, the Organization rises up against the

219

'Jewish–liberal–democratic–equalitarian plague,' murdering Jews and Blacks to establish an all-white 'New Era' (*ibid.*: A10). First published in 1978, it has become a cult classic amongst the militia movement, being used as a form of guide to action. In the early 1980s, for example, a group called the Order, which consciously modeled itself on the Organization, robbed armored cars and assassinated a Jewish radio host called Allan Berg in Denver before being broken up by the FBI in a 1984 shoot-out on an island in Puget Sound. McVeigh, it appears, followed another *Turner Diaries* model with his bomb plot. Early in the story of the Organization, a similar explosion to that in Oklahoma occurs when the hero and his group pack a truck with a homemade bomb mixed from fertilizer and fuel oil and set it off at FBI headquarters, killing 700 people. They hope it will serve to inspire others also to revolt against the government.

After the real explosion in Oklahoma, after the arrest of McVeigh, and after it became clear that he had sympathies with the militias, Pierce himself – having previously celebrated his book as 'too strong a dish for any reader who has not thoroughly prepared himself for it' – expressed less conviction in the value of such propaganda acts. He said he was 'shocked' by the bombing. However, he went on to say that President Clinton and Janet Reno, the Attorney General, were 'the real terrorists.'

> When a government engages in terrorism against its own citizens, it should not be surprised when some of those citizens strike back and engage in terrorism against the government. . . . There are many Americans . . . who have come to consider the U.S. government their worst enemy [because it caters to] the politicians and bureaucrats. And the homosexuals and the 'career' women. And the minorities. . . . They can't imagine why anyone would want to go back to the bad, old days when this was a White country, and men were men, and women were women, and the freaks stayed in the closet, and everyone worked for his living.
>
> (quoted in Stern 1996: 206)

Writing novels like *The Turner Diaries*, Pierce has successfully disseminated his imagination of nation, enlisting Patriot system products like McVeigh and molding their patriotism into his fascistic worldview. Despite its bold black and white outlines, this view of inside and outside is thoroughly convoluted, easily allowing for a displacement in fantasy from outsiders like Saddam Hussein and Iraq on to insiders like Clinton, Reno and the federal government. In this fantasy world, the inside enemy becomes metaphorized as a cancer that needs to be eradicated. Thus runs Earl Turner's apologetics for the bombing in *The Turner Diaries*:

> All day yesterday and most of today we watched the TV coverage of rescue crews bringing the dead and injured out of the building. It is a

heavy burden of responsibility for us to bear, since most of the victims of our bomb were only pawns who were no more committed to sick philosophy or the racially destructive goals of the System than we are.

But there is no way we can destroy the System without hurting many thousands of innocent people – no way. It is a cancer too deeply rooted in our flesh. And if we don't destroy the System before it destroys us – if we don't cut this cancer out of our living flesh – our whole race will die.

We have gone over this before, and we are all completely convinced that what we did was justified, but it is still very hard to see our own people suffering so intensely because of our acts. It is because Americans have for so many years been unwilling to make unpleasant decisions that we are forced to make decisions now which are stern indeed.

(quoted in Kifner 1995a: A12)

The real bombing, the real death and the real tragedy is, of course, occluded by this pathetic logic of patriotic action. Nevertheless, it remains a logic that must be considered as a likely enabling logic for McVeigh. For a young man traumatized by his failure to become a green beret, alienated from loving relationships with other people, and instead in love with guns, military routine and the cult of the patriotic renegade, such an argument may well have made sense. He may thus have identified a cancerous outside inside the nation and sought, as a patriot, to rip it out. One year later, the official patriot system responded with the Anti-Terrorism and Effective Death Penalty Act of 1996, an Act that fully foreclosed analysis of the connections between the bombing and the growth of white male paramilitary patriotism. However, by retracing McVeigh's doubled-up odyssey from Gulf War 'outside' to heartland 'inside' and his own identification of an 'outside' 'inside' the nation, it has been possible here to highlight a series of displaced identifications that themselves in turn displace any easy inside/outside geopolitics of the nation. Identifying the outside inside the nation may thus be transformed from fascist alibi for violence into a way of remembering the secret truths of patriotism: that it is precarious, personal, shifting, exclusionary and always potentially violent. Or, to reuse the chorus from the theme song of the first Rambo movie: 'The real war is outside your frontdoor.'

References

Abraham, N. (1992) 'The Gulf crisis and anti-Arab racism in America,' in Cynthia Peters, (ed.) *Collateral Damage: The New World Order at Home and Abroad*, Boston: South End Press, pp. 255–278.

Andrews, E. (1992) 'Agency to dismiss analyst who estimated Iraqi dead,' *The New York Times*, March 7th, Y5.

Appadurai, A. (1993) 'Patriotism and its futures,' *Public Culture*, 5, 411–429.

Bachelard, G. (1969) *The Poetics of Space*, translated by Maria Jolas, Boston: Beacon Press.

Bernstein, N. (1997) 'Defense's portrait of political outrage may have backfired,' *The New York Times*, June 14th.

Bhabha, H. (1994) 'DissemiNation: time, narrative and the margins of the modern nation,' in *idem*, *The Location of Culture*, New York: Routledge, 139–170.

Cooper, M. (1995) 'Montana's mother of all militias,' *The Nation*, May 22nd, 714–721.

Daniels, J. (1997) *White Lies: Race, Class, Gender and Sexuality in White Supremacist Discourse*, New York: Routledge.

Dedman, B. (1997) 'Death sentence for McVeigh would put him in rare group,' *The New York Times*, June 4th, A15.

Gibson, J. W. (1994) *Warrior Dreams: Violence and Manhood in Post-Vietnam America*, New York: Hill & Wang.

Jeffords, S. (1989) *The Remasculinization of America: Gender and the Vietnam War*, Bloomington: Indiana University Press.

Jeffords, S. (1993) 'The patriot system, or managerial heroism,' in Amy Kaplan and Donald E. Pease (eds) *Cultures of United States Imperialism*, Durham: Duke University Press, 535–556.

Jeffords, S. (1994) *Hard Bodies: Hollywood Masculinity in the Reagan Era*, New Brunswick: Rutgers University Press.

Kellner, D. (1992) *The Persian Gulf T.V. War*, Boulder, Colo.: Westview Press.

Kifner, J. (1995a) 'Oklahoma bomb suspect: an angry man, with an obsession for guns,' *The New York Times*, December 31st, A1 and A12.

Kifner, J. (1995b) 'Bomb suspect felt at home riding the gun-show circuit,' *The New York Times*, July 5th, A1 and A10.

Labaton, S. (1996) 'New limits on prisoner appeals: major shift of power From U.S. to states,' *The New York Times*, April 19th, C18.

Lacayo, R. (1995) 'Rushing to bash outsiders,' *Time*, May 1, 7–8.

Limbaugh, R. (1995) 'Why I'm not to blame,' *Newsweek*, August 5th, 1995.

Mitchell, A. (1996) 'President signs bill on terrorism and death row appeals,' *New York Times*, April 25th, A6.

Morganthau, T. and P. Annin (1997) 'Should McVeigh die?' *Newsweek*, June 16th, 21–30.

Naureckas, J. (1995a) 'The Oklahoma City bombing: the Jihad that wasn't,' *EXTRA! The Magazine of FAIR*, 8, 4, 6–10.

Naureckas, J. (1995b) 'Talk radio on Oklahoma City: don't look at us,' *Extra*, July/August, 17–18.

Oklahoma Today (1995) *9:02 a.m. April 19, 1995: the historical record of the Oklahoma City bombing*, compiled by *Oklahoma Today* magazine, Oklahoma City.

Pfeil, F. (1995) 'Sympathy for the devils: some white guys in the ridiculous class war,' *New Left Review*, 213, 115–124.

Renan, E. (1990) 'What is a nation?' in Homi Bhabha (ed.) *Nation and Narration*, New York: Routledge, 8–22, p. 11.

Ridgeway, J. (1990) *Blood in the Face, The Ku Klux Klan, Aryan Nations, Nazi Skinheads, and the Rise of a New White Culture*, New York: Thunder's Mouth Press.

Said, E. W. (1993) *Culture and Imperialism*, New York: Alfred Knopf.

Schmitt, E. (1996), 'Antiterrorism bill provision cuts rights of illegal aliens,' *The New York Times*, April 19th, C19.

Sparke, M. (1999) *Negotiating Nation-States: North American Geographies of Culture and Capitalism*, Minneapolis: University of Minnesota Press.

Stern, K. S. (1996) *A Force Upon The Plain: The American Militia Movement and the Politics of Hate*, New York: Simon & Schuster.

Theweleit, K. (1987) *Male Fantasies*, Volume 1: *Women, Floods, Bodies and History*, Minneapolis: University of Minnesota Press.

Thomas, J. (1996) 'Somber days of remembrance,' *Seattle Post Intelligencer*, April 20, A8.

Thomas, J. (1997a) 'McVeigh guilty on all counts,' *The New York Times*, June 3rd, A1.

Thomas, J. (1997b) 'McVeigh jury decides on sentence of death in Oklahoma bombing,' *The New York Times*, June 14th, A1.

Thomas, J. (1997c) 'Friend says McVeigh wanted bombing to start an uprising,' *The New York Times*, May 13th, A1.

Thomas, J. (1997d) 'McVeigh team depicts a man, not a "monster",' *The New York Times*, June 11th, A14.

Thomas, J. (1997e) 'A tarnished case: flaws at FBI lab offer latest setback to prosecutors in Oklahoma City bombing,' *The New York Times*, 17th April.

Thomas, J. (1997f) 'Bomb trial judge warns both sides against "lynching",' *The New York Times*, June 4th, A1 and A15.

Thomas, J. (1997g) 'No one offered an alibi to McVeigh,' *Seattle Post-Intelligencer*, June 3rd.

Thomas, J. (1997h) 'McVeigh's lawyers cite Waco,' *New York Times*, June 7th, A1.

Thomas, J. (1997i) 'Somber day of remembrance,' *Seattle Post-Intelligencer*, April 20th, A1 and A8.

Ó Tuathail, G. (1996) *Critical Geopolitics: The Politics of Writing Global Space*, Minneapolis: University of Minnesota Press.

Tyler, P. (1991) 'Disease spirals in Iraq as embargo takes toll,' *The New York Times*, June 24th, A6.

Walker, R. B. J. (1993) *Inside/Outside: International Relations as Political Theory*, Cambridge: Cambridge University Press.

West, C. (1993) *Keeping Faith: Philosophy and Race in America*, New York: Routledge.

10

WHAT IS IN A GULF?

From the 'arc of crisis' to the gulf war

James Derrick Sidaway

> Some nights we climbed to the upper floors [of Al Rasheed hotel
> in Baghdad where the remaining 34 foreign journalists were stay-
> ing] for a better view of the show. Pointed out the targets to each
> other. Front row seats at a live snuff movie, except we never saw
> any blood....There were flashes, red stains that crept past the
> censor, but mostly it was fun: stealth missiles, smart bombs, mind-
> reading rockets, flashes of tracer fire in the night. F16s and war
> games.
>
> On the day the war ended, at a bus station south of Baghdad,
> dusk was falling and the road was covered with weeping women.
> The Iraqi survivors of the 'turkey shoot' on the Basra road were
> crawling home with fresh running wounds. Their women were
> throwing themselves at the battered minibuses and trucks, pulling,
> pleading, begging: 'Where is he, have you seen him? Is he not with
> you?' Some fell to their knees on the road when they heard the
> news. Others kept running from bus, to truck, to car, looking for
> their husbands, their sons or their lovers – the 37,000 Iraqi sol-
> diers who did not come back. It went on all night and it was the
> most desperate and moving scene I have ever witnessed.
>
> (O'Kane 1995: 13)

In the West, the horrors that Maggie O'Kane[1] writes of have generally been justified in a variety of ways: defense of sovereignty, preservation of order, access to
oil, containment of danger, democracy and the rule of international law. On
closer examination, these justifications reveal profound contradictions. Whilst
most media and official presentations were black and white in terms of right-
eousness (the West) against wickedness (Iraq), there are many ways to disrupt
such claims. A number of critical accounts have shown how this is the case.[2] In
these 'dissident' accounts, the Gulf War emerges as much more complex than a
simple 'just cause.' The moral and political issues that it raises are not black and
white, but grey. One of the most sustained critical engagements and explorations
of such 'grey' areas is the monograph by David Campbell (1993) on *Politics*

without Principle: Sovereignty, Ethics, and the Narratives of the Gulf War.
Reflecting on the official rationales for the war, Campbell (*ibid.*: 79) details how

> Each of these issues [containment of danger, access to oil, etc.] can be
> rendered in a manner that demonstrates how – contrary to the admin-
> istration's [and mainstream media] rhetoric – the United States cannot
> be divorced from the problem in a fashion that locates the responsibil-
> ity for evil with the Other.

In his references to 'the Other,' Campbell is writing about a form of *orientalist*
logic: but a very particular one.[3] To be sure, the Iraqi other is represented as different and inferior: part of the orientalist stereotype denounced by Edward Said.
However, what was particularly evident in the Gulf War was the way that Iraq
became the sole bearer of responsibility. In this, Iraq's independent and heinous
action (the occupation of Kuwait) was presented as *the* cause of the war. And in
some versions an individual (demonized) Saddam Hussein became the simple
source of evil. In these scriptings not only was the complex historical, social and
political geography of the Gulf erased (see, Ó Tuathail 1993; Sidaway 1994),
but so too, the West and its local allies were removed from any form of historical or contemporary responsibility for or complicity in the situation there. It was
necessary to forget that Western imperialism had drawn the boundaries in the
first place, that more recently Iraq had been armed in part by the West, which in
turn had remained relatively quiescent in the face of Saddam Hussein's first use
of poison gas in the war against Iran and similar genocidal actions against the
Kurdish population of the north. In this ignoring or side-stepping of Western
complicity, Iraq was also marked as different from all the other states (amongst
them prominent Western allies such as Israel,[4] Turkey, Indonesia and South
Africa) that have invaded and/or illegally occupied adjacent territories.

And unlike illegal invasions or occupations perpetrated by these allies of the
West, Iraq's action was not allowed to stand. This time, the 'security' of an
entire 'vital strategic region' was held to be at stake. Furthermore, no diplomatic
solution – we were told – was available. The only solution lay in the latest smart
weapons and strategic practices of electronic and mechanized command, control
and information. All this to wage a war in the name of democracy and freedom,
in the name of justice and in the name of a new world order, yet which at the
same time restored a reactionary anti-democratic monarchy (the al-Sabah family), allied itself with another (the al-Saud family) and left contained, but more
or less intact, a quasi-fascist regime that anyway had been substantially armed
and bankrolled by those now allied against it.

But where did all this Western concern with the security of the Gulf come
from? This chapter will chart aspects of the complex emergence of the Gulf as a
place of Western strategic anxiety. It therefore seeks to bring out and reflect on
some of the key historically determined conditions of possibility for the war of
1990–1991. This is done through examining some of the ways in which the
Gulf was coded and scripted as 'vitally' geostrategic in the United States in the

late 1970s and early 1980s. In this context, the chapter will interrogate representations of the Gulf by influential American policy makers (specifically the
national security advisor to the Carter administration, Zbigniew Brzezinski) and
how these were taken up and disseminated by aspects of the print media. In particular, *Time* magazine's January 1979 cover story on 'The Crescent of Crisis'
(Figure 10.1) is read as an example of the way that the notion of the Gulf in
danger was constructed. *Time* has been chosen because it is the foremost and
pioneering example of the genre of the US news magazine. At a moment when
critical analysis (including that of Baudrillard) focuses on the televisual nature of
representations of US strategy, it is important to restate the role of the written
media in the production of popular geopolitical knowledge. Publications such as
Time, Newsweek, US News and World Report and the monthly *Readers Digest*
(see Sharp 1993, 1996) seek to provide a more distanced, 'historically informed'
discussion of events that are played out on TV screens each day (Wark 1996).
Soon after its establishment in 1923, *Time* came to be seen as *the* authoritative
American news magazine. As Peterson (1975: 330) noted in his history of magazines in twentieth-century America:

> The two major national magazines of general news that followed Time
> – Newsweek and US News – both appeared in 1933. By then Time had

Figure 10.1 Cover of *Time*, January 1979: 'The Crescent of Crisis' (© Time Inc., reproduced by permission).

forcibly demonstrated the appeal of the news weekly. As the nation struggled out of the depression and then went along the road to war, the uncertainties and tensions, the complexities of the times perhaps made Americans more needful than ever of publications that tried to sift the news and explain its significance.

Its status reinforced in the Cold War,[5] *Time* in particular became an important signifier of American (global) culture. It became a symbol that something is important. By the 1950s, movies were using fake front pages in *Time* to symbol that the protagonist has 'made it'. The rise of television only served to reinforce the authority of the more measured analysis of the news magazine, and documentaries began to show real covers of *Time* to underscore the importance of the topic they covered. And as Peterson (1975: 329) pointed out: 'the most distinctive characteristic of Time's writing was its overall omniscience which hinted that one person had studied all the events of the week and was passing along the real story with its true significance.' This was to play a significant role in representing a geopolitics of 'danger concern' in the Gulf to an 'informed' American public.

The Gulf emerges – inside an arc

There may be considerable ambiguity and arbitrariness concerning the lines that divide it up into 'sovereign' states, but on a general level, the space of the Gulf is fairly clearly demarcated. It has a more or less agreed physical beginning and an end – unlike other geopolitical designations such as 'Europe'[6] or the 'Pacific Rim.'[7] It has however different names, the Persian Gulf, the Arabian Gulf or – as American strategists usually elect to say (preferring neither to admit its *Persian* genealogy nor its contemporary *Arabness*) – just 'the Gulf.'[8]

But just how and when did this *Gulf* come into the Western and specifically the US geopolitical vision? The first point to note here is that in earlier imperial scripts, notably those of Britain, the Gulf emerges as of vital interest and deep concern (see Adelson 1995; Rich 1991). For the British, this part of what came to be called the 'Middle East' was in turn part of a wider 'Eastern Question' triggered by the weakening of the Ottoman Empire and the competition between France, Germany, Russia and Britain for the spoils. It was also a vital *connection* (literally in the case of Aden, a staging post) en route to the 'Jewel in the (Imperial) Crown' of India. Although India was independent by 1947, the Gulf was one of the last parts of the 'post-colonial' world to experience a hegemonic succession from a European colonial power to the USA. Whilst this had happened in the 1940s (at the start of the Cold War) in the Balkans and the Levant and Iran, in Southeast and South Asia by the mid-1950s and to much of Africa in the 1960s, it was not until 1971 that the British were fully replaced by the USA as *the* dominant foreign power in the Gulf. US intelligence and military forces and petroleum companies had been in evidence in parts of the Middle

East since the 1940s, for example, forging a special relationship with the Saudis.[9] However, Britain continued to rule Kuwait until 1961, South Yemen (Aden) until 1968 and Bahrain, Qatar and the United Arab Emirates (Trucial States) until 1971. Perhaps it was only in southern Africa that a later transition from European colonial hegemony was completed, following the collapse of the Portuguese empire and Zimbabwe's independence.

Yet once the British had withdrawn and the Gulf was firmly and directly rooted within the vision of US strategy and the USA extended overall responsibility for the 'security' of the region, so anxiety escalated. Earlier periods of Cold War confrontation had seen the Middle East scripted as a buffer of containment, formalized in British and US-mediated security pacts (the Baghdad Pact and Central Treaty Organisation), but it was not a *primary* focus of Cold War strategic concern to the degree that Europe or Northeast Asia was. However, writing a critical account of Western strategy at the start of the 1980s, Fred Halliday noted how the Gulf in particular had recently come to be a core focus of *renewed* Cold War confrontation in which:

> restatement of an earlier theme [the danger of the USSR] is accompanied by a significant geopolitical shift, one that highlights the originality of this New Cold War. For the focus of the First Cold War was distinctly Europe, and, in the second instance, the Far East; while the focus of the Second Cold War is South-west Asia, and in particular the Persian Gulf. It is here that the West is now said to have its major strategic interests, and it is here that the full blast of the Soviet threat is alleged to be most evident.
>
> (1982: 14–15)

Halliday went on to chart something of the representation of this 'threat' and what went into the making of it, seeing these as a composite response reflecting a variety of causes and purposes, chief of which were a wave of successful Third World revolutions in the 1970s:

> The Persian Gulf became a particularly apt place in which to respond to this wave of revolutions . . . First, it was geographically near to some of the most important social upheavals of the period – in Ethiopia, Iran and Afghanistan. Ethiopia was the site of a large-scale and successful Cuban intervention, in support of the Ethiopian government. Iran was the site of the most humiliating individual incident in the whole process of Third World revolutions – the hostages affair. Afghanistan was the site of large-scale Soviet military intervention.
>
> (*ibid.*: 18)

These events were woven into a script of danger–instability–threat.[10] They were given purchase in combination with a second material factor:

namely the fragility of the West's remaining allies in that area. . . . All the West's allies round the Gulf were monarchies, ruling without the consent of their people and with enormous corruption and inequality of wealth. The events of Iran showed, moreover, that apparently secure regimes could be rapidly overthrown once a popular movement started to grow.

(ibid.: 18)

Halliday also refers to the developing notion of the strategic significance of Persian Gulf oil. In the context of the apparent challenge of OPEC power, this related to growing concern about Western difficulty/vulnerability in securing an adequate supply of 'strategic' minerals. This was related to a geopolitical discourse of 'lifelines,' 'chokepoints' and 'arteries' – a vocabulary deployed to good effect with respect to other regions too, such as Central America and southern Africa (Ó Tuathail 1986, 1992). Certain places were deemed vital to the supply and open uninterrupted flow of the raw materials (particularly oil) on which Western societies depended. In terms of the Gulf, all this was given a particular twist when fed into a story that the USSR was running out of domestic oil supplies – and looking to capture or control new ones beyond its traditional sphere of influence.

It is in these contexts that in the frame of the Middle East, a so-called 'arc of crisis' entered Western geopolitical discourse. The phrase was coined by President Carter's national security advisor Zbigniew Brzezinski, who had referred to:

An arc of crisis [that] stretches along the shores of the Indian Ocean, with fragile social and political structures in a region of vital importance to us threatened with fragmentation. The resulting political chaos could well be filled by elements hostile to our values and sympathetic to our adversaries.

(Time, 15 January 1979: 6)

The geopolitical metaphor of the 'arc' was enthusiastically adopted by the US media, including *Time* magazine, which placed a graphic of what it termed a 'Crescent of Crisis' on the front cover of its 15 January 1979 edition. In the *Time* graphic, the Soviet bear was pictured looming over a crescent (rather like the ones on the top of mosques and on the flags of Pakistan, Turkey and Algeria).[11] The growing mobilization against the US-backed regime of the Shah in Iran was fed into this 'danger–instability' script. Inside, the magazine devoted eight pages to the cover story. Interchanging 'arc' with 'crescent', *Time* set out how:

In the broadest and grandest of measurements, this crisis crescent . . . reaches all the way from Indochina to southern Africa. In practical terms, however, what Brzezinski is really speaking of are the nations that stretch across the southern flank of the Soviet Union from the Indian subcontinent to Turkey, and southward through the Arabian Peninsula to the Horn of Africa. The center of gravity of this arc is Iran. . . . Regardless of what kind of regime comes to power in this immensely strategic land, the politics of the region, and indeed the geopolitics of the entire world will be affected.

(*ibid.*)

The magazine went on to chart a version of this geopolitics, allowing that:

The conditions that make for instability along the arc vary greatly from country to country, and it would be imprudent to apply the Cold War domino theory to the area . . . Nonetheless, it is also apparent that what happens next in Iran could have an impact on the whole region. The international rivalry that Rudyard Kipling once described as 'the great game' for control of warm-weather ports and lucrative trade routes between Suez and the Bay of Bengal is still being played, except that the chief contestants today are not imperial Britain and czarist Russia but the U.S. and the Soviet Union, and the big prize is not trade but oil. Former Secretary of State Henry Kissinger . . . [pictured in front of a map and described later as 'a judicious analyst of geopolitics' and someone who could fulfil 'the need for a cold blooded assessment,' p. 15] long has argued that in a situation of what he called 'rough parity' between Moscow and Washington, the global balance could be profoundly affected by events at the regional level – and, in recent years, the tide throughout the crescent of crisis could be construed to have been running in Moscow's favour . . . governments that were strongly pro-Western have either fallen or been weakened in Iran, Turkey and Pakistan. Pro-Moscow regimes have come to power in Ethiopia, Afghanistan and South Yemen. The collapse of the Portuguese colonial empire gave the Russians new opportunities in southern Africa. Soviet naval vessels now call at ports from Mozambique to Viet Nam.

(*ibid.*: 6)

What should the West do? An interview with Kissinger followed, in which he deployed his characteristic objectifying language of 'geopolitics.' 'Geopolitically,' we are told, 'this area has been a barrier to Soviet expansion.' Now, however: 'The issue is not only formal Soviet exploitation, but the geopolitical momentum which in that area has turned against us' (Ogden 1979: 15). The Soviet Union must be taught that it cannot pursue 'a systematic attempt to overturn the geopolitical equilibrium.' 'Detente should not become a tranquillizer,' or else, US 'weakness' 'will accelerate the geopolitical decline' (*ibid.*: 15).

Statements by Kissinger could still carry authority despite the fact that he no longer occupied a formal office of state. In *Time*, they were put forward to rein-force and confirm the diagnosis and analysis of danger in the arc. Kissinger's 'expertise' confirmed the seemingly objective importance and need for appropri-ate Western responses in the 'arc of crisis' that was designated by Brzezinski.

Yet whilst Kissinger's references to 'geopolitics' have been the subject of a number of good critical accounts (e.g. Hepple 1986; Ó Tuathail 1994), Zbigniew Brzezinski (who has been almost as prolific as a source of geopolitical discourse) has not received the same critical scrutiny.[12] In addition to their fre-quent references to 'geopolitics,' there were many similarities between them. Like Kissinger, Brzezinski was an immigrant from Europe. He was born into a diplomatic family in Warsaw in 1928. The family fled to Canada in 1938 and, like Kissinger, Brzezinski first made a career in an academic environment. As a policy intellectual he directed an Institute of Communist Affairs at Columbia (1961–), while advising the Kennedy administration and then the State Department (1966–1968). From 1973 to 1975 he chaired the Trilateral Commission, an elite international think tank dedicated to problems of gover-nance and Western strategy that was to have some impact on conservative–liberal internationalism in the 1970s (Gill 1990; Sklar 1980). By the standards of the Carter administration, Brzezinski was at first something of an unrecon-structed cold warrior. His resolute realist characterization of the Soviet Union (also shared with Kissinger) seemed rather at odds with the apparent relaxation of Cold War tensions and the spirit of *détente* of the mid–late 1970s.[13]

It was Brzezinski who had coined the metaphor of the arc and insisted on the danger posed by events there, when juxtaposed with Soviet ambitions. However, his real moment came after the Soviet invasion of Afghanistan in early January 1980. Following the débâcle in Iran, the Soviet intervention was then fed into a script of danger, warning and admonishment for the advocates of *détente* and peaceful coexistence. The Soviet intervention was allowed to confirm the realist scenarios of Soviet threat to and designs on the 'crescent' that Brzezinski and others had been sketching. In the 'Crescent of Crisis' article, *Time* had cited Brzezinski: 'I'd have to be blind or Pollyannish not to recognize that there are dark clouds on the horizon,' before declaring that 'within a decade, according to intelligence reports, the Soviet Union will be running short of the oil it needs to fuel an expanding economy. Thus the region could easily become the fulcrum of world conflict in the 1980s' (*ibid.*: 6). This analysis was blended with a story about a Russian and Soviet search for 'warm water ports,' often (re)invoking British expertise on the matter first codified during the nineteenth century 'Great Game.' *Time* magazine took the trouble to dig out quotes from Tsar Peter the Great on this, citing his transhistorical geopolitical authority, and explaining that as far as present-day Soviet conduct was concerned: 'Things have not changed much since czarist times' (*ibid.*: 9).

Once the Soviet Union made the intervention to prop up the ailing and faction-ridden revolutionary regime in Kabul, so all the intelligence and analy-sis concerning Soviet designs could re-emerge as *confirmed* prediction and therefore as objective geopolitical 'fact.' After the Soviet invasion, President Carter appeared on US television networks on the evening of Friday 11 January 1980 for a 13-minute speech largely devoted to Soviet conduct. In its subsequent elaboration and analysis, *Time* was approving, describing it as 'the toughest speech of his presidency':

> Warned Carter: 'Aggression unopposed becomes a contagious disease.' He denounced the Soviet invasion of Afghanistan as 'a deliberate effort by a powerful atheistic government to subjugate an independent Islamic people' and said that a 'Soviet-occupied Afghanistan threatens both Iran and Pakistan and is a stepping-stone to their possible control over much of the world's oil supplies.'
>
> (14 January 1980: 6)

Again, *Time* cited those British diplomatic experts about the true motives of the Soviets, their quest for warm-water ports and the Soviet design to be in 'a position of being able to turn off the oil tap for Western consumers almost at will' (*ibid.*: 9).

Carter duly prepared his State of the Union address – devoting it mostly to 'foreign policy' issues. According to *Time*, his speech was redrafted over the weekend prior to its delivery, reputedly with an input from his wife Rosalynn, whom *Time* approvingly described as 'a "Brzezinski-liner" because she has long shared the security advisor's hawkish views' (4 February 1980: 7). Therefore: 'By the time Carter returned to Washington on Monday, he had a new speech ... It was a hard, anti-Soviet address that largely reflected Brzezinski's views, rather than those of [Secretary of State for Defense] Vance. Said a senior State Department official: "Zbig's finally got his cold war"' (*ibid.*: 8).

Carter's half-hour speech set out a 'doctrine' – in the manner of Truman, Eisenhower and Nixon before him. Beginning with an invocation of America *as* the world and *of* the world: 'As we meet tonight, it has *never* been more clear that the state of our union depends on the state of the world. And tonight, as throughout our generation, freedom and peace in the world depend on the state of the American union' (President Carter's State of the Union Address, 23 January 1980, cited in *Kessing's Contemporary Archives* 1980: 30245–30246). In the midst of the address, following an account of Soviet aggression and the expansion of Soviet 'military power far beyond its genuine security needs, using that power for colonial conquest,' came reference to the security of the Gulf:

> The region now threatened by Soviet troops in Afghanistan is of great strategic importance. It contains more than two-thirds of the world's exportable oil. The Soviet effort to dominate Afghanistan has brought Soviet military forces within 300 miles of the Indian Ocean and close to the Strait of Hormuz [i.e. at the mouth of the Gulf, between Iran and Oman] – a waterway through which much of the free world's oil must

flow. The [USSR] is now attempting to consolidate a strategic position that poses a grave threat to the free movement of Middle East oil . . . This situation demands careful thought, steady nerves and resolute action – not only for this year, but for many years to come . . . Meeting this challenge will take national will, diplomatic and political wisdom, economic sacrifice and, of course, military capability. We must call on the best that is in us to preserve the security of this crucial region. Let our position be absolutely clear: Any attempt by any outside force to gain control of the Persian Gulf region will be regarded as an assault on the vital interests of the United States. It will be repelled by the use of any means necessary, including military force.

(*ibid.*: 30245–30246)

The Gulf had come to occupy center-stage in the US strategic gaze. This was the first new commitment to deploy US troops abroad since Vietnam. It was followed up by the development of a rapid deployment force and the upgrading of a range of permanent strategic facilities in neighbouring states – in Oman, Saudi Arabia, Turkey, Pakistan and Kenya and on the British Indian Ocean island colony of Diego Garcia (whose indigenous population had been summarily expelled and relocated – with minimal compensation).[14] It also served the quest for higher appropriations by the US military and the account books of the US avionics, armaments and petroleum industries (see Bichler and Nitzan 1996). In this context, the cruise missile, which had been a floundering Cold War project, found a new rationale (see Hayes *et al.* 1987). This was much to the delight of the Boeing Corporation, which was selected to manufacture it. On the award of the contract, *Time* featured Boeing on its front cover, noting in a leading article (under the rubric of 'Economy and Business' but titled 'Masters of the Air') that:

It was the biggest single Air Force contract since the Viet Nam war, and it is almost certain to be remembered as the arms deal that propelled the U.S. into the weapons-bristling decade of the 1980s. The prize was the $4 billion air-launched, cruise missile system that the U.S. is counting on to maintain a strategic edge over the Soviet Union for much of the remainder of the 20th century. The victor, not unexpectedly, was the Boeing Co. of Seattle, one of the nation's most successful corporations.

(7 April 1980: 38)

In demonstrating US 'resolve' and playing on Western European anxieties concerning energy security, the commitment to the Gulf (with associated debates for an 'out of area' role for NATO) was also a significant means of seeking to re-establish a general discipline within the Western alliance. It served to demonstrate who would – in the last instance – protect Western (oil) interests in the Gulf.

Under Carter's successor, the Gulf strategy was given some new twists but essentially followed the course set out by Carter and in particular his national security advisor. In 1983, the rapid deployment force was upgraded and renamed Central Command (CENTCOM), putting it on a formal par with the US military commands in the Pacific and Atlantic. In addition, weapons were poured into Afghanistan as part of a 'Reagan Doctrine' of support for counter-revolutionary forces around the world. And the Afghan *mujahadeen*, one of the most reactionary (and faction-ridden) movements in the world, were scripted as 'freedom fighters' (UNITA in Angola and the Contras in Nicaragua were other beneficiaries). In due course, Reagan's obsessive support for such 'freedom fighters' was to lead the USA into the further imbroglios of the 'Irangate' affair, in which weapons parts were secretly supplied to Iran and the profits passed illegally to the Nicaraguan Contras. The complexities and geopolitical discourses offered to justify the illegal (in US law) Iran–Contras affair are beyond the scope of this chapter (see Marshall *et al.* 1987), as must be the wider destructive consequences of the Reagan Doctrine. What this chapter has set out to show is how a set of representations and commitments made in hysterical Cold War times have had unforseen but profound (post-Cold War) consequences. The conclusions will reflect at greater length on this. For beyond the moment when the Gulf occupied center-stage in an 'arc of crisis,' it remained filled with geopolitical scripts of instability–threat–danger that continued to demand – as Carter had explained early in 1980 – 'resolute action.'

Conclusions

The Gulf War preparation took around six months following the Iraqi invasion of Kuwait. It took this time to hammer the alliance together, gain United Nations approval, whip up anxiety concerning the intentions of Saddam Hussein (including a later forgotten and never verified claim that Iraq was about to invade Saudi Arabia)[15] and ship in further US troops and material. But the network of bases, the weapons systems and the commitment to and capacity for fighting in the Gulf were over a decade in the making. In short, the Gulf War's historical preconditions were directly linked to Brzezinski's and Kissinger's geopolitical rhetoric of the 1970s. Ironically, Brzezinski (1991) himself, in testimony before the Senate Foreign Relations Committee in December 1990, called for *diplomatic* means to force Iraqi compliance. But the path to war was already set in the logic of commitment–containment into which he had sought to place the Gulf whilst in office.

In this respect, what is striking is how easily the Gulf was re-inscribed and represented as a place of vital interest and/or danger when the notion of Soviet threat (of which Brzezinski had made so much) was no longer relevant. And as it had in the Cold War confrontation, the most influential US national magazine repeated (and elaborated) the official view. A few days after the Iraqi invasion, in a column headed by a graphic of Saddam depicted as an octopus, *Time* explained that: 'The threat to US interests is not some distant danger. It is very real, and not only because of the region's oil reserves. Does America really want to let the Saddams of the world shape the new global power structure?' (Beyer 1990: 9).

The following week, the magazine opened with a 'letter from the publisher.' In this, *Time*'s 'Deputy Chief of Correspondents' was pictured holding a telephone in front of a map of the world. Readers learned that he 'deploys TIME troops to the Gulf.' *Time*'s reporters on the ground joined the US military sanctioned and escorted press pool. Their version of what must be done for the 'new world order' was resolute and did not allow for alternatives. In the months that followed only rare examples of anything other than a more-or-less official view turned up in the pages of the magazine.

Right and wrong – black and white – good and bad. What had changed since the Cold War first began in 1947? When *Time* had also:

> shared the president's view of the new divisions of the world, the antagonists, and the ideological content of the worldwide struggle. Equally important, the magazine blithely noted that if people did not like the way that some described the policy (*imperialism*), then another more congenial description (containment) should be adopted. It was as simple as that. Change the language and one changes the way people view the policy.
>
> (Hinds and Windt 1991: 157, my emphasis)

Today there is, we are told, a 'new world order.' Indeed, the Gulf War was one of its inaugural moments. This new order is full of recycled and new scripts of threat–danger. As in the Cold War, (geopolitical) language and representation matters. The wives, mothers or lovers of those '37,000 Iraqi [conscript] soldiers who did not come back' could vouch for that. If they had a chance.

Acknowledgements

I am very grateful to Simon Dalby, Klaus Dodds, Robina Mohammad, Jo Sharp and Gearóid Ó Tuathail for helpful comments on earlier drafts.

Notes

1 For a useful study of what he terms Maggie O'Kane's 'antigeopolitical' journalism in the context of the Bosnian war, see Ó Tuathail (1996).
2 Examples include: Falk (1991), Kellner (1992), Melling (1995), Mueller (1995), Taylor (1992), Thompson (1992) – all good reviews; Scarry (1993) – written in a genre perfected by Baudrillard (1991, 1995); O'Conner (1991) – an early reflection on the positions taken by US peace and ecological movements. Useful collections can be found in Sifry and Cerf (1991), Walsh (1995) and in Mowlana, Gerbner and Schiller (1992) – which contains contributions on media coverage from 35 authors in 18 countries. See Enloe (1993) for an analysis of the militarization of gender difference during the war. *Cf.* Mernissi (1983) – a critical response to the war by a Moroccan feminist sociologist.
    At first sight, it seems that an element of farce or irresponsibility appears in the forms

of claims by an avant-garde French intellectual, Jean Baudrillard, that '*la guerre du)golfe n'a pas eu lieu*' (the Gulf War has not had a place). Yet Baudrillard's arguments demand a response by anyone who would seek to develop a critical history of the Gulf War. Here is France's leading scholar of postmodernism declaring in a leading left-wing daily (*Libération*) that the Gulf War would not, was not and did not take place. Such an argument was bound to earn some notoriety. Indeed, such claims seem farcical or irresponsible when the unnumbered dead Iraqi troops and civilians are taken into account. However, Baudrillard's apparently disconcerting claims can (and most probably should) be read on a number of levels. They are not simply a straightforward reactionary and nihilistic degeneration of post-structuralism/postmodernism that denies truth and facts – as Norris (1992) has claimed. Other readings claim Baudrillard as a morality story and radical denunciation of (televised) high-technology violence (Merrin 1994); as a genre of science fiction response to the evident fiction of a new world order and of clean smart weapons (as Baudrillard 1993a, 1993b, has claimed for himself); or part of a French intellectual response to or dis-orientation in a late capitalist dystopia (Mathy 1993); and as a parody of orientalist tales of savage Arabia.

The value of working through Baudrillard's polemics has been demonstrated particularly well by Merrin (1994) and Patton (1995), who point to the way that they reveal the limits of taking a position for or against the war without *first* interrogating the nature and type of reality proper to events such as those that unfolded in the Gulf.

3 On other orientalist logics, specifically the idea that the 'Middle East' is culturally unsuited for democracy, see Sadowski (1993), who cites Saudi despots, CIA chiefs and an array of academic and media regional 'experts.' Sadowski critiques this discourse as 'neo-orientalism.' It should be noted, however, that parallel (and equally contestable) arguments have been made by the right (and sometimes by anti-democratic factions of the left) with respect to Southern Europe, East Asia and Latin America.

4 Even if we are now being told that East Jerusalem and the West Bank are 'disputed' rather than (Israeli) occupied (Neff 1994).

5 On the significant roles of *Time* (and *Newsweek*) in disseminating and reinforcing official discourse, particularly at the beginning of the Cold War, see Hinds and Windt (1991: 155–177).

6 Amongst others, see Neumann and Walsh (1991) and Taylor (1991).

7 See the excellent collection of essays *What is in a Rim? Critical Perspectives on the Pacific Region Idea* in Dirlik (1993), whose title I have drawn upon for this chapter.

8 In British imperial texts, it seems to have usually been called the 'Persian Gulf,' the epithet of 'Arabian' having become more widely used in the 1960s heyday of Arab (Nasserist) nationalism. It should be added that in Texas and the south of the USA more generally, 'Gulf' first and foremost signifies the Gulf of Mexico.

9 Writing about Truman's guarantees to Ibn Saud, which reiterated Franklin Roosevelt's commitments during the Second World War, Yergin (1991: 428) notes that: 'The special relationship that was emerging represented an interweaving of public and private interests, of the commercial and the strategic. It was effected both at the governmental level and through Aramaco, which became a mechanism not just for oil development, but also for the overall development of Saudi Arabia – though insulated from the wide range of Arabian society and always within the limits prescribed by the Saudi state. It was an unlikely union – Bedouin Arabs and Texas oil men, a traditional Islamic autocracy allied with modern American capitalism. Yet it was one that was destined to endure.'

10 In turn, other conservative security 'experts' in the USA were busy painting the USSR as an expansionist power and therefore requiring a revival and reinvigoration of US-directed containment (see Dalby 1990).

11 It should be added that this also prefigured a subsequent concern with the 'threat' of

radical Islam – a theme that would develop in the 1980s and after the Cold War. The front cover of *Time* of 15 June 1992 juxtaposes the shadow of a minaret and a rifle with the question 'ISLAM: SHOULD THE WORLD BE AFRAID?'

12 Although there is no scope in this chapter for a detailed study, Brzezinski's (1985) memoirs of his term as national security advisor (1977–1981) and his subsequent text on the 'geostrategic' conduct of the US–Soviet contest (1986) are replete with references to geopolitics and geostrategy.

13 For a careful consideration of Brzezinski's role in the administration, see Dumbrell (1995), especially pp. 111–115 and 194–209.

14 The population of Diego Garcia in the 1960s consisted of about 2,000 Ilois people. Once the islands had been leased to the USA for 50 years from December 1966, the people were progressively removed. As the Minority Rights Group subsequently documented: 'Between 1965 and 1973 the British government went about the systematic removal of its own subjects from Diego Garcia; it deposited them in exile in Mauritius without any workable resettlement scheme; left them in abject poverty, gave them a tiny amount of compensation and later offered more on condition that the islanders renounced their rights ever to return home' (Madeley 1982: 4). The report goes on to contrast the fate of the Ilois with those of a roughly equal number of Falkland Islanders.

15 In the days after the Iraqi invasion of Kuwait, the claim that Saudi Arabia was now threatened was used to launch the idea of 'Operation Desert Shield' on a rather sceptical UN and Western public. Although the 'evidence' for this was limited (O'Kane 1995), what matters in the context of subsequent events was the anxiety that it could happen.

References

Adelson, R. (1995) *London and the Invention of the Middle East: Money, Power and War 1902–1922.* New Haven, Conn.: Yale University Press.

Baudrillard, J. (1991) *La Guerre du Golfe n'a pas eu Lieu.* Paris: Éditions Galilée, Paris [English translation by P. Patton, 1995, Sydney: Power Publications.]

Baudrillard, J. (1993a) '"This beer isn't a beer" interview with Anne Laurent,' in M. Gane (ed.) *Baudrillard Live – Selective Interviews,* London and New York, Routledge, 180–190.

Baudrillard, J. (1993b) '"Baudrillard: the interview" interview with Monique Arnaud and Mike Gane,' in M. Gane (ed.) *Baudrillard Live – Selected Interviews* London and New York: Routledge, 199–207.

Beyer, L. (1990) 'Iraq's power grab,' *Time,* 13 August, 8–12.

Bichler, S. and Nitzan, J. (1996) 'Putting the state in its place: US foreign policy and differential capital accumulation in Middle East "energy conflicts",' *Review of International Political Economy,* 3(4) 608–661.

Brzezinski, Z. (1985) *Power and Principle: Memoirs of the National Security Advisor, 1977–1981,* New York: Farrar, Straus and Giroux.

Brzezinski, Z. (1986) *Game Plan. A Geostrategic Framework for the Conduct of the U.S.-Soviet Contest,* New York: The Atlantic Monthly Press.

Brzezinski, Z. (1991) 'The drift to war. Testimony before the Senate Foreign Relations Committee, December 5, 1990,' in M. L. Sifry and C. Cerf (eds) *The Gulf War Reader: History, Documents, Opinions,* New York: Random House, 251–254.

Campbell, D. (1993) *Politics Without Principle: Sovereignty, Ethics, and the Narratives of the Gulf War,* Boulder, Colo.: Lynne Rienner Publishers.

Dalby, S. (1990) *Creating the Second Cold War: The Discourse of Politics*, London: Pinter.

Dirlik, A. (ed) (1993) *What is in a Rim? Critical Perspectives on the Pacific Region Idea*, Boulder, Colo.: Westview Press.

Dumbrell, J. (1995) *The Carter Presidency: a Re-evaluation*, Manchester: Manchester University Press.

Enloe, C. (1993) *The Morning After: Sexual Politics at the End of the Cold War*, Berkeley: University of California Press.

Falk, R. (1991) 'Reflections on democracy and the Gulf War,' *Alternatives*, 16(2), 263–274.

Gill, S. (1990) *American Hegemony and the Trilateral Commission*, Cambridge: Cambridge University Press.

Halliday, F. (1982) *Threat from the East? Soviet Policy from Afghanistan and Iran to the Horn of Africa*, Harmondsworth: Penguin.

Hayes, P. Zarsky, L. and Bello, W. (1987) *American Lake: Nuclear Peril in the Pacific*, Harmondsworth: Penguin.

Hepple, L. (1986) 'The Revival of Geopolitics,' *Political Geography Quarterly*, 5, 21–36.

Hinds, L. O. and Windt, T. O. (1991) *The Cold War as Rhetoric. The Beginnings, 1945–1950*, New York: Praeger.

Kellner, D. (1992) *The Persian Gulf TV War*, Boulder, Colo.: Westview.

Kessings Record of World Events (previously *Kessings Contemporary Archives*), various issues.

Madeley, J. (1982) *Diego Garcia: A Contrast to the Falklands*, Minority Rights Group Report Number 54, London: Minority Rights Group.

Marshall, J., Scott P. D. and Hunter, J. (1987) *The Iran Contra Connections: Secret Teams and Covert Operations in the Reagan Era*, Montreal: Black Rose Books.

Mathy, J.-P. (1993) *Extreme-Occident: French Intellectuals and America*, Chicago: University of Chicago.

Melling, P. (1995) 'Burial party: the Gulf War as epilogue to the 1980s,' in J. Walsh. (ed.) *The Gulf War Did Not Happen: Politics, Culture and Warfare Post-Vietnam*, Aldershot: Arena, 63–86.

Mernissi, F. (1993) *Islam and Democracy: Fear of the Modern World*, London: Virago Press.

Merrin, W. (1994) 'Uncritical criticism? Norris, Baudrillard and the Gulf War,' *Economy and Society*, 23(4), 433–458.

Mowlana, H., Gerbner, G. and Schiller, H. I. (eds) (1992) *Triumph of the Image: the Media's War in the Persian Gulf – A Global Perspective*, Boulder, Colo.: Westview Press.

Mueller, R. (1995) 'The perfect enemy: assessing the Gulf War,' *Security Studies*, 5(1) 77–117.

Neff, D. (1994) 'Embracing Israel's claims at the UN,' *Middle East International*, 1 April 3–4.

Neumann, I. B. and Walsh, J. (1991) 'The Other in European self-definition: an addendum to the literatures on international society,' *Review of International Studies*, 17, 327–348.

Norris, C. (1992) *Uncritical Theory: Postmodernism, intellectuals and the Gulf War*, London: Lawrence and Wishart.

O'Conner, J. (1991) 'Murder on the Orient Express: the political economy of the Gulf War,' *Capitalism, Nature, Socialism*, 2(2) 1–17.

Ogden, C. (1979) 'An interview with Kissinger,' *Time*, 15 January 1979, 15–16.

O'Kane, M. (1995) 'Bloodless words, bloody war,' *The Guardian Weekend*, 16 December 12–18.

Ó Tuathail, G. (1986) 'The language and nature of the "new" geopolitics: the case of US–El Salvador relations,' *Political Geography Quarterly*, 5, 73–85.

Ó Tuathail, G. (1992) 'Foreign policy and the hyperreal: the Reagan administration and the scripting of "South Africa",' in T. J. Barnes and J. S. Duncan (eds) *Writing Worlds: Discourse, Text and Metaphor in the Representation of Landscape*, London: Routledge, 155–175.

Ó Tuathail, G. (1993) 'The effacement of place: US foreign policy and the spatiality of the Gulf Crisis,' *Antipode*, 25: 4–31.

Ó Tuathail G, (1994) 'Problematising geopolitics: survey, statesmanship and strategy,' *Transactions of the Institute of British Geographers*, NS, 19(3), 259–272.

Ó Tuathail, G. (1996) 'An anti-geopolitical eye: Maggie O'Kane in Bosnia 1992–93,' *Gender Place and Culture*, 3(2), 171–186.

Patton, P. (1995) 'Introduction in Baudrillard', J., *The Gulf War Did Not Take Place*, Sydney: Power Publications, 1–21.

Peterson, T. (1975) *Magazines in the Twentieth Century*, 2nd Edition, Chicago: University of Illinois Press.

Rich, P. (1991) *The Invasions of the Gulf: Radicalism, Ritualism and the Shaiks*, Cambridge: Allborough Press.

Sadowski, Y. (1993) 'The New Orientalism and the Democracy Debate,' *Middle East Report*, 183, 14–40.

Scarry, E. (1993) 'Watching and authorizing the Gulf War,' in M. Garber, J. Matlock and R. L. Walkowitz (eds) *Media Spectacles*, London: Routledge, 57–73.

Sharp, J. (1993) 'Publishing American identity: popular geopolitics, myth and the Readers Digest,' *Political Geography*, 12, 491–504.

Sharp, J. (1996) 'Hegemony, popular culture and geopolitics: the Readers Digest and the construction of danger,' *Political Geography*, 15(6/7), 557–570.

Sidaway, J. D. (1994) 'Geopolitics, geography and "terrorism" in the Middle East,' *Environment and Planning D: Society and Space*, 12, 357–372.

Sifry, M. L. and Cerf, C. (1991) (eds) *The Gulf War Reader: History, Documents, Opinions*, Toronto: Random House.

Sklar, H. (ed.) (1980) *Trilateralism*, Montreal: Black Rose Books.

Taylor, P. J. (1991) 'A theory and practice of regions: the case of Europes,' *Environment and Planning D: Society and Space*, 9, 183–195.

Taylor, P. M. (1992) *War and the Media: Propaganda and Persuasion in the Gulf War*, Manchester: Manchester University Press.

Thompson, A. (1992) *Smokescreen: The Media, the Censors and the Gulf*, Saffron Waldon: Laburnham and Spellmont.

Time, various issues (Time Life, Washington).

Walsh, J. (ed.) (1995) *The Gulf War did not Happen: Politics, Culture and Warfare Post-Vietnam*, Aldershot: Arena.

Wark, B. (1996). *Virtual Geography: Living with Global Media Events*, Bloomington: Indiana University Press.

Yergin, D. (1991) *The Prize: The Epic Quest for Oil, Money and Power*, New York: Simon & Schuster.

11

GOING GLOBILE

Spatiality, Embodiment, and mediation in the Zapatista insurgency

Paul Routledge

The storm

It will be born out of the clash between two winds, it will arrive in its own time, the coals on the hearth of history are stoked up – and ready to burn. Now the wind from above rules, but the one from below is coming, the storm rises . . . so it will be.

(EZLN communiqué 1994a: 5)

On 1 January 1994, media vectors around the world carried the dramatic news that ski-masked guerrillas had captured the town of San Cristobal de las Casas (in the Mexican state of Chiapas) and declared war on the Mexican state. As the drama unfolded, it became apparent that this was not simply another Latin American *foco*[1] movement, in the tradition of previous revolutionary uprisings. The EZLN (Ejercito Zapatista Liberacion National) or Zapatistas, as they became known, differed from the recent FMLN (Farabundo Marti Liberacion National) movement in El Salvador, and the FSLN (Frente Sandanista Liberacion National) in Nicaragua. Unlike them, the EZLN did not see itself as the vanguard directing a struggle to seize state power. Rather, it demanded the democratic revitalization of Mexican civil and political society, and autonomy for, and recognition of, indigenous culture.[2]

Perhaps more than any other recent guerrilla insurgency (e.g. FMLN, FSLN) the Zapatistas have consciously sought to use the media as an integral part of their overall political strategy, and have used an articulate, humorous, and poetic spokesperson in the masked figure of Subcommandante Marcos. As a result, the Zapatistas have received enormous national and international media attention. Thousands of column inches have been devoted to them in the print media; a Zapatista web site – including information updates, Zapatista communiqués, and analyses of the movement – exists on the world wide web, and journal articles and books about the Zapatista rebellion have appeared in both academic and activist publications (e.g. Ross 1995; Harvey 1994; Burbach 1994). In this chapter, I wish to discuss the importance of the Zapatistas' strategy to a critical geopolitics. In order to do so, let me first consider the importance of social movement practices for critical geopolitics.

Critical geopolitics and social movements

An important area of investigation within critical geopolitics is the role of social movements[3] in challenging the power of both the state and international institutions to enact particular economic and political programmes. Indeed, as Dalby notes, social movements are important within the 'concomitant reconstruction of political community at the global, local and regional scales' (1991: 277).

Social movements are usually located within the political boundaries of a state, and affected by the actions and policies of the state (as in the effects of the development project). However, it is important to note that, increasingly, certain resistances are becoming regional and international in focus and organization, such as the recent protests against the 50th anniversary of the World Bank (see Brecher and Costello 1994). Also, within states, social movements may address their actions to groupings or institutions other than the state, for example ethnic groups, or the media. Hence, while social movements are frequently in perpetual interaction with states (e.g. via negotiations or conflicts), they are not reducible to them. They tend to pose political, cultural, discursive, and economic alternatives to the state and state policies.

Attention to the actions of social movements such as the Zapatistas contributes to a critical geopolitics in at least two ways. First, it (de)centers analytical focus away from an exclusive concern with the machinations of the state. Second, it enables critical geopolitics to investigate how different types of social movement challenge state-centered notions of hegemony, consent, and power and contest the colonization of the 'political' by the state (Routledge 1994; Dodds and Sidaway 1994; Falk 1987). The consequences of such an approach may include the broadening of political geography to encompass more radical understandings of the political; an understanding of how place is central to particular terrains of resistance and the creation and articulation of alternative knowledges; and how local contexts of resistance may interplay with global processes (Routledge 1993; Dodds and Sidaway 1994).

Also of interest to a critical geopolitics are the strategic practices of social movements – how resources and actions are coordinated to pursue desired outcomes in a fluid process of interaction with opponents, including those tactical vectors of dissent in specific encounters that express a social movement's ingenuity, and deployment of forces (Routledge 1996a). Strategic mobility within social movement practice allows for myriad perspectives to be explored, which may reveal possibilities of action unimagined from other positions and collectivities. Such an itinerary of resistance is important for at least two reasons: first, because the opponents of social movements (be they states, international institutions, private interests, etc.) occupy constantly changing positions regarding policies, strategies, and relations and thus social movements must be able to maneuver in relation to these shifting sands of power; and, second, because mobility enables social movements to avoid entrapment by those options offered by dominating discourses (Rose 1993). Perhaps nowhere is this potential for mobility more apparent than in the use of contemporary media within resistance, and it is to a discussion of these themes within social movement practices in general, and the Zapatista rebellion in particular, that I will now turn.[4]

Media-tion and social movements

The use of telecommunications has the potential to alter the balance of power in social struggles. This is in part effected by the refusal of social movements to accept the boundaries of communication taken for granted by established systems of domination (e.g. states). Through their use of media vectors, social movements can escape the social confines of territorial space upon which much of the legitimacy of the state is predicated. Indeed, the globalization of communications provides new opportunities for decentralized political practices, as many social movements increasingly locate their strategies within local and translocal spaces as well as national and transnational spaces (Adams 1996: 419). Moreover, political conflicts frequently involve struggles over the representation of events and whether the conflict should be bounded in particular spaces, or publicized and traverse those spaces. It is usually in the interests of governments to restrict the bounds of a conflict in order to effect containment (if not total control) of events, whereas social movements frequently wish to publicize their struggles in order attract the attention of as wide an audience as possible to their aims and grievances. I refer to the uses of media by social movements as 'going globile': effecting strategic mobility within the increasingly global space of contemporary media.

Within an increasingly media-ted world, Wark (1994: 62–63) has argued that contemporary subjectivities are formed within two sets of exterior relations, the map and the territory. The former constitutes broadcast areas, satellite and telephone networks, and the signs and images that accumulate through the interactions in this space of media vector fields. The latter constitutes the physical space of interactions, social relations of production and reproduction, and the places of work and habitation. He argues that the occupation of time in the information network is an important aspect of contemporary struggle, the occupation of space in the symbolic landscape being a means to that end.

Referring specifically to practices of resistance, Melucci (1989) argues that, since collective action frequently focuses on cultural codes, the forms of contemporary social movements are themselves messages, operating as signs, representing a symbolic challenge to dominant codes. This leads Melucci to argue that social movements are a kind of new media, acting to transmit messages to society, frequently as symbolic challenges, that attempt to make power visible and thus negotiable. By confronting power, contemporary social movements aim to challenge the symbolic order of what constitutes permissible thinking and action on specific issues. They aim to force power to take differences into account by articulating alternative ideas and practices in space.

In doing this, social movements engage in what I term 'event-actions' (Routledge 1997), these being specifically symbolic and media-oriented, attempting to create spectacular images that may catalyse political effects. Indeed, Baumann (1992: 33) has argued that the contemporary media present the real world as a drama, a staged spectacle. In most strategic sites of the 'real world,' events happen because of their potential fitness to be televised. Under such circumstances, both politicians and activists 'act' for television, hoping to elevate their private actions into public events.[5] As Baudrillard observes, 'the manipulation of media images constitutes the continuation of politics by other means' (1988: 16), indeed,

> the vocabulary of resistance must be expanded to include the means of electronic disturbance. Just as authority located in the street was once met by demonstrations and barricades, the authority that locates itself in the electronic field must be met with electronic resistance. Spatial strategies may not be key in this endeavour, but they are necessary for support, at least in the case of broad spectrum disturbance.
>
> (Critical Art Ensemble 1994: 24)

Media-tion and the Zapatistas

The Zapatistas have been represented as 'shadow warriors' (Guillermoprieto 1995; Ó Tuathail 1997) globalizing their resistance through electronic vectors, fighting a netwar (of words) against the Mexican state and its international investors. Indeed, it is tempting to consider the Zapatistas primarily as a media event, since for the many onlookers, the *experience* of the Zapatistas has been entirely a media-ted one. However, I want to argue that it is more appropriate to consider the Zapatistas as an example of *imagineered resistance* that exists both as embodied and media-ted (Routledge 1997), for the Zapatistas articulate challenges to both the *material* political and economic power of the Mexican state, and to the monopoly of *representations* – imposed by political elites – of Mexico as an 'emerging market,' economically and politically stable for foreign investment. In so doing, they attempt to 'nurture the seeds of their own forms of representation' (Calderon et al. 1992: 25), which they have articulated to national and international audiences through various media. Hence, the Zapatistas are located within a terrain of resistance that includes both the jungles and towns of Chiapas and what Appadurai (1990) calls the 'mediascape': the global distributions of electronic and other mediated information (e.g. newspapers, e-mail, television) and the signs and images of the world created by these media.

The latter are important because, as Ó Tuathail (1997) argues, contemporary geopolitical evaluations of international events have become primarily geo-financial in character. A dispersed informational network, which Ó Tuathail terms the 'geo-financial panopticon' (*ibid.*), has developed beside the infrastructures of the techno-military panopticon, which imagines and produces the surface of global affairs as a world of 'emerging markets.' Hence, political events that may previously have been viewed as local or national in significance – such as the Zapatista rebellion – now reverberate globally as sign wars with potentially dramatic consequences. Such a process is deeply entangled with the increasing importance of multivariate media in the lifeworlds of business, news, politics, and culture.

Through material and media-ted strategies, the Zapatistas have attempted to create public spaces in order to render power visible, and thus negotiable. By staging an insurgency, the Zapatistas' aim has been to challenge the symbolic, political, and economic order within Mexico in an attempt to force the Mexican state to take differences into account by articulating alternative ideas and practices in both embodied and virtual space. Following some traces of the Zapatista rebellion, I will consider the Zapatistas' strategies of resistance and their importance for a critical geopolitics.

Traces of the Zapatista rebellion

The appearance of the Zapatistas in the political life of Mexico and on the international media vectors can be traced, in part, to the ongoing material economic and political conditions of Chiapas in particular, and of Mexico in general. The economic and political marginalization of Chiapan peasants has been extensively discussed elsewhere (Burbach 1994; Collier and Quaratiello 1994; Guillermoprieto 1995; Harvey 1994; Ross 1995; Wollock 1994) and has led to various forms of resistance in the state. Many of these owed their origin, in part, to broader political struggles that have taken place within Mexico. Indeed, resistances rarely have precise times of birth or death. Memories of rebellion live on in people's imaginations, stories and dreams, and in the tactics and strategies of subsequent struggles. As such we speak of traces of resistance – those moments of rebellion, the passing of social movements – that inspire, nurture, and inform subsequent upsurges of people's power. Hence, although the Zapatistas 'appeared' on 1 January 1994, they were the outcome of many years of political organizing and resistance.

> We are the product of 500 years of struggle; first against slavery, then in the insurgent-led war of Independence against Spain; later in the fight to avoid being absorbed by North American expansionism, next to proclaim our Constitution and expel the French from our soil; and finally after the dictatorship of Porfirio Diaz refused to fairly apply the Reform laws, in the rebellion where people created their own leaders. In that rebellion Villa and Zapata emerged – poor men, like us.
>
> (EZLN communiqué 1994b: 5)

The Mexican Revolution, and particularly the figure of guerrilla leader Emiliano Zapata (who was murdered in 1919), provide the Zapatistas with both their name and their claim to popular legitimacy. The revolutionaries fought an ongoing war against successive Mexican governments from 1911 to 1917 in order to secure land reforms and social justice for the peasantry. Although the revolutionaries were ultimately defeated militarily by constitutional forces, some of their goals were, nevertheless, enshrined in the constitution of 1917 and institutionalized under populist reformer General Lazaro Cardenas and his Party of Institutionalized Revolution (PRI) when he became president in 1934. The same party has ruled Mexico uninterrupted until the present day, while the egalitarian goals of the revolution have bled away in the PRI's particular brand of privilege, political rhetoric, and authoritarianism.

Closer to the present, a variety of political upheavals sowed the seeds of contemporary rebellion. During the student unrest at the National Autonomous University of Mexico in 1968, economics professor Adolfo Orive organized Emiliano Zapata-allied brigades to protest against the military occupation of the campus. All decision making was by popular assemblies of students, workers, and *campesinos*. These assemblies became the cornerstone of Orive's later 'Politica Popular' movement. Following the student rebellions, which culminated in the massacre of protesting students in the Plaza de las Tres Culturas on 2 October 1968 in Mexico City, the Politica Popular movement moved to different parts of Mexico where leftists organized squatters, workers, and *campesinos* around a variety of land, wage, and housing issues. This resulted in massive, violent land takeovers and the creation of *Tierra y Libertad* (Land and Liberty) encampments in various parts of Mexico, including Monterrey and the Laguna region. The movement focused upon creating associations of *ejidos*[6] in different parts of the country. In Chiapas, peasant agitation (via Politica Popular, union activity, etc.) brought 54 per cent of the state's land under *ejidos*, albeit land of the poorest quality (Ross 1995; Harvey 1994; Burbach 1994).

Meanwhile, during the late 1960s and early 1970s, fifteen urban and rural guerrilla *focos* flowered in different parts of Mexico, including the Liga Communista 23rd de Septiembre, the Armed Forces of National Liberation (FALN), and the Party of the Poor. In the early 1970s, the pro-Cuba Fuerzas Armadas de Liberacion National set up a short-lived training camp near Ocosingo in Chiapas. As Major Mario of the EZLN commented: 'We have no one with us from that time, but they are our examples' (quoted in Ross 1995: 273).

Agitation and organizing was aided by the liberation theology of the Catholic Church. In Chiapas, Bishop Ruiz set up a human rights office, and organized the First Congress of Indigenous Peoples in San Cristobal in 1974. The focus of the Church's activities were the empowerment of peasant and Indian communities in their struggle against poverty and exploitation. The work of the leftists gradually found sympathy among the Mayan communities of Chiapas, aided by their notion that no one person should be above another. Throughout the 1970s and 1980s, various independent peasant organizations emerged in

Chiapas, challenging evictions of peasants from the lands that they occupied, the complicity of the government-controlled National Peasant Confederation (CNC) in this process, and government intransigence in acting upon land claims by Indians against *ladino* ranchers and *latifundistas* whose holdings exceeded legal limits. There were also many *tomas de tierras* (invasions of ranches) by *campesinos* who had been abused by plantation and ranch owners. Some of the peasant organizations were organized by *brigadistas*[7] who emerged from the resistances of the previous decades.[8] The most important of the peasant organizations were the Emiliano Zapata Peasant Organization (OCEZ), the Independent Confederation of Agricultural Workers and Peasants (CIOAC) and the Union of Ejido Unions of Chiapas (Union de Uniones) (Collier and Quaratiello 1994; Montejo 1994; Weinberg 1994; Harvey 1995).

The Union de Uniones was the largest of the peasant alliances formed under the joint leadership of the Church and leftists. The union consisted of two sides: one radical, demanding land; the other reformist, concerned with marketing and credit mechanisms. Both sides, however, appear to have been infiltrated by EZLN cadres, who proposed armed self-defense against the ranchers, paramilitary patrols, and state government. In 1988, the Union de Uniones split, one side becoming the Rural Association of Collective Interest (ARIC–Union de Uniones), concerned with credit issues. The radical wing, which included 6,000 families out of the 10,000 living in the region, decided to join the Zapatista formations (Guillermoprieto 1995). In 1989, the Emiliano Zapata Independent Peasant Alliance (ACIEZ) emerged in the Chiapan towns of Altamirano, Ocosingo, San Cristobal, Sabanilla, and Salto de Agua. In 1992, it changed its name to ANCIEZ (adding National to its title), claiming membership in six central and northern states. It was especially strong among the indigenous Tzotil, Tzetal, and Chol communities of Chiapas. In a march in San Cristobal on 12 October 1992 to mark 500 years of indigenous resistance, 50 per cent of the 10,000 Indian participants were members of ANCIEZ. In early 1993, ANCIEZ went underground, presumably to commence training for armed rebellion (Guillermoprieto 1995; Harvey 1994).

> As far as the peasants are concerned, the EZLN arose as a self-defense group to defend against the ranchers' hired gunmen, who try to take their land and mistreat them, limiting the social and political advance of the Indians. So they took up arms so as not to be defenseless. Then later, the comrades saw it wasn't enough to do self-defense of a single ejido or community but rather to establish alliances with others and to begin to make up military and paramilitary contingents on a larger scale, still for the purpose of self-defense.
>
> (Subcommandante Marcos, quoted in Collier and
> Quaratiello 1994: 84)

Interestingly, in May 1993, the Mexican army raided a EZLN garrison in Corralchen, and, according to initial reports, discovered a guerrilla camp with underground bunkers and a scale model of Ocosingo. However, the government subsequently downplayed the reports of a guerrilla movement in Chiapas so as not to endanger the chances that the US Congress would accept NAFTA (Collier and Quaratiello 1994).

Embodiment and media-tion: the interwoven strategies of the Zapatistas

As noted above, the Zapatista uprising was not the first peasant uprising in Chiapas. However, it was the first to gain international attention. Indeed, the movement developed in counterpoint to numerous other peasant organizations operating in the state. During the past 20 years, grassroots organizations in the Lacandon forest have experimented with different forms of cooperative community development; innovative projects to protect natural resources, networks of intercommunity ties between different ethnic groups; combining ethnic, regional, and class consciousness; and the increased participation of indigenous women (who now comprise a third of the EZLN ranks) (Toledo 1994). Indeed, the Zapatistas articulate a third way beyond the demand for land (agrarian movements) and the control over the means of production (economic movements). In their calls for indigenous rights and ecological preservation (i.e. an end to logging, a programme of reforestation, an end to water contamination of the jungle, preservation of remaining virgin forest), they articulate an ecological and cultural struggle that forms part of what Gedicks (1993) terms the 'resource wars': the struggle over the remaining natural resources between indigenous and traditional peoples and the government and multinationals.[9]

The extent to which the insurgency in Chiapas commenced with an eye for international media vectors – beginning on 1 January to coincide with the inauguration of NAFTA – has been put into question by Marcos himself. Although EZLN communiqués first identified NAFTA as having inspired its rebellion, it appears that the designation of NAFTA was a pretext:

We had not planned to rise up on 1 January . . . We thought about various dates, taking various factors into account. For example, we needed to show clearly that we weren't drug traffickers. We had to do something related to cities, the pretext couldn't be just about rural conditions so that they couldn't write us off in the jungle, as they did in the incident in the mountains of Corralchen, that garrison of ours they discovered, when they decided we were [a fringe group], marijuana growers, or Guatemalans. We didn't say anything, because we were watching to see what would happen . . . Then there were logistical questions, apart from political ones. For example, when would our food reserves be greatest, given that the war would be long, that we would be surrounded, that they would drive us into the mountains, so [the uprising] had to begin just after the harvest, when we could get

money together ... The truth of the matter is that decision about
when to rise up didn't take national politics into account. That's not so
important to the comrades, not so much as not being able to stomach
things any longer, regardless of national or international conditions.
(Subcommandante Marcos, quoted in Collier and
Quaratiello 1994: 86–87)

However, this does not mean that the Zapatistas did not consciously use the
media. They staged their uprising in a spectacular manner to ensure maximum
media coverage and thus gain the attention of a variety of audiences – including
civil society, the state, the national and international media, and international
finance markets. Indeed, the Zapatistas' use of the media formed part of a political strategy that consisted of three *interwoven* facets: (1) the physical occupation
of space; (2) the media-tion of images (the movement as a form of media); and
(3) the manipulation of discourse (a war of words). What differentiates the
Zapatistas, in part, from many of the earlier political struggles that have pep-
pered the history of Latin America is not so much their waging of guerrilla war
on both the landscape and the mediascape, but rather the strategic importance
of images and words in their struggle.

Embodied spaces of resistance, images of resistance

The initial uprising in Chiapas was staged as an event-action by the protagonists,
articulating indigenous and *campesino* resistance to NAFTA and the reform of
Article 27.[10] The Zapatistas staged a spatial struggle in their occupation of San
Cristobal and other Chiapan towns. The initial occupation of San Cristobal was
a masterstroke of public relations, a 'poem' Marcos styled it (Ross 1995). It
served to awaken the Mexican government and the international media to what
at first may have appeared like a traditional guerrilla insurrection. While the
EZLN communicated its Declaration of War to the Mexican government (see
EZLN communiqué 1994b: 5), it also attacked symbolic targets. The Zapatistas
destroyed town halls, police stations, and the municipal palace of Altamirano.
The main towns of conflict in Chiapas were Margaritas, Altamirano, Rancho
Nuevo, Comitan, Ocosingo, and San Cristobal (M-A-R-C-O-S). The Zapatistas
mobilized 1,200–1,500 fighters for the New Year's Day offensive (400 in San
Cristobal, 300 in Ocosingo). The rebels moved on to local military barracks
after San Cristobal, where they were successfully turned back. Other rebel
columns attempted to take the towns of Ocosingo, Altamirano, and Las
Margaritas and the villages of Chanal, Huixtan, and Oxchuc: in every conflict
they were defeated by the army (see Ross 1995).

However, the national and international media conveyed images of ski-
masked guerrillas engaged in what appeared to be the creation of a liberated
zone. In reality, the 'liberation' was brief, lasting 30 hours, before the Zapatistas
melted back into the Lacandon jungle. It appears they were more concerned

with the occupation of time on the information network than permanently securing control of Chiapas' major towns. Rather, their material occupation of space was symbolic, staged to gain access to media vectors. This focused upon the images of a guerrilla war rather than the undertaking of a protracted armed struggle.

Having said this, two important caveats are to be noted. First, the Zapatistas *are* a guerrilla force – embodied rather than virtual – which has engaged in several armed conflicts with the military. Their tactics of armed struggle were developed over many years and have their own culturally specific history:

> We learned out tactics from Mexican history itself . . . from resistance to the Yankee invasion of 1846–1847, and from popular resistance to the French intervention, from the heroic deeds of Villa and Zapata and from the long history of indigenous resistance in our country.
>
> (EZLN communiqué, quoted in Collier and
> Quaratiello 1994: 64)

More specifically, Marcos mentions

> Pancho Villa, as far as regular army tactics are concerned, and Emilio Zapata, with respect to the interchange between guerrilla and peasant. We got the rest out of a manual of the Mexican army that fell into out hands, and a small manual from the Pentagon, and some work by a French general whose name I can't remember.
>
> (Subcommandante Marcos, quoted in Collier and
> Quaratiello 1994: 83)

Armed struggle was also important in enabling the Zapatistas to place themselves in opposition to the reformist tactics used by other peasant groups in Chiapas (Collier and Quaratiello 1994).

Second, the practices and discourses of resistance require some form of co-ordination and communication, which, in turn, require (however temporarily) material social spaces and socio-spatial networks that are insulated from control and surveillance. These spaces act as sources of self-dignity and agency, sites of solidarity in which and from which resistance can be organized and conceptual-ized. Such sites are created, claimed, defended, and used (strategically or tactically) in the practice of resistance. For the Zapatistas, the terrain of the Lacandon jungle had served as a material base for the creation, organization, and subsequent emergence of their resistance. This 'homeplace' of resistance (Hooks 1990: 41) was created out of an intimate knowledge of the local physi-cal terrain established by the guerrillas from years of living (and organizing) in the jungle and mountains. It provided a material sanctuary for the guerrillas when they were faced with the military might of the Mexican army and air force.[11] Moreover, it also served as a base from which the Zapatistas were able to articulate their grievances and demands to the media, via their communiqués, and engage in their war of words with the government.

Throughout the conflict, the EZLN has emerged sporadically from the jungle, taken over small villages and towns, then retreated back into the jungle when the army appeared. Since it declared itself in the face of power, the rebel army has entrenched itself in and around Las Canadas (in the Lacandon jungle), a *de facto* zone of rebel control. Since the cease-fire (see below), it has been surrounded by 15,000 government troops (Ross 1995). However, even when surrounded, the Zapatistas have managed to filter out via remote mountain trails, as they did on 19 December 1994.[12] In addition, the Zapatistas locate their General Command in mobile encampments.

This itinerancy has not only frustrated the government in its attempts to curtail the insurgency, it is also symptomatic of the symbolic character of the insurgency. The Zapatista rebellion presents little military threat to the Mexican state. The Mexican army boasts 175,000 troops spread through 36 regions. The air force includes 80 US-supplied helicopter transports and gunships (HEUYs), and a front-line strike force of 75 Swiss-made P–7 jet trainers. Meanwhile, the largest Zapatista military parade comprised 400 people in arms (which included those with wooden rifles). While Zapatista troops number anywhere between 3,000 and 15,000, their spatial control has extended to only four municipalities out of the 2,000 into which the country is divided (Ross 1995).

Ski-masked, armed guerrillas occupying Chiapan towns then 'disappearing' back into the Lacandon jungle was a visually arresting event-action with a deep cultural and political resonance throughout the Americas. The tactic of armed struggle and the discourse of insurgency has an important place both within Mexican society – whose revolution was secured by force of arms – and within a region that had witnessed numerous armed rebellions, including the Cuban and Nicaraguan revolutions, and armed insurgencies in El Salvador, Colombia, and Guatemala. As noted above, such event-actions are specifically symbolic and media-oriented, attempting to create spectacular images that attract the attention of a variety of 'publics' in order to catalyse political effects. In the case of the Zapatistas, they also created an image out of proportion to the (military) threat they posed to the Mexican state.

While the EZLN threatened to march on the nation's capital, its takeover of Mexico City was symbolically achieved by a demonstration of 100,000 people in the city's Zocalo Square in solidarity with the EZLN struggle. Alongside solidarity demonstrations in other Mexican cities, these were staged in response to the Mexican army's attacks on the EZLN following its occupation of San Cristobal. Following demonstrations of popular support for the EZLN, President Salinas called a unilateral cease-fire on 12 January, which was followed by an EZLN agreement to enter into peace talks with the government.

Discourses of resistance

Out of the mouths of the guns of the faceless men and women spoke the voices of the landless campesinos, the agricultural workers, the small farmers, and the indigenous Mexicans. The voice of those who have nothing and deserve everything.

(EZLN communiqué 1994e: 16)

The Zapatistas have made use of a variety of media – including the Internet and news print media, and radio – to give voice to their struggle, and to organize their resistance: 'they have . . . organized the whole radio situation, which is how the entire jungle stays in touch' (Mardonio Morales, quoted in Collier and Quaratiello 1994: 83). Much of their demands have been articulated through communiqués, which were initially sent to four media that the Zapatistas thought would publish their words.[13] The eloquence of the communiqués was partly responsible for the success of the Zapatistas. These open letters helped to mobilize civil society to put pressure on the federal government to end its early military attempts to destroy the EZLN. They also helped to create deep sympathy for the Zapatistas throughout Mexico. Several US publications published interviews with Marcos (e.g. *The New York Times*, *Vanity Fair*), and these contributed to international public opinion calling for a peaceful settlement in Chiapas.[14]

The EZLN communiqués were cast in different voices. Many of the early ones were in a diplomatic language as the EZLN struggled to be formally recognized as a belligerent force by the federal government, and to denote the official nature of the EZLN organization. Meanwhile, many of Marcos's 'personal' letters were written in the idioms of Mexico City's streets and universities. Marcos used many voices: self-mocking and wisecracking; those of a letter writer, storyteller, and poet; the voice of a visionary community, of the indigenous people of Chiapas; and as the voice of the Comite Clandestino Revolucionario Indeigena-Coordinadora General (CCRI-CG), of the EZLN. The communiqués articulated the amalgamated injustices, grievances, and demands of the peasants and indigenous people of Chiapas (and Mexico). Such demands had little to do with ideology and everything to do with *campesino's* needs, and hence their resonance with civil society. In addition to the print media, EZLN communiqués were also posted on numerous Mexican news bulletin boards on the Internet, then downloaded and photocopied by groups in other countries (e.g. Spain) to be handed out in demonstrations in solidarity with the Zapatista cause.[15]

However, the Zapatistas' use of media-tion for strategic purposes was certainly not unique in the history of guerrilla politics. For example, Fanon (1965: 82–85), writing about the Algerian revolution, discussed the importance of the radio in enabling information about the revolution to spread amongst the population, and in mobilizing the Algerians against the French colonizers. Moreover, Fidel Castro invited news reporters and film camera crews to the

Cuban guerrilla bases in the Sierra Maestra mountains in order to conduct interviews, and made extensive use of the mimeograph to print propaganda communiqués. The difference between these particular resistances and the Zapatistas is the *primacy* of guerrilla war within their struggle (to which media-ted forms of resistance were a support). In contrast, the Zapatistas' insurgency comprised a series of event-actions, designed to create images of rebellion that would attract public attention towards their discourses of resistance: i.e. those demands, critiques, and analyses that were contained within the Zapatistas' communiqués.

We might argue that this strategy was armed propaganda rather than armed struggle. However, I want to stress that the *materiality* of armed struggle – the embodiment of resistance *par excellence* – within the Zapatista rebellion is an integral part of their strategy, and one *without which* the political effect of the Zapatistas' war of words would have been negligible. As Marcos noted:

> We did not take up arms to appear in the newspaper. We took up arms so as not to die of hunger. If the Government does not heed our demands, the problem is going to continue and the war is going to continue.
>
> (Subcommandante Marcos, quoted in Ó Tuathail 1997)

The Zapatistas articulated a counter-discourse to that of Mexican government, which as Ó Tuathail (1997) has argued had attempted to create the appearance of Mexico as a newly emerging market, stable for foreign investment, and an economically viable partner within NAFTA. This counter-discourse has served to (1) make people aware of the unequal distribution of land, and economic and political power in Chiapas; (2) challenge the neoliberal economic policies of the Mexican government (particularly the reform of Article 27 of the Mexican constitution) and articulate the imagined long-term detrimental effects of NAFTA upon the peasant economy; (3) articulate an alternative to *indigenismo* – the politics of assimilation of indigenous peoples into the fabric of the Mexican state – by advocating the politics of *indianismo*, which articulates the indigenous worldview and promotes Indian political autonomy (Ewen 1994: 34, 39); and (4) articulate a call for the democratization of civil society, which in turn has enabled the creation of political space for numerous indigenous and peasant organizations to articulate their own political and material challenges (see below). This discourse is a form of 'semiotic resistance' that articulates the 'desire of the subordinate to exert control over the meanings of their lives, a control typically denied them in the material social conditions' (Fiske, quoted in Escobar, 1992: 75).

Effects of the Zapatistas' appearance

The Zapatista rebellion has had various effects within Mexican civil society, the economy, and the corridors of international finance. Since the uprising, village-based groups under the banner of the State Council of Indian and Campesino Organizations (CEOIC), which includes representatives of 280 state *campesino* and indigenous organizations, have launched a campaign of land occupations in Chiapas. They have declared five 'pluriethnic autonomous regions' in Chiapas, and seized over 50,000 hectares of land (Harvey 1995: 67). Moreover, at the end of 1994, a Rebel Government in Transition was inaugurated in Chiapas, headed by Amado Avendano,[16] in opposition to the official state government (headed by Eduardo Robledo). While the latter was located in Tuxtla Gutierrez, the state capital, the former was located in San Cristobal de las Casas, the capital of indigenous Chiapas (Kampwirth 1996).

Meanwhile, the Zapatistas have announced their presence in 38 of Chiapas' 111 municipalities, pronouncing these municipalities in rebellion and loyal to the rebel government of Avendano.[17] While PRI municipal governments have administrative control over the village centers, real territorial control lies with independent Indian and *campesino* organizations, which generally support Avendano and the EZLN. Indeed, Avendano is supported by the CEOIC, the state PRD, the Women's Group, and Civil Society (a grassroots political formation independent of political parties) (Weinberg 1995). Solidarity actions have also occurred including land takeovers, road blockades of government vehicles, and blockades of government offices. In March 1994, there were 200 land occupations in Chiapas alone (Weinberg 1994). In the states of Oaxaca, Veracruz, and Tabasco the left opposition is challenging the installation of PRI state governors. Solidarity actions have taken place in other states, including Pueblo, Mixtec Sierra, Oaxaca, Juchitan, Michoacan, and Guerrero. The Zapatista rebellion has provided the political space for hitherto independently operating community-based indigenous and *campesino* organizations to begin working together in a state-wide network (Ross 1995; Weinberg 1995). However, many guerrillas and *campesinos* have been killed in combat with the Mexican army, while Chiapan peasants also face daily harassment, torture or even murder by the military, in addition to the ongoing lack of health facilities, adequate nutrition, etc.

Regarding the Mexican economy, by 10 January 1994 the Mexican stock market was down by 6.2 per cent of its total worth, the biggest fall since October 1987 (Ross 1995). In March alone, investment in the stock market had declined by US\$5.9 billion. The collapse of the peso on 20 December 1994 was blamed on rebels for undermining investor confidence and dashing expectations that Mexico would transform itself into a modern economy.[18] As a result, President Clinton organized a US\$47.8 billion emergency loan to steady the peso and allow Mexico to restructure its short-term debt (Guillermoprieto 1995). In February 1995, when President Zedillo ordered the army against the Zapatistas, those in Chiapas were convinced it was linked to loan guarantees from the USA. Indeed, in a report in *The Independent on Sunday*, Chase Manhattan Bank issued a memorandum to investors on 13 January stating: 'The government will need to eliminate the Zapatistas to demonstrate their effective control of the national territory and of security policy.' It continued, while the insurrection in Chiapas 'does not pose a fundamental threat to Mexican political stability, it is perceived to be so by many in the investment community' (Doyle 1995: 14). Wark has argued that when an event causes political time (with its struggles) to intersect with economic time (with its investments) then a crisis arises (1994: 203). The Zapatistas' appearance seems to confirm this, since the political event of the rebellion, while predominantly local and national in character and significance, has caused global geo-financial effects.[19] Since that time, an uneasy truce has developed between the guerrillas and the government, with the peace talks stalled.

Critical geopolitics, mediation, and the Zapatistas

Media images are increasingly seen by social movements as an essential aspect of organization, a tool for changing attitudes, raising public awareness, and relaying movement views to a wider public. The use of images has become an important strategy in the conflict over (re)presentations of events between activists, governments, private corporations, and the public. As I have shown, the Zapatistas articulate material and representational challenges to the Mexican state and its elites. Eschewing the capture of state power, the Zapatistas nevertheless articulate challenges, both to the material political and economic power of the Mexican state – through the mobilization of civil society – and to the monopoly of representations of Mexico as an 'emerging market,' economically and politically stable for foreign investment. In order to effect these challenges, the Zapatistas have undertaken a political strategy that focuses upon the appearance of a guerrilla war rather than the undertaking of a protracted armed struggle. This strategy has comprised the waging of guerrilla war on both the physical landscape and the mediascape, with strategic importance given to the images and discourses of their struggle. As such, the Zapatistas can be considered as a form of media, engaging in event-actions and the articulation of counter-discourses that pose symbolic challenges to geopolitical and geo-financial power.

Calderon *et al.* (1992: 27) are pessimistic about the ability of social movements to oppose national and global power relations in order to create new spaces of struggle and debate within societies. They argue that collective action is relatively powerless in the face of the forces of international domination, the cultural homogenization of the market, the increased pervasiveness of the mass media, and the central role of financial capital in the world economy. However, while sites and methods of resistance have traditionally been defined in terms of physical space, recent technological advances (and the associated power of the geo-financial panopticon) suggest that we should begin to rethink spatial disturbance as the *only* productive form of resistance (Critical Art Ensemble 1994).

Indeed, the use of various forms of media can alter the balance of power in social struggles effecting the refusal of established (and frequently state-centered) boundaries of communication, and enabling social movements to

254

escape the social confines of territorial space. Through the use of communications technology, social movements can articulate resistance on an international stage, and can 'go globile' by effecting a strategic mobility within the increasingly global space of contemporary media. However, going globile can be an ambiguous process. The media-tion of resistance practices can lend legitimacy to a campaign, serve to inform the public, and attract new recruits to a particular struggle. It can also serve to deepen the commitment and empower those who are participating in the campaign. Such coverage may also contribute to changing the long-term climate of public opinion at the local, national, and global scales (Burgess 1990). The use of the media can lend itself to a strategic mobility within social movement practices that enables political and discursive maneuverability in relation to their opponents (be they states, international institutions, private interests, etc.). The ability to be seen and heard by national and international audiences lends legitimacy to resistance and enables counter-hegemonic discourses to circulate within the public/global realm.

However, the use of the media is still dependent upon the overall narrative framework of the news; and media coverage, how ever extensive, does not guarantee the success of social movement campaigns. As Herman and Chomsky (1988) argue in their 'propaganda model,' news in the mainstream US media passes through successive filters (e.g. corporate control of the media; the profit orientation of these media; the dependence of news corporations on information supplied by the government and industry) that 'cleanse' material of its subversive, counter-hegemonic content. Moreover, the media frequently creates personalities out of the spokespeople of particular social movements, thereby potentially individualizing the collective character of political action (Gitlin 1980).

As social movements engage increasingly with the media – as they become a form of media – so there is a tendency for them to perform what Truett-Anderson (1990) terms 'theatrical politics.' This is particularly the case with news media, which get consumed as one entertainment among many. As the practice of resistance becomes increasingly dramaturgical, there is a danger that politics may become more about appearance than effect, more about symbolic protest than material change. While social movements may achieve a short period of media coverage, it is far more difficult for them to maintain positions as 'established,' authoritative, or legitimate actors regarding political decision making.

Having said this, however social movement messages are received – and there are, of course, multiple possible readings that a public may attribute to social movement 'texts' – a media vector leaves traces in people's imaginations and memories that exist beyond the experience of a particular event-action or campaign. Memories of previous resistances can be aided by a vectoral record (whether spoken, written, recorded, or filmed), which may provide an archive of tactics and strategies for future use – what I would term the *ghosting* of resistance. Moreover, although the Zapatistas imagineer their resistance, it would be a mistake to characterize such protests merely as a politics of gesture. For, as I have discussed above, the Zapatistas have engaged in armed propaganda and event-actions in order to appear (and as guerrillas, to just as suddenly disappear) on media vectors and thus symbolically and materially challenge the Mexican state. Such *spectacular* acts of resistance have become the sites of media-tion between social relations and between the embodied and virtual worlds – *within and between* materiality and representation, the immediate and media-tion.

This has important implications for a critical geopolitics that seeks to engage critically with social movements such as the Zapatistas. While the immediate is the world of the Chiapan peasants who face displacement and impoverishment from the *ejido* reforms, the Mayan Indians who face pauperization and cultural ethnocide, and the Zapatista guerrillas hiding deep in the Lacandon jungle, it is the media-ted images and words that are consumed and appropriated by academics. Hence *from where we watch it*, the media-ted event is not embodied, or *lived*. Rather, such events become subjected to intellectual post mortems, which frequently feel completely *disembodied*. There is then a temptation, perhaps, to romanticize the struggle and its principal actors, and in representing them from a position of solidarity, make a critical geopolitics feel good about itself. Moreover, there is a danger that theories of resistance may tend to nullify the physical, embodied acts of resistance, taming them macropolitically, or teleologically (see Routledge and Simons 1995) or appropriating them, and then marketing them as cultural capital.

This is not to suggest that these theories and discourses do not have merit, or interest (predominantly to the academic caste that produces, consumes then recycles these words). Nor is it to argue that the written word does not have explanatory and potentially liberatory power. Indeed, e-mail messages of solidarity, book chapters, articles celebrating the Zapatistas, newspaper reports, etc. all have a role to play in the discursive terrain of resistance surrounding the EZLN challenge within Mexico. Rather, it is to locate a critical geopolitics *within and between* the discursive and embodied terrains of resistance.

Such a critical engagement effects a 'politics of articulation' involving an interactive process of collaboration between critical theorists and social movements as subjects working together to understand the questions under examination, the heterogeneous accounts of the world (see Routledge 1996a, 1996b). Such a process implies social relations of conversation rather than discovery, the creation of political formations between social movements and critical theorists as actants.[20] It would necessitate a constant interrogation of the positionality of the critical intellectual in relation to the sites of resistance under study, acknowledging the myriad differences that exist between people, while constructing 'flexible, practical relations of solidarity' through various forms of dialogue and struggle (Pfeil 1994: 225).

Through listening to, learning from, and participating in conflicts (i.e. embodying resistance), critical engagement can attempt to tell stories of resistance that traverse between and within discursive and material sites, opening up spaces for practical actions that are as heterogeneous and multiplicitous as our imaginations. The actualization of such situated theory becomes a participatory and collaborationist process *within* the terrain of resistance and raises crucial issues of representational, ethical, and political practice within and without critical geopolitical enquiry. For, as Marcos argues, referring to the struggle of the indigenous peoples of Chiapas:

> they have made their own a word that cannot be understood with the head, that cannot be studied or learned by memorization. It is a word that lives in the heart . . . This word is dignity, respect for ourselves, for our right to better ourselves, for our right to struggle for what we believe, for our right to live and to die in accord with our ideas. Dignity is not to be studied, it is to be lived and died for.
>
> (Subcommandante Marcos, quoted in *Z Magazine* 1995: 5)

Acknowledgements

I would like to thank Gearóid Ó Tuathail for his comments on an earlier draft of this chapter.

Notes

1 The *foco* – a mobile strategic base – involves launching armed struggle without preparatory political work amongst the peasantry. The theory of the *foco* was formulated by Che Guevara (1968). See also Chailand (1982, 1989).
2 The Zapatistas also represented the first Latin American revolutionary organization with an entirely Indian command.
3 A social movement is a heterogeneous formation that comprises myriad (and at times conflicting) interests and identities (e.g. of gender, race, class, sexuality, etc.) that constitute an analytical and political–cultural terrain of contestation in which the hegemonies of state, the development project, aspects of modernity (economic growth, progress) can be explored, defined, and challenged.
4 I will not attempt, here, to give a full and accurate picture of the Zapatista rebellion. I have not visited Mexico. Nor am I a 'Latin American expert.' Even those who have visited Chiapas in order to report, or do research, have found that rather than Spanish, many *campesinos* (peasants) speak only indigenous languages, and those who speak Spanish are shy and distrustful of foreigners (see Guillermoprieto 1995). Indeed, the 'real' story of the EZLN is complex and tangled, essentially unknown to outsiders. Hence, I can only add my interpretation of events to others, and acknowledge that this interpretation is but a shadow of the lived reality of the Zapatista struggle. Further, I do not, in this paper, seek to become the Zapatistas' representative, or have the arrogance to speak for the injustices suffered by the Mayan peoples during the 500-year war waged against the indigenous peoples of the Americas. I recognize that it is all too easy to appropriate these complex events into an intellectual terrain that is far removed – physically, politically, culturally – from the everyday life and struggles of the people of Chiapas. However, as a politically engaged individual, I do write from a position of solidarity – that small piece of my heart that is Zapatista. ('We only ask that a small piece of your heart be Zapatista. That it never sell out. That it never surrender.

That it resist.' From a Zapatista speech to the Native Forest Network in Montana, USA, November 1994).

5 The use of the media or video technology by activists may also represent the strategy of 'media witnessing' of events that might otherwise remain hidden from the public eye.

6 Individual plots and communal property which cannot be legally sold, rented or used as collateral.

7 Volunteers, working in political brigades.

8 Indeed, Major Mario of the EZLN argues that the Zapatistas had their inception in 1983, when six idealists from Mexico's north arrived to join forces with dissident peasants and Indians in a movement that immediately went underground to commence military and political organizing.

9 The EZLN has an eleven-word program: 'Trabajo, Tierra, Techo, Pan, Salud, Educacion, Democracia, Libertad, Paz, Independencia, and Justica' (work, land, shelter, bread, health, education, democracy, liberty, peace, independence, and justice). See Ross 1995.

10 The reform of this Article of the Mexican Constitution in 1991 freed all ejidos for sale to the highest bidder, potentially leading to the displacement of campesinos by wealthier mining, cattle, and logging interests.

11 During the first months of the rebellion, aerial bombings were conducted by the Mexican military upon civilian populations in Chiapas (Ross 1995).

12 After Robledo (PRI) became governor of Chiapas on 8 December 1994, the EZLN halted its cease-fire with the government. The army surrounded the Zapatistas in their Lacandon stronghold, but on 19 December, the guerrillas filtered out unseen via remote mountain trails.

13 These were *Tiempo*, a San Cristobal de las Casas-based magazine; *La Journada* and *El Financiero*, Mexico City-based newspapers; and *Proceso*, a Mexico City-based weekly magazine (EZLN communiqué 1994f: 11).

14 The Zapatistas also hosted an 'Intercontinental Gathering for Humanity and Against Neoliberalism' in the Lacandon jungle during the summer of 1996.

15 Internet postings for medical aid to Commandante Ramona met with massive offers within hours of being posted (Wood 1995).

16 The gubernatorial candidate for the Party of Democratic Revolution, PRD.

17 But see above, concerning their *spatial* control.

18 Presumably defined as one without the instabilities associated with a rebel insurgency.

19 See Ó Tuathail (1997) for a fuller analysis of the geo-financial effects of the Zapatista rebellion.

20 Collective entities in action (see Haraway 1992).

References

Adams, P. C. (1996) 'Protest and the scale politics of telecommunications,' *Political Geography*, 15(5), 419–441.

Appadurai, A. (1990) 'Disjuncture and difference in the global cultural economy,' *Theory, Culture and Society*, 7, 295–310.

Baudrillard, J. (1988) *The Evil Demon of Images*. New South Wales, Australia: Power Institute Publications No. 3, University of Sydney.

Baumann, Z. (1992) *Intimations of Postmodernity*. New York: Routledge.

Brecher, J. and Costello, T. (1994) *Global Village or Global Pillage*. Boston: South End Press.

Burbach, R. (1994) 'Roots of the postmodern rebellion in Chiapas,' *New Left Review*, 205, 113–124.

Burgess, J. (1990) 'The production and consumption of environmental meanings in the mass media: a research agenda for the 1990s,' *Transactions of the Institute of British Geographers*, 15, 139–161.

Calderon, F., Piscitelli, A., and Reyna, J. L. (1992) 'Social movements: actors, theories, expectations,' in Escobar, A. and Alvarez, S. E. (eds) *The Making of Social Movements in Latin America*. Boulder, Colo.: Westview Press. 19–36.

Chailand, G. (1982) *Guerrilla Strategies*. London: Penguin.

Chailand, G. (1989) *Revolution in the Third World*. London: Penguin.

Collier, G.A. and Quaratiello, E. L. (1994) *Basta! Land and the Zapatista Rebellion in Chiapas*. Oakland, Calif.: The Institute for Food and Development Policy.

Critical Art Ensemble (1994) *The Electronic Disturbance*. New York: Autonomedia.

Dalby, S. (1991) 'Critical geopolitics: discourse, difference, and dissent,' *Society and Space*, 9, 261–283.

Dodds, K. J. and Sidaway, J. D. (1994) 'Locating critical geopolitics,' *Society and Space*, 12(5), 515–524.

Doyle, L. (1995) 'Did US bank send in battalions against Mexican rebel army?,' *The Independent on Sunday*, March 5, p. 14.

Escobar, A. (1992) 'Culture, economics, and politics in Latin American social movements theory and research,' in Escobar, A. and Alvarez, S.E. (eds) *The Making of Social Movements in Latin America*. Boulder, Colo.: Westview Press. 62–85.

Ewen, A. (1994) 'Mexico: the crisis of indentity,' *Akwe:kon*, 11(2), 28–40.

EZLN (1994a) 'Chiapas: the southeast in two winds, a storm and a prophecy,' in *Anderson Valley Advertiser*, vol. 42, no. 31, 1–5.

EZLN (1994b) 'Declaration of the Lacandon jungle: today we say "Enough",' in *Anderson Valley Advertiser*, vol. 42, no. 31, 5–6.

EZLN (1994c) 'Here we are, the forever dead,' in *Anderson Valley Advertiser*, vol. 42, no. 31, 6.

EZLN (1994d) 'The sup will take off his mask, if Mexico takes off its mask,' in *Anderson Valley Advertiser*, vol. 42, no. 31, 8.

EZLN (1994e) 'Zapata will not die by arrogant decree,' in *Anderson Valley Advertiser*, 42, no. 31, 16.

EZLN (1994f) 'Reasons and non-reasons why some media were chosen,' in *Anderson Valley Advertiser*, vol. 42, no. 31, 11.

Falk, R. A. (1987) 'The state system and contemporary social movements,' in Mendlovitz, S.H. and Walker, R.B.J. (eds) *Towards a Just World Peace: Perspectives from Social Movements*. London: Butterworths. 15–48.

Fanon, F. (1965) *A Dying Colonialism*. New York: Grove Press.

Gedicks, A. (1993) *The New Resource Wars*. Boston: South End Press.

Gitlin, T. (1980) *The Whole World is Watching*. Berkeley: University of California Press.

Guevara, E. Che (1968) *Guerrilla Warfare*. New York: Vintage Books.

Guillermoprieto, A. (1995) 'The shadow war,' *The New York Review of Books*, March 2, 34–43.

Haraway, D. (1992) 'The promises of monsters: a regenerative politics for inappropriated others,' in Grossberg, L., Nelson, C. and Treichler, P. (eds) *Cultural Studies*. London: Routledge. 295–337.

Harvey, N. (1994) 'Playing with fire: the implications of ejido reform,' *Akwe:kon*, 11(2), 20–27.

Harvey, N. (1995) 'Rebellion in Chiapas: rural reforms and popular struggle,' *Third World Quarterly*, 16(1), 39–72.

Herman, E. S. and Chomsky, N. (1988) *Manufacturing Consent: The Political Economy of the Mass Media*. New York: Pantheon Books.

Hooks, B. (1990) *Yearning: Race, Gender, and Cultural Politics*. Boston: South End Press.

Kampwirth, K. (1996) 'Creating space in Chiapas: an analysis of the strategies of the Zapatista army and the rebel government in transition,' *Bulletin of Latin American Research*, 15(2), 261–267.

Melucci, A. (1989) *Nomads of the Present*. London: Radius.

Montejo, V. (1994) 'Border dynamics: a Mayan perspective,' *Akwe:kon*, 11(2), 49–52.

Ó Tuathail, G. (1997) 'Emerging markets and other simulations: Mexico, the Chiapas revolt and the geofinancial panopticon,' *Ecumene*, 4(3), 300–317.

Pfeil, F. (1994) 'No basta teorizar: in-difference to solidarity in contemporary fiction, theory, and practice,' in Grewal, I. and Kaplan, C. (eds) *Scattered Hegemonies*. Minneapolis: University of Minnesota Press. 197–230.

Rose, G. (1993) *Feminism and Geography*. Cambridge: Polity Press.

Ross, J. (1995) *Rebellion from the Roots*. Monroe, Maine: Common Courage Press.

Routledge, P. (1993) *Terrains of Resistance: Nonviolent Social Movements and the Contestation of Place in India*. Westport, Conn.: Praeger.

Routledge, P. (1994) 'Backstreets, barricades and blackouts: Nepal's urban terrains of resistance,' *Society and Space*, 12, 559–578.

Routledge, P. (1996a) 'Critical geopolitics and terrains of resistance,' *Political Geography* 15(6/7), 509–531.

Routledge, P. (1996b) 'The third space as critical engagement,' *Antipode*, 28(4) 397–419.

Routledge, P. (1997) 'The imagineering of resistance: Pollok Free State and the practice of postmodern politics,' *Transactions of the Institute of British Geographers*, (forthcoming).

Routledge, P. and Simons, J. (1995) 'Embodying spirits of resistance,' *Society and Space*, 13, 471–498.

Subcommandante Marcos (1995) 'A letter,' *Z Magazine*, September, 5–6.

Toledo, V.M. (1994) 'The ecology of Indian campesinos,' *Akwe:kon*, 11(2), 41–46.

Truett-Anderson, W. (1990) *Reality Isn't What It Used To Be: Theatrical Politics, Ready-to-Wear Religion, Global Myths, Primitive Chic, and Other Wonders of the Post-Modern World*. San Francisco: Harper & Row.

Wark, M. (1994) *Virtual Geography*. Bloomington: Indiana University Press.

Weinberg, B. (1994) 'Zapata lives on: a report from San Cristobal,' *Akwe:kon*, 11(2), 5–12.

Weinberg, B. (1995) 'Chiapas one year later: rumbles of war, rumors of peace,' *The Nation*, 260(5), 164–166.

Wollock, J. (1994) 'Globalizing corn: technocracy and the Indian farmer,' *Akwe:kon*, 11(2), 53–66.

Wood, D. (1995) 'Net wars. Chiapas: the revolution will not be televised (but it will be on line)' *Index on Censorship* no. 3. June. http://www.oneworld.org/index-oc/wood.html.

12

'ALL BUT WAR IS SIMULATION'

James Der Derian

On Rigor in Science

In that Empire, the Art of Cartography reached such Perfection that the map of one Province alone took up the whole of a City, and the map of the empire, the whole of a province. In time, those Unconscionable Maps did not satisfy and the Colleges of Cartographers set up a Map of the Empire which had the size of the Empire itself and coincided with it point by point. Less Addicted to the Study of Cartography, Succeeding Generations understood that this Widespread Map was Useless and not without Impiety they abandoned it to the Inclemencies of the Sun and of the Winters. In the deserts of the West some mangled Ruins of the Map lasted on, inhabited by Animals and beggars; in the whole Country there are no other relics of the Discipline of Geography.

(Jorge Luis Borges, 'Viajes de Varones Prudentes,' *Dreamtigers*)

As a society, we're leaving the landscape and moving onto the map, without paying much attention to the process or the destination.

(Mitch Kapor, founder of Lotus, *Mondo 2000*)

The evidence lies, as it were, in the images. Flash back once more to the Berlin Wall, taking its first hammer blows, President George Bush and Secretary of State James Baker appearing at a televised press briefing with a map of Germany in front of them, seeking in cartography what they could no longer locate in reality: the fixity of former borders and former times. For those with an investment in the status quo, when events were moving too quickly and too unexpectedly, the map became a more appealing, more plausible home than the world itself. Now flash forward to the twenty-first century, long after the End of History has been remaindered, The Coming Anarchy is the name of a retro-punk band, and the *X-Files* are under subpoena by a special prosecutor: What maps will we inhabit? In search of some answers, I undertook a trip to the Simulation Triangle.

But first, a word on cautionary road signs. In times of 'phase transitions'

between order and disorder – the best definition so far for the indefinable phenomenon of complexity – the main task of the critical geopolitician (crit-geo) is
to challenge the arrogance of the Empire's cartographers as well as the mapping
imperatives of the social sciences that would colonize the present, reduce the
other to the same, even confuse the map for the 'real thing.' Whether it is in the
name of abstraction, parsimony, or tradition, there is a scientific predilection in
mapping that favors the global reach over the grasp of the local, the thin over
the thick description, the revisionist over the visionary perspective. This is why
the crit-geo must go where the signs say not to: the edge of the map. There we
might find the dangerous complexities that the mapped world effaces. And since
this is usually where the sea monsters lurk, we might also gain some insight into
the dark allure of the edge that keeps drawing us on.

> This idea of the indirect approach is closely related to all problems of
> the influence of mind upon mind – the most influential factor in human
> history. Yet it is hard to reconcile with another lesson: that true con-
> clusions can only be reached, or approached, by pursuing the truth
> without regard to where it may lead or what its effect may be – on dif-
> ferent interests . . . In strategy, the longest way round is often the
> shortest way home.
>
> (B.H. Liddell Hart 1967: 5)

'All but war is simulation': this is the slogan that took me over the edge of the
map. I first heard it at the annual Interservice/Industry Training Systems and
Education Conference in Orlando (I/ITSEC), where it kept popping up like a
bad mantra. I had ambushed a colonel for a hallway interview after he had finished a briefing on the virtues of virtual simulations to a packed room. At the
end of the interview, he handed me a standard-issue business card with the slogan as its banderole. When I asked him what it meant, he gave me a quick
history of his current base, STRICOM.

The slogan originated in 1992 with the activation of STRICOM (Simulation,
Training, and Instrumentation Command), the newest and – as I was to find out
– the most unusual command post in the military. Tasked to provide the US
Army's 'vision for the future', STRICOM chose a bold motto to go with the
command post logo of a 'land warrior' bisected by a lightning bolt in the middle of a bull's-eye. In the tone of instruction, he told me what the phrase means:
'Everything short of war is simulation.' But, he then added hastily, 'we don't
really look at it that way, because you can't manage that properly.' Sensing my
confusion, he offered an analogical assist: 'When you think about it, well, it's
kind of like your love-life: everything short of it is simulation.'

An officer of lesser rank, someone who knew a dodgy soundbite when he
heard one, cut in to remind the colonel that he had a plane to catch. I was left
standing in the hallway, the potted palm with a frozen half grin. What did he
mean by 'love-life'? Did this mean war was to simulation as 'love' was to

stimulation? Was STRICOM into some kind of William Gibson *Neuromancer*, 'sim/stim' thing?

These were not the kind of questions that had originally brought me to I/ITSEC. I came to Florida to bear witness to an auspicious alignment of the military, new media, and Mickey. In one corner of Orlando, I/ITSEC was occupying the Marriot World Center for three days, with over 60 panels, 180 exhibition booths, and enough uniforms and suits to grid-lock the Beltway. Gathering under one convention roof for this year's theme, 'Information Technologies: The World Tomorrow,' the conference included an impressive list of special events, keynote speakers, and a who's who of industry CEOs, Defense Department higher-ups, and officers from all branches of the military. And truth be told, I was drawn to the prospect of hearing Tom Clancy as the banquet speaker (a no-show, as it turned out). At the other corner of Orlando, 40 minutes up the Central Florida Greeneway, STRICOM was setting the stage for an award ceremony for the $69 million 'JSIMS' contract. According to the press release, JSIMS (Joint Simulation System) was 'a distributed computerized warfare simulation system that provides a joint synthetic battlespace . . . to support the 21st century warfighter's preparation for real world contingencies.' And making up the third leg, a few miles down International Drive through the pink arches and under a pair of mouse ears, Disneyworld was celebrating its twenty-fifth Anniversary with a paroxysm of imagineered (copyrighted) fun.

I entered the Orlando Triangle as one might enter a paradox, where slogans like 'everything but war is simulation,' 'prepare for war if you want peace,' 'the land where the fun always shines,' quickly enhance the appeal of tour guides who do not rely on linear reasoning and conventional cartography. My intention was to ask a few questions, make some observations, and get in, get out quick. For guidance, I drew from some thinkers who well understood the seductive powers of simulations. To jump the monorails of spectacle, I borrowed from Guy Debord the subversive power of the 'psychogeographic drift' (the preferred situationist method of studying the psychological effects of a geographical environment on inhabitants as well as the transient observer) to tour a world where 'everything that was directly lived has moved away into a representation' (Debord 1983). To counter the hazards of hyper-real simulacra, I relied on the hyperbole of Jean Baudrillard, who could well have had in mind the military–industrial–media–entertainment (MIME) complex when he warned of 'a group which dreams of a miraculous correspondence of the real to their models, and therefore of an absolute manipulation' (1983: 43). And to avoid becoming one more casualty of 'the war of images,' I planned to take seriously Paul Virilio's advice that 'winning today, whether it's a market or a fight, is merely not losing sight of yourself' (1988: 7).

> The problem of security, as we know, haunts our societies and long ago replaced the problem of liberty. This is not as much a moral or philosophical change as an evolution in the objective state of systems.
>
> (Jean Baudrillard 1990: 37)

This indirect approach was prompted by my first pilgrimage to I/ITSEC, five years earlier in the wake of the Gulf War. Back then there was a real Patriot missile in the lobby, flanked by two looped videos extolling its virtues through a series of blatantly phallic images. Many of the military still seemed to be shaking the sand out of their boots. This year, however, with the kill ratio of the Patriot dramatically downgraded, Kurds in refugee camps in the no-fly zone, and Saddam Hussein still playing the rogue, the victorious aura of the Gulf War had somewhat faded. Moreover, the poisonous snakes that emerged from the belly of the dead dragon – the post-Cold War metaphor and prophecy of former CIA director Admiral Woolsey – had since morphed into multi-headed hydras, in the former Yugoslavia, Somalia, Chechnya, Rwanda, and in other expanding pockets of the new chaos.

The pride and patriotism of I/ITSEC '91 still flared on occasion into imperial hubris and technological hype, but this year's model was more a meld of corporate steel and glass with infotainment show-and-tell. Envisioning the future was still the goal, but enriching yourself and entertaining the stockholders *en route* made for a burgeoning of concessions on the way to Tomorrowland. Nowhere was this more apparent than in the war of signs itself, with the self-help vocabulary of management consultants giving the acronymic, ritualized language of the military a run for its fiscal allocations. 'Synergy' was the conference buzzword. Synergy between the high-flyers in the military and top players in defense industries, to make those thinner and thinner slices of the budgetary pie go that much further. Synergy in the form of alliances or outright mergers among the major defense industries. But also synergy at the advanced technological level, to imagine and engineer a new form of virtual warfare out of networked computer simulation (SIMNET) and Distributed Interactive Simulation (DIS), a command, control, communication, computer, and intelligence system of systems (C4I), and complete interoperability through a common high-level architecture (HLA). Perched at the top of this synergy pyramid was the endgame of all war games, JSIMS, the macro-, mega-, meta-simulation of the twenty-first century. Or so they said.

When I arrived at the convention, the synergy wave was making its way through the Grand Ballroom, where the Flag and General Officer Panel was in full session. On a podium at one end of the vast room, against a projected backdrop of the American flag, multiplied and magnified by two oversized video screens, the top brass and officials from the Department of Defense presented their views on the role of information technologies for the military. Deputy Undersecretary of Defense Louis Finch warned of a return to a post-Vietnam 'hollow army' if new information technologies were not harnessed 'to manage a massive transition.' Vice Admiral Mazach called for a post-Cold War strategy that could deal with more complex, multiple threats in a time of military downsizing, declaring that 'We must walk down the information highway – or be run down.' Vice Admiral Patricia Tracey endorsed the use of 'infomercials' in boot camp to train our troops in issues like drug and alcohol abuse as well as in new sensitive areas like gender relations: 'Disney has used it for years, we're ready to use it now.' Major General Thomas Chase of the Air Force, citing the displacement of traditional battlefields by a digitized 'battlespace,' endorsed a global linking up of 'synthetic environments.'

Not everyone was so eager to jump on the cyber-wagon. Wearing battle ribbons from two tours in Vietnam, unaccompanied by snappy graphics or intricate flow charts, Major General Ray Smith of the Marine Corps took a more cautious approach to simulations. No luddite, he acknowledged the need for new skills and training techniques for the soldier, offering the story of a lance corporal abroad, who in a single day might rehydrate a starving child, mediate between members of warring clans, handle the media, and use a global positioning system with a satellite link-up to call in a gunship attack. Simulations, while useful, are not sufficient to train such a range of complex and compressed duties: only experience in the field would do. When asked from the floor what industry can do to help, he paused, then bluntly said: 'Make it cheap.' After the panel I probed him for the source of his guarded skepticism. 'In war you fight people not machines. We're training to beat computers, instead of training to beat the enemy. You cannot model the effects of confusion and surprise, the friction and fog of war.'

> This makes the decisive new importance of the 'logistics of perception' clearer, as well as accounting for the secrecy that continues to surround it. It is a war of images and sounds, rather than objects and things, in which winning is simply a matter of not losing sight of the opposition. The will to see all, to know all, at every moment, everywhere, the will to universalised illumination: a scientific permutation on the eye of God which would forever rule out the surprise, the accident, the irruption of the unforeseen.
>
> (Paul Virilio 1994a: 70)

Smith's view ran against the grain of an emergent technological imperative to manage uncertainty, unpredictability, and worst-case scenarios of chaos through superior simulation power. All the major corporate players were making the pitch in force – Lockheed Martin, McDonnell Douglas, Boeing, Hughes, Evans and Sutherland, Raytheon, and Northrop – along with the rising stars of the simulation business, like SAIC, Silicon Graphics, Reflectone, and Viewpoint DataLabs. They had come to sell the hardware and software of the future. Human wetware was more problematical. Indulged as a consumer, it otherwise took on the look of an expensive add-on, or a plug-in with compatibility problems. In most instances, the human component added a bizarro effect to the synergy mix. Consider an excerpt from one of the papers presented in the 'Modeling and Simulation' section, called 'Human Immersion into the DIS Battlefield':

> Recent advances in human motion capture and head mounted display technologies, coupled with Distributed Interactive Simulation (DIS)

capabilities, now allow for the implementation of an untethered, fully-immersable, DIS-compliant, real-time Dismounted Soldier Simulation (DSS) System. The untethered soldier, outfitted with a set of optical markers and a wireless helmet-mounted display, can move about freely within a real-world motion capture area, while position and orientation data are gathered and sent onto a DIS network via tracking cameras and image processing computers.

Fortunately, for those who cannot tell DIS from DSS, there was a demonstration on hand to cut through the techno-babble. Occupying some prime real estate at the entrance to the exhibition hall, the STRICOM booth was running a looped version of the 'dismounted soldier' – the 'dis' saying it all about the level of respect for a grunt without wheels. Tracy Jones, lead engineer of individual combatant simulations at STRICOM, gave me the blow-by-blow:

> We are trying to prove the principle of immersing the individual soldier in a virtual environment and having him interact with other entities in real time. What we've got is a wireless optical-reflective marker system developed by the entertainment industry about ten years ago in movies like *Batman* and *Aliens*. It consists of a series of four camera systems with spotlights, 16 markers on the soldier's body and three on this M–16. These markers will pick up exactly where he is in real time and render it into a 3-D model for a virtual database. He's wearing a wireless virtual head display so he can see where he is in the virtual environment.

She directed me to the back of the display. Lifting an edge of the camouflage netting, she revealed the *deus ex machina* of DSS. 'This is a MODSAF SGI station.' She translated for me: 'Modulated Semi-Automated Forces, Silicon Graphics Images.' It was a program developed by the Army to construct computer-generated forces, because, as she put it, 'you're never going to have enough men – uh, *people* – in the loop to populate a simulated battlefield, so we have computer-generated forces that are smart and intelligent, that can fight against our men in the loop.' When I asked why '*semi*-automated', she admitted that 'they're not completely smart, you can't just push a button and let them go.' I was going to ask her if she knew about SKYNET and the semi-automated sentinels in *The Terminator* that synergized into a very nasty Arnold Schwarzenegger. But I feared she might find that condescending.

I asked the wired soldier instead, a big guy in camouflage who looked more like Sly with a mustache than Arnold in shades. 'Isn't this getting close to the Terminator? Aren't you afraid of the machines getting smarter than the soldiers and taking over?' He gave me the narrow-eyed Clint look – or maybe it was just the camera lights: 'Uh. . .' Tracy intervened: 'They're not that smart yet.' Not sure who wasn't that smart, I asked who usually wins in the simulations. No hesitation from Sly this time: 'I do.' Is that programmed in? 'Well, they can't kill me. Otherwise we'd have to stop and restart the program.' So you're immortal? 'No, I'm Rambo.' Before I can get him to elaborate on this distinction, Tracy announces that it is time to start the demonstration.

At the front of the booth I recognized the new commander of STRICOM, Brigadier General Geis, from the front cover of the recently launched magazine *Military Training and Technology*. He is surrounded by some VIPs but he amicably agrees to a quick interview. What I get is a verbal version of a press release on the cost-effectiveness of simulations in a period of military draw-down – which is understandable, given his short tenure of the job. But there is a pay-off: after I confess to continued confusion about JSIMS, he invites me to come out to the base the next day to witness the signing ceremony of the contract award. There I could get a first-hand account from the architects and builders of JSIMS.

> The spectacle is the map of this new world, a map which exactly covers its territory. The very powers which escaped us show themselves to us in all their force.
>
> (Guy Debord, 1983)

I left the STRICOM booth and plunged into the belly of the beast. The exhibition hall was vast, full of simulated gunfire, flashing computer monitors, and reps who varied in style from barkers at a freak show to the Zen *haiku* of a Nissan ad. There were simulated cockpits of jets and helicopters, tanks and spaceships. You could fire a simulated M–16 at 'terrorists' (all looking like cousins of Arafat), throw simulated grenades (you smell the post-traumatic stress with each flash-bang), tear up some turf in a simulated M1A2 tank (no German farmer to complain), take out a bad guy in a simulated drug raid (in a curious fashion-lag, the Miami Vice look prevails), or blow up a building with a simulated truck bomb (essential viewing for every militia member). In this electromagnetic maelstrom of simulation, patriotism, and profit, I thought a seizure was more likely than synergy. I drift, heading nowhere, searching everywhere for a psychogeography that might provide a map of meaning for the sound and light show.

I found a familiar landmark immediately behind the STRICOM booth, where a small group of marines was using the synergy to make simulation fun. Compared with the surroundings, theirs was a low-tech operation: cordoned off by black curtain, there were four monitors with keyboards, a projection screen, and a sound system all hooked up to a minicomputer. I had stumbled upon *Marine Doom*. On a tight budget, and always looking for off-the-shelf technology, the Marine Corps Modeling and Simulation Office had decided to appropriate rather than innovate, to simulate what marines do best: to fight independently in squads with small arms. There wasn't a smart weapon in sight, just a computer-generated four-man fire team in a re-tooled game of *Doom*. The monsters had been replaced by distant, barely visible forces that kept popping up out of foxholes and from around bunker walls to lay down some lethal fire.

After giving a history and description of *Marine Doom* – 'a mental exercise in command and control in a situation of chaos' – the lieutenant wanted to know if I was ready to walk the walk. Having spent some time in the video arcade, I thought it couldn't be too tough, especially since I would be playing with the lieutenant and two kids barely in their teens, who seemed to have acquired squatter rights. That was my first mistake. The plan was simple enough: with mouse and keyboard strokes controlling speed and direction, we were to head out of our foxhole, traverse the road, go around some bunkers, and clear a building of bad guys. In eight attempts, I was killed seven times. The single time I made it all the way to the building, I killed the lieutenant in a burst of 'friendly fire.' I was not sure if you had to say you were sorry in simulations, but I apologized nonetheless. The high-quality graphics, sounds of gunfire and heavy breathing, and the sight of rounds kicking up in your face, as well as the constant patter of the lieutenant ('Save your ammo. Point man, take that bunker. You're taking rounds. I'm going up, cover me. Ahh! I'm down'), gave the 'game' a pretty high dose of realism, especially if accelerated heartbeat is any measure.

The appropriation of *Doom* by the Marine Corps was significant for another reason. Usually the technology transfer goes in the other direction, with military applications leading the way in research and development, from the earliest incarnations of the computer in simulation projects like 'Whirlwind' at MIT's Servomechanisms Laboratory during World War II, to 'SAGE', the first centralized air defense system of the Cold War. We could say there has been from the very first a close 'Link' between military simulations, the development of the computer, and the entertainment industry. In 1931, the Navy purchased the first aircraft simulator from its designer, Edward Link. By 1932, the military still had only one Link Trainer; the amusement parks had bought close to 50. Now the developmental lag between the real thing and its simulation has just about disappeared. From the F–16 to the F–117A, the M1A2 tank to the Bradley armored vehicle, the Aegis cruiser to the latest nuclear aircraft carrier, the video game version arrives on the shelves almost as soon as the weapon system first appears. Indeed, a Pentium chip and a joystick will get you into the Comanche helicopter, the F–22, and the newest *Seawolf* SSN–21 submarine – which is more than a real pilot or sailor can currently claim as these projects suffer delays and budget cuts.

> For it is with the same Imperialism that present-day simulators try to make the real, all the real, coincide with their simulation models. But it is no longer a question of either maps or territory. Something has disappeared: the sovereign difference between them that was the abstraction's charm.
>
> (Jean Baudrillard 1983: 2)

My drift was interrupted by an invitation to attend a lunch laid on by Lockheed Martin. Over a catered meal in a hotel suite, Stephen Buzzard, vice-president for business development at the company walked a group of journalists – mainly from the military and defense industry journals – through a series of organization flow charts that seemed to be in constant need of verbal revisions. Merger mania had outstripped the capabilities of the graphics and public relations departments. Lockheed, having barely digested Martin Marietta, added Loral in July and Quinitron in August, and has since reorganized 40 subsidiary companies into 'virtual organizations' to 'create a mix of cultures.' And just in case the assembled press missed the point, Buzzard concluded by stating in the sovereign voice usually reserved for statesmen, 'We have alliances with various other companies.'

The first-name basis of the journalists and the corporate executives, the inside jokes, and the closest thing to investigative reporting appearing to be a vying for stock tips, all combined to make 'synergy' a continuation of monopoly capitalism by other means – only this time the highest stage was not Lenin's vaunted imperialism but Baudrillard's hyperbolized simulation. This suspicion was supported the following week by Boeing's announcement of a $13 billion takeover of McDonnell Douglas, creating one more aerospace colossus.

But the smaller industries were not waving any white flags. Silicon Graphics, for some time the David among the simulation Goliaths, had developed the most powerful slingshot yet, the Onyx2, with a memory capacity of 256 gigabytes, a memory bandwidth of 800 megabytes per second, and, most importantly for simulation graphics, the capability to generate 20 K polygons at 60 hertz. Watching one of these generate a simulation of a helicopter on the deck, down to the details of its reflections in the water and cows stopping in mid-rumination as it passes overhead, was a reality check that everyone seemed eager to cash. A hierarchy of booths could be drawn from those that did and those that did not have one (or even two or more) of the sleek, black Onyx2; obvious from their placement that they were there not just to run displays but to *be* the display of the simulation edge. Other firms were compensating by making synergy work at the organizational level. Highly visible – and offering the best food and drink at its reception – was 'The Solution Group,' a consortium of close to 20 industries formed by Paradigm Simulation in 1994 to integrate product, services, and support for the simulation consumer. Judging from current trends, one could imagine two, maybe three enormous booths filling the hall at I/ITSEC 2000: if you are not part of the solution, you are part of Lockheed Marietta, or Boeing McDonnell. And even if there are no more enemies in sight by the year 2000, one could surmise that there would still be a solution in search of a problem.

Niche synergy was another way to go. One member of the Solution Group was leading the way, infiltrating the military–industrial–entertainment nexus by creating an ever-expanding database of hyper-real, real-time 3-D simulations. Viewpoint DataLabs might not have high name recognition, but anyone who has viewed a commercial, a television show, a hit film, or a video game with computer-generated graphics over the last year has probably sampled Viewpoint's product. Its booth's promo video was riveting and revealing, for the eclecticism of the content as well as for the monotony of the style. It opens with the memorable scene of the alien foofighters swarming the F-18s in *Independence Day*, which buzz-cuts into a pair of attacking mosquitoes in a Cutter insect repellent commercial, then to spaceships attacking in *Star Trek Voyager*, followed by some requisite mega-explosions, a simulation of a missile launch from two helicopters, the dropping of a fuel–air dispersal bomb from *Outbreak*, and a trio of Eurofighter 2000s doing maneuvers that are aerodynamically impossible (a case of wishful flying, since the problem-plagued real Eurofighter has yet to make it into the air). Interspersed is a whimsical scene of a museum-bound *T. Rex* doing a little chiropracty for a MacDonald's ad and, to my émigré eye, an offensive ad of Lady Liberty plucking an Oldsmobile Aurora off the Staten Island ferry lady (give her your riches, your muddled mind, and she'll make the right car choice for you – that's freedom). Big Bang backed by Bang-Bang, especially when it comes in 3-D with a techno-rave sound track, is a big seller.

That night, I made the rounds of the receptions, hosted mainly by the larger defense industries. I learned a lot about the field from ex-fighter jocks turned corporate VPs, ex-artists turned graphic designers, ex-hackers turned software developers. After a few drinks, nearly everyone was eager to let me know about their former lives. I suppose making a living making the machines that help to stop others living does not make for cocktail chatter. Nonetheless, it was there, in all the stories about what they once did. I did not dwell on it for fear of sounding sanctimonious, but also because I too was in the triangle, collecting data to entertain/train others in the ways of war, making war fun for the consumer/reader. I took some notes on what was being said, but I lost the cocktail napkin on which they were written.

> An idol capable of realizing exactly what men's faith has been unable to accomplish. . . . A utopia of *technical fundamentalism* that has nothing at all to do with the religious variety that still requires virtues of men instead of advantages to 'machines'?
>
> (Paul Virilio 1994b: 202)

The previous night was not the only reason I was late the next morning for the awards ceremony at STRICOM. I could not find the place. When you drive up to most military bases, there is a perimeter, a guard booth, at the very least a recognizable headquarters with flags flying out front. Here there were just row after row of sleek steel and glass buildings, interrupted by nicely landscaped parking lots. This was military base as corporate research park, with all of the major defense industries represented on the base. I finally located the right building and room, and joined a circle of dark suits and a mix of army khaki, air force blue, and navy white, standing around a large conference table. At the front, naval Captain Drew Beasley, Program Manager of JSIMS, was just getting into the background of the program. It began with a memorandum of agreement between the leadership of the armed forces and the Department of Defense, signed in 1994, to develop 'an interoperable training simulation capable of combining warfighting doctrine, command, control, communication, computer and intelligence (C4I), and logistics into a team event.' It would replace, said Beasley, older war games devised 'for the dreaded threat of the great Russian hordes coming over the tundra.' Since the end of the Cold War, 32 military operations ranging from famine relief to armed conflict, have demonstrated that 'we need a different paradigm that allows us to work cooperatively and jointly.'

JSIMS would make it possible to combine and distribute three forms of simulation: *live simulations* (conducted with soldiers and equipment in 'real' environments), *virtual simulations* (conducted with electronic and mechanical replications of weapons systems in computer-generated scenarios), and *constructive simulations* (the highest level of abstraction, where computer-modeled war games play multiple scenarios of conflict). Advances in microprocessor speed, interactive communication, and real-time, high-resolution video mean that military exercises will be able to mix and match live, virtual, and constructive simulations not only in Synthetic Theaters of Wars (STOW) but also on commercially available computers and networks. Experiments have already been conducted where a group of colonels at Fort Leavenworth in Kansas introduce an electronic 'OPFOR' (enemy or 'opposing forces') battalion into an actual training exercise at the National Training Center in the Mojave Desert, while soldiers in Martin Marietta tank simulators at Fort Knox 'ride along' in real time with either side as part of a distributed Battle Lab simulation. But by 2003, JSIMS would make it possible for 'all the services to play together' with 'just-in-time' mission rehearsal, and 'a worldwide terrain database.'

With the flash of cameras and a round of applause, Captain Beasley and Lane Arbuthnot, program manager at TRW of 'JSIMS Enterprises,' put pen to the $69 million contract. A very efficient public affairs officer had arranged an interview for me with Beasley, Arbuthnot, and Kurt Simon, also from TRW, who was actually in charge of the technical aspect of building the simulation. The captain once more deployed his demise of the Russian hordes metaphor to emphasize the external motivation for a new macro-simulation but spent most of the time going over the internal factors, like the need to standardize the disparate models of the different services (some based on hex-systems, others on Cartesian coordinates) and to globalize our preparation for future threats. Sounding like a modern-day Francis Bacon ('knowledge is power'), he made JSIMS sound as glorious as the founding of the library of Alexandria: 'We are building a synthetic environment that can be used to pull down objects and representations out of our electronic libraries, objects that other services have placed there . . . as part of an overall streamlining process to bring a joint focus, commonality and collaboration within government and with industry.' The captain moved to a white marker board to draw a series of circles representing live and constructive simulations, which increasingly overlap as JSIMS goes through its stages of development: in his schema, the constructive had engulfed the live by 2003.

> The Disneyland imaginary is neither true nor false; it is a deterrence machine set up in order to rejuvenate in reverse the fiction of the real.
>
> (Jean Baudrillard 1983: 25)

Compared to Disneyworld, the military and industry were open laptops when it came to the role of simulations. My efforts to set up interviews with the architects of Imagineering and Audio-Animatronics (always with superscripted trademarks affixed), or better yet, to get a glimpse behind the technology of simulators like 'Star Tours' or 'Body Wars,' were met by some very polite, very efficient stonewalling. People were in meetings, on vacation, in California. Getting into STRICOM was a piece of cake compared with the obstacles I faced at Team Disney's po-mo headquarters. A series of abstracted mouse-ear arches, a formidable defense-in-depth of receptionists, multiple mazes of cubicles, and a sundial atrium that looked like a nuclear cooling tower, did not invoke a sense that this was a place where the fun always shines. When I finally reached the right cubicle, I was told that my designated handler was in a meeting. Further efforts produced meager results. After a couple of phone calls, clearance was finally reached from higher up: I was given a copy of the '25th Anniversary Press Book and Media Guide' and sent on my way.

The guide was full of noteworthy information, like the fact that Eddie Fisher and Debbie Reynolds had a flat tire and showed up late for the 1955 opening of Disneyland (but no mention that film star Ronald Reagan emceed the event), and that Walt Disney did not live to see the opening of Walt DisneyWorld (but no mention that his vision of the future as a frozen past included a cryogenic funeral for himself). The chronology provided for the opening year of Disneyworld is even stranger. In the Disney version, in 1971 astronauts take the lunar buggy for a spin, George C. Scott wins an Oscar for *Patton*, 18-year-olds get the right to vote, and President Nixon fights inflation. And just about saying it all, 'Everyone was wearing smile buttons,' and 'Charles Manson was convicted of murder.' Others might have different memories, like President Nixon set up the 'Plumbers,' 18-year-olds drafted to fight in Vietnam, the Pentagon Papers leaked to *The New York Times*, and 31 prisoners and nine hostages killed at Attica State Prison. Simulations of the future sometimes require a re-imagineering of the past.

> The society signs a sort of peace treaty with its most outspoken enemies by giving them a spot in its spectacle.
>
> (Guy Debord 1978)

My trip to the Simulation Triangle yielded more anxieties than insights. If anything, STRICOM's motto, 'all but war is simulation,' had taken on an even denser, fractal ambiguity. Through technical reproduction, repetition, and regression, proliferations of simulation nuked any sense of an original meaning to war – or fun. Indeed, with increasing orders of verisimilitude, the simulations displayed a capability to precede and replace reality itself: Borges' nightmare again. Design and desire partially explain the spread of simulation. At the abstracted level of deterrence, simulations can and have 'worked.' Total transparency through surveillance (at the airport or by satellites), combined with the occasional direct application of simulations (*COPS* or the Gulf War), have proven to be powerful, if not always democratic, cyber-deterrents. It is understandable why some might desire the virtual security of simulation (JSIMS or Main Street USA) to the risks of the real (conflict overseas and crime in the cities), even if it puts liberty as well as the reality principle at risk.

But there remains an irony if not a danger lurking at the edge of the map, where it comes up hard against the contingencies of life. As superior computing power and networking increase in representational power and global reach, simulation leaves little room to imagine the unpredictable, the unforeseeable, the unknowable, *except* as accident. Will God's will, nature's caprice, human error seem puny in effect as simulation becomes more interactive, more complex, more *synergistic*? In the context of industrial accidents, organizational theorists have already identified a 'negative synergism' in complex systems that can produce unpredictable, worst-case failures. In the technological drive to map the future – to deter known threats through their simulation – are we unknowingly constructing new, more catastrophic dangers? Conversely, will the 'new' only be construed, and feared, as the unmapped event? Or, worse, if the map does become truly, hyper-really global, without the edge beyond which lies the unmappable, where will the monsters go?

References

Baudrillard, J. (1983) *Simulations*. New York: Semiotext(e).
Baudrillard, J. (1990) *Fatal Strategies*. New York: Semiotext(e).
Debord, G. (1978) *In girum imus nocte et consumimur* (We go around in circles in the night and are consumed by fire), film.
Debord, G. (1983) *Society of the Spectacle*. Detroit: Black and Red.
Liddell Hart, B.H. (1967) *Strategy*. New York: Signet.
Virilio, P. (1988) *Block 14* (Autumn).
Virilio, P. (1994a) *The Vision Machine*. Bloomington: Indiana University Press.
Virilio, P. (1994b) *Bunker Archeology*. New York: Princeton Architectural Press.

13

RUNNING FLAT OUT ON THE ROAD AHEAD

Nationality, sovereignty, and territoriality in the world of the information superhighway

Timothy W. Luke

An introduction

The digerati tell us that a new world order, or one composed of 'digital nations,' is being born in cyberspace, and it will bring a revolutionary end to politics as we know it. While much of this talk is just talk, the wired writings of these info-insurrectionists circulate some intriguing conjectures about nationality, sovereignty and territoriality in the contemporary world system. This chapter re-examines these digital discourses to poke and prod at the still indistinct expanses of cyber-space, whose n-dimensionalities stretch away from the on-ramps to the telematic infobahns that the digerati tell us are 'the road ahead.' For all the pilgrims turning onto these telematic roads ahead, new kinds of productive power, working within the datasphere's network of networks, are reshaping subjectivity and society at the (wo)man/machine interface. Such digital discourses, with their own supercon-ductive disciplinarities, are warping the Euclidean spaces and Newtonian times of the modern sovereign state, where relative wealth, power or culture has been mea-sured by the movements of physical material across landscapes or seascapes, against clock and calendar time.

Negroponte (1995) asserts that being digital requires us to rethink the condi-tions of human existence, because the information revolution is not really about moving matter, or 'atoms,' as much as it is actually centered upon coping with flows of information, or 'bits.' This shift from the rhetorical register of atoms to that of bits provides a primary pivot point for this chapter. It wonders if a world where bits displace atoms can remain one in which modern sovereign states play realist political games around the old rules of an 'international' politics. Codes of nationality, sovereignty and territoriality concocted out of the 'real life' (RL) of the atom-state may not access 'virtual life' (VL) in the bit-state. Therefore, we need to head down information superhighways to survey the secondary,

274

unanticipated consequences of building these digital roadbeds and telematic roadhouses as they branch out alongside the infobahn.

The information superhighway represents a vast grid of multimedia/multilingual/multivalent VL spaces tunneling through a diverse array of mono-media/monolingual/monovalent RL settings. At this juncture, the construction of information networks seems to be a blockbuster, breaking apart hitherto more culturally integrated, socially homogeneous or politically organized blocks of human settlement into many more diverse and divided informational fragments. Anglo-American infotelcos, like the would-be MCI/BT combine, incessantly beam TV promos at their customers, asserting that 'the Internet is utopia,' where 'there is no race, no gender, no class, no age, only minds in communication,' and then begging the basic question: 'Is this a great time, or what?' The greatness of this time, however, is unclear. Indeed, it is an era characterized by tremendous contradictions between VL possibilities and RL actualities. To be informatic about it, many North Americans swap data packets on dial-up connections at 56,000 baud, while phone lines in Madagascar lose most messages at 300 baud – the slowest possible infotransmission rate (Wresch 1996: 130). The United States and the United Kingdom publish nearly 40 per cent of all scientific articles, South Africa and Zimbabwe, which are the leading African centers of scientific research public 0.439 per cent (*ibid.*: 84). North America now publishes annually 461 book titles per million population, and Africa publishes 29 titles per million population (*ibid.*: 39). Kenya has one phone line for every 100 people, while Canada and the United States have 54 (United States Bureau of the Census 1993: Wresch 1996: 125). Finally, North America has over 5,400 on-line databases, or 71 per cent of the world's total, while Africa has eight on-line databases, or less than 1 per cent, a figure that South America, Eastern Europe and Asia also do not exceed (*ibid.*: 97). The Internet is still essentially a very Anglophone world in which IBM, Compaq, Digital, Apple, Hewlett-Packard, Microsoft or Oracle produce almost all of the hardware and software, and North American or European telcos support most of the connectivity arrangements. ASCII implicitly sums up the digital planet at this time; it is one dominated by American Standard Codes for Information Interchange.

MCI is mistaken: the Internet carries many signs of race, gender, class, age and nationality in its everyday operations, yet it still might be right about this moment being 'a great time.' The Net constitutes an extraordinary utopian space, or a nowhere/everywhere, which is populated by millions of 'netizens,' whose collective sense of their common possibilities and actualities are corroding many existing principles of nationality, territoriality and sovereignty, which are pegged to the assumptions of atom(ic) existence. On one level, the Internet appears to be a vast intranet for the still quite powerful advanced industrial West, whose cybernetic influences are felt as the effects of exocolonizing extranets by the very weak, arrested, non-industrial Rest, carrying the cyberporn of 'Occidentalosis' everywhere at microwave or laser light speed. There are signs of these tendencies, but, on another level, the Internet, and all of its many parallel or successor networks, also constitute new supranational, post-territorial, anti-sovereign social formations, whose destabilizing effects can be experienced everywhere and anytime its on-line VL practices pervade off-line RL existences.

Consequently, we need to re-examine the same question that Susan Strange addresses in *The Retreat of the State*, namely, how and why 'the territorial boundaries of states no longer coincide with the extent or the limits of political authority over the economy and society' (1996: ix). We should, however, push beyond Strange's incomplete *tour d'horizon* to ask a more fundamental question, because she essentially looks only at how atoms are bundled and bounded by territorial borders in RL. The new authors of political power are rewrighting states, societies and economies as bits whose VL telemetries no longer necessarily coincide with old atom-based territorial boundaries. Who then is/are the new author(s) of post-atomic, embitted power/knowledge in the economies and societies of VL?

A digital ontology: cyberspace as third nature

Cyberspace is not another fantasy about things to come; its bits are embedded within the material condition of many things already here and at work today. A census of Internet users needs to be updated daily or weekly, not monthly or yearly, to keep accurate track of its exponential growth rates. Many millions of computers – some 40 million at this time – are linked into this network, directly or through other, smaller networks. How many human users actually utilize the Internet from these multiple points of entry is less clear. Numbers can be cited, but they become inaccurate even as they are reported. Most of the world's money, almost all of its communications, much of its transportation system, and many forms of data archiving now move by means of operations in cyberspace. In perhaps too many ways, it is already the age of the smart machine (Zuboff 1988). Cyberspaces might be understood as the latest iteration in Nature's anthropogenic pluralization through commodification, reification or technification. Human beings have always reshaped their biophysical environmental settings, or a terrestrial 'first nature,' through purposive–rational action, as illustrated traditionally by the territorialized 'second nature' of technological artifacts fabricated as part and parcel of human industrial and agricultural activity in the vast arcologies of post-Neolithic civilization (Lukács 1971). Informationalization goes one more iteration beyond these technical artifices of second nature, creating the ultimate imagined community (Anderson 1991) from the hyper-real domains of the digitized 'third nature' generated within cybernetic telemetricalities.

Many prevailing concepts of power, subjectivity and community may not, however, fully capture the atomic-level changes happening in both the industrial technosphere of modern second nature and the ecological biosphere of premodern first nature as these elaborate human nature constructs become overlaid, interpenetrated and reconstituted in the embitted postmodern 'third nature' of an informational cybersphere/telesphere (Lyotard 1984; Jameson 1991; Jones 1995). Vattimo argues: 'the society in which we live is a society of generalized communication. It is a society of the mass media' (1992: 1). With informationalization, power shifts focus, speed overcomes space, orders become disordered, time moves standards, community loses centers, values change denomination as the atom settings of industrialized human agency are being shaken completely into bits. Third nature expresses its bit forms on the cyberscape/infoscape/mediascape of telemetricality. It too is an anthropogenic domain, but built on-line out of 'bits' (Lucky 1989). If, as Smith contends, 'it is in the production of nature that use-value and exchange-value, and space and society, are fused together' (1984: 32), then embitted third nature is now recombining society with space to render these fusions in electronic and photonic forms, by producing exchange-values in unprecedented ways from the use-values of the electromagnetic spectrum, the industrial era's telecommunication infrastructures, and the digitized restructuring of labor and leisure (Luke 1989).

'As a social product,' the embitted spatiality of third nature remains, like first and second nature, 'simultaneously the medium and outcome, presupposition and embodiment, of social action and relationship' (Soja 1989: 129). Digitization can shift the sites of human agency and social structure into registers of informational bits from those of manufactured matter. Most importantly, the setting of agency, the character of power and the structure of meaning can change significantly in the n-dimensionality of this emergent third nature as what could be described provisionally as telemetrical 'bit-states' displace territorialized 'atom-states' as sites of power/knowledge.

From geography to infography

While such digital ontologies might invite us to reflect upon digital beings and virtual times (Luke 1996), third nature is drawing many microsoft explorers to its digital domains in droves – all eager to navigate its netscapes so that its mosaics of meaning might be better managed by all operating systems. As Ó Tuathail asserts, 'geography is about power. Although often assumed to be innocent, the geography of the world is not a product of nature but a product of histories of struggle between competing authorities over the power to organize, occupy, and administer space' (1996: 1). This chapter upgrades Ó Tuathail's insights about geography and terrestrial space by shifting our focus to the informational writing and wrights of cyberspace. There one finds many new eyes and voices, all seeking to envision its expanses and make claims about its qualities as they eagerly grab for access, domains and connections within its n-dimensional ambit. As the worldwide webs of informationalized cyberspace proliferate at near paralyzing speed – in January 1993, for example, there were only 50 known web servers, but there were over 100,000 by January 1996 (Tapscott 1996: 21) – a new class of digital intellectuals performing symbolic analyst labor for the wired elites (Brockman 1996; Lasch 1995; Reich 1991) is hustling to define or articulate its sense of info-power in new intellectual discourses, which one might label 'infography.'

Like geography, infography might be mostly a foreign imposition, but it is not, as Ó Tuathail aptly describes geography, 'a form of knowledge conceived in imperial capitals and dedicated to the territorialization of space along lines established by royal authority' (1996: 2). It too is a verb, but its information writing mostly parallels the disparate agendas held by the data wrights of informational/telematic/digital power in major transnational computer, software and telecom corporations. These ambitious enterprises are seeking to create/seize, make/master or generate/organize cyberspace, as Ó Tuathail says of geographers, 'to fit their own cultural visions and material interests' (*ibid.*). So just as the medieval vision of space was religious, and its maps pivoted around Jerusalem, Rome or Mecca to represent the known expanses of Eurasia, and a new vision of geographic space had to re-envision that world space instead as the territorially bordered, imagined communities of nation-states seen from London, Paris or Madrid, infographers are now remapping the world of print capitalist nationalism around new hypertextual capitalist transnationalisms, showing corporate intranets, government extranets and global internets spinning new webs of infographed telemetrical dominion over/through/around the sublated boundaries of geographed territorialized space. Trade no longer follows the flag, it also comes on the Net; hence, the rewrighting of space through infography simply articulates the rewriting of power/knowledge with the closed codes of informationalization.

Telecommunicative cartographies of telematic nodes and data links are rewriting/rewrighting older imperial map-making practices, which scrupulously strove to document all contiguously containerized lands and waters for some sovereign authority. Of course, infography does not erase geography, just as geography could not efface medieval theographies: one can map the world from Jerusalem or Mecca, and so too are the maps from London and Paris still in use. Nonetheless, fresh maps of newborn(e) cyberspaces, whether they plot new paths of some telco's broadband backbones, points of presence of Internet uplink ports, dynamics of the Wintel OS *vs* Mac OS population growth among world PCs, numbers of WWW servers, or frequency and source of new netcasts, are giving us infographies of life out on the road ahead.

Atom-states against bit-states

Booting up in RL to head out into VL on the Net poses one of today's most challenging philosophical questions: '"The *splitting of viewpoint*," the sharing of perception of the environment between the animate (the living subject) and the inanimate (the object, the seeing machine), which leads, in turn, to (con)fusion of "the factual" (or operational, if you prefer) and the virtual; the ascendancy of the "reality effect" over a reality principle already largely contested elsewhere'

(Virilio 1994: 60). Splitting sight, then, also splits sites, creating fresh reality effects in many new spaces beyond, behind and beneath the domains occupied principally by atoms. Computer images of the real space of material objects or data profiles of the real characteristics of animate subjects (re)/(dis)place atomic observables with non-observable bits. Human agency and social structure, at the same time, jump *'from the actual to the virtual'* (*ibid.*: 67) to roam through hypertopic terrains on hyperchronic timelines. Out on the infobahn and down in the central processors, human agents must coexist through 'the *automation of perceptions*' and 'the innovation of artificial vision, delegating the analysis of objective reality to a machine' (Virilio 1995: 59), because machinic formations arrayed in telematic mediascapes are the objective reality that automatic perception must scan as virtual clusters of bits. A comparison of the atom-state and bit-state – as ideal types in this preliminary conceptual comparison – might lend more substance to Negroponte's schema.

In cyberspace, RL time warps and VL space distorts, leaving zones of communicating lasers and wireless transmissions as the virtual spaces that human beings must traverse. On-line communication, which is mediated quite often through the blank screens and dead connections of the World Wide Web's wildly worsening wait, can approach light speed in megabytes of fiber-optic blitzkrieg. As Virilio claims, virtual reality is running on warp drive, causing it to lose real-life dimensionality (1991). Because of 'the principle of instantaneous emission and reception' of bits through the ether, '*change-over* has already superseded the principle of *communication* which still required a certain delay,' leaving cyberspatialized virtualities recalibrating all conventional notions of 'the real and the figurative, since the question of reality would become the PATH of the light interval, rather than a matter of the OBJECT and space–time intervals' (Virilio 1994: 74). Chronopolitics in states of bits, then, remediate power as hyper-kinetic speed effects, whose co-acceleration and asynchronous modes of transferring influence sublate geopolitics in states of atoms where power mediates matter filling spatial expanses, energy traversing territorially objectified space–time intervals, or information communicating content as its encoded matter diffuses across space and advances in time.

Cyberspace problematizes a geography of space and place. What has been central to living beings, transpiring through their biotic times at metabolic speed, is place. Now, Virilio and Lotringer suggest, 'in some way, place is challenged. Ancient societies were built by distributing territory. Whether on the family scale, the group scale, the tribal scale or the national scale, memory was the earth; inheritance was the earth. The foundation of politics was the inscription of laws, not only on tables, but in the formation of region, nation or city. And I believe this is what is now challenged, contradicted by technology' (1983: 142). Politics was once the polis enscribing its jurisdiction across the atoms of lands, peoples, and their settlements. The bottom line of informationalization may be this deterritorialization and, by extension, its disembodiment, disinheritance and dismemberment of terrestrialized social emplacements.

How much of this reasoning, however, rests upon moving atoms instead of bits? Is place totally challenged, or is it simply shifted into new terrains? Contemporary infographic societies may be built by distributing network bandwidth, routing data packets or colonizing electromagnetic spectra, but physical infrastructures are still quite necessary. The foundation of cyberpolitics could then be found in systematizing operations, fixing code routines and printing circuits. After all, one can inscribe laws on silicon, fiber optics and disk drives as well as on stone tablets, and thereby generate informatic regions, nations, cities. Indeed, cybertechnics repositions living beings in neobiotic times and spaces, where they seem to transpire telematically at extrametabolic speeds and diffuse into postmetabolic spaces. Some memories may remain tied to the Earth, but one need not leave their containers behind to pass through Heaven's Gate to see how permanent web memorials are proliferating on-line, where the multimediated memories of dead and departed souls or entire lost peoples can be pulled down by interactive browsers *ad infinitum*. On a family group, tribal or national scale, one discovers that cyberspaces also provide flexibly scalable memorializations that are as reliable and durable as the Earth, while having the added virtues of flexibility and accessibility to serve members of mobile, distributed and remote users. The fact that 'none of the so-called great politicians today is able to approach' (Virilio and Lotringer 1983: 142) this sort of cybermodernity from fixed sites in atom-states does not mean that any great cyberpoliticos are not working on their 'killer applications' in bits and bytes.

What do Negroponte's digital ontologies for the atom- and bit-states imply for the bodies politic of international relations? To keep his crude distinctions, most of what we understand as international politics applies to territorially arrayed nation-states that contain and channel the energies of their physically constrained inhabitants and lands in the realm of an 'atom'-state rather than the ambit for a state of 'bits.' Yet, what if the atom-state can be displaced by bit-states? What if the locus of governmentality can be upgraded as a portfolio of power applications from, as Virilio foresees, 'that of an inert territorial body' to one of 'the centralized and miniaturized control of a constantly active, yet invisible and unknowable, body of communications' (Virilio 1990: 94)? What new social code might bring power out of its solid state of atom agency and structure into much more amorphous conditions of bit-borne agency and structure? Perhaps Negroponte's naive narratives about world civilization making another great transition – this time from off-line/nondigitized atoms to on-line/digitized bits – reveal the engines of governmentality undergoing a digital refitting. Out on the Net, government can become, to paraphrase Foucault, the right informational disposition of atomic things telematically arranged so as to lead to a convenient digitized end (Foucault 1991).

The density of net-based dromological structures in the bit-state, then, acquires its own quiddity. In the last analysis, digital beings share time and reparse space in their Internet protocol packets composing the mediascapes of third nature. For Virilio, the built environments of atomic second nature – cities and towns – have not expanded as profusely as the fractalizing domains of VL out on the Net:

> If you want proof, you need only look at a map of the physical geography of France . . . this one showing the totality – visible and invisible – of communication networks: canals, railways, airways, highways and, from the visual path of Claude Chappe's ocular telegraph to the electronic age, radar. We immediately realize that during the last two centuries of our history, the physical geography of France has completely disappeared under the inextricable tangle of different media systems; that *not only does delocalization occupy more territory than does* localization, *but it occupies it in totalitarian* fashion . . . make the whole thing entirely techno-logistical; then you will have before your eyes *the true physical body of the modern totalitarian state, its speed-body.*
>
> (1990: 91–91)

Dromological net-centric existence, then, is delocalized, mobilized and instrumentalized living as a speed-body within the hyperchronic flow and hypertopic domain of speed. The totalizing reach of the media – electronic and machinic – represents for Virilio the inversion of Clausewitzian war reasoning, because the speed-body of bit-states must endocolonize their actual territoriality with virtual telemetricalities. Politics is now war carried on by other means, and the doctrine of security founded upon this recognition leads to 'the saturation of time and space by speed, making daily life the last theater of operations, the ultimate scene of strategic foresight' (*ibid*.: 92). And embitted victory in these digitized internal wars comes in fully mediatized on-line form. Indeed, '*beating an enemy involves not so much capturing as captivating them*' (Virilio 1995: 14). So the heavy artillery of the bit-state's post modernizing totalitarian regime fires advertorial rhetorics and infomercial rhetorics down all of its circuits and conduits in commodified imageries of communion, desire and power (Schwartau 1994).

As John Perry Barlow asserts, 'cyberspace is naturally anti-sovereign,' and loosely linked netizen leagues, like those allegedly represented by the Electronic Frontier Foundation, must defend civil liberties in cyberspace against 'hegemonic incursions by various power sources from the terrestrial world' (Brockman 1996: 13). Here, Barlow and others see the cyberspace of the bit-state as a digital domain whose populations and frontiers need to be defended, planned or expanded by their inhabitants. These netizens, in turn, do not necessarily share the same 'imagined community' with their atomic neighbors. Instead, they populate new on-line societies as speed-bodies whose internetworking populations are strings of bits. Their liberties, meanings and loyalties are 'a separate matter' from those incarnated as civil affairs in the physical world of the atom-state. In turn, the diverse and divided number of atom-states virtually guarantees that no single sovereign power ever will be able to enforce its local, regional, national or supranational community standards upon its citizens.

As David Bunnell notes, 'if you want to sell dirty pictures on the Internet to the U.S. market in this post-Communications Decency Act era, all you have to do is set up your file server outside the border. . . . Although it was created by the government, the technology of the Internet is now beyond the control of the government. Hackers will always figure a way to get around laws that attempt to censor the Internet' (*ibid.*: 36). Sovereignty in atom-state terms, then, can only be realized temporarily or tenuously as either power effects on intranets, where state controls can be imposed on all coded traffic, or as coercive controls over access into or out of any nation's Internet switches, where all abnormal messages can be blocked, cached or eradicated, if they can be detected.

The cyberporn issue in the United States can be seen as simply a cat's paw for insecure atom-state politicos, who are anxiously anticipating their loss of sovereign authority, to threaten otherwise secure populations with new insecurities.

> In Germany, they want to control cyberspace to keep the skinheads from using it. In Iran, they want to control it to keep people from having infidel conversations or having inappropriate contact between the sexes. Every culture is going to try to use its primary bogeymen to give it an excuse to go into cyberspace and ride roughshod over it. The pornography issue in the United States is nothing but a stalking horse for control.
>
> (*ibid.*: 14)

According to the digerati, these RL moves against VL liberties are doomed to fail. They are actions taken in the off-line world of atoms to regulate bit behaviors in the on-line world. VL will always outpace RL, exerting the heteronomous anarchy of the Net against the monocratic autonomy of the atom-state.

On one level, as the cyberporn question illustrates, the inside/outside dynamic of sovereignty might well change as the otherizing discourses of difference and identity open up new notions of community and exchange that could have quite open architectures. Closed architectures of sovereign power typically presume exclusive circuits for the circulation of authority to guarantee that only one approved type of embodied subjectivity can operate within its domains. Authority accepts embodied atoms in slow-bodies that will act only in the ways and forms that the jurisdictive atom-state approves – it is monological. A bit-state, however, could accept open architectures and multiplatform diversities, acknowledging any and all alternative interaction opportunities that flow through, with, beneath all other speed-bodies like encoded bits. Shared access instead of contiguous aggregation could well bound the scope of bit-state preserves, making possible the coherent organization of mobile, distributed and nonterritorialized communities of governance. Bit-states could be embitted object-oriented mobilities for speed-bodies, and no longer atomic subject-centered fixities for slow-bodies. In other words, serving some delimited functionality, as with special-purpose governance bodies now operating intranationally, might supplant the creation of some fixed identity, as with modernized territorial nation-states today (Rheingold 1993). Bodies politic, if telematic/electronic/cybernetic, might enact far more limited and much more expansive powers through such object orientations (Edwards 1996). Their populations would sort automatically by speed and access in dynamic functionalities instead of searing inactivation by homogenizing all forms and flavors of atoms to accord with one jurisdictive standard of interaction as massed bodies or embodied masses in fixed places.

Likewise, 'the work of nations' in the atom world, which Reich (1991) anticipates must change, can be undercut by the workings of the bit world as people (de)port themselves as bits into domains where laws made for atomic personae do not pertain. The deliberative democracy of face-to-face embodied politics is being challenged by the discursive dromocracies of many embitted communities in user.groups, list.serves and network.communities. Esther Dyson argues that these fast, disembodied, textualized exchanges of networked communities will soon exceed on-line commerce and socializing in importance. In fact,

> it's magical how it's going to affect people and relationships. A new kind of community, not a culture, is coming. The difference between a culture and a community is that a culture is one-way – you can absorb it by reading it, watching it – but you have to invest back in a community. Absent this return investment, it's not really a community. People will be investing in sharing content and spending measures to each other, in spending time together, and, in part, that's what builds these communities.
>
> (cited in Brockman 1996: 86)

The magic that Dyson celebrates, of course, might also be seen as secession, insurrection or independence from fixed territorialized national cultures and governments. To the extent that people invest their time, energy and spirit in cybercommunitarian contacts, they must rob any embodied face-to-face groups of their human presence.

Bit-driven states could well be populated by such magical speed-bodies, who regard one another as co-accelerants, time sharers or telepresent gatherings. A census taking of shared files, multi-user domains or threaded interactions in dedicated data streams or stand-alone networks could reveal possibilities for new cyberbiopolitical regimes that could easily coexist with atom-states or other bit-states. Netscapes can be closed to outside traffic; but, ironically, their virtual ecologies may wither and die, unlike natural landscapes, if everyone cannot browse across them. Being everywhere and nowhere, the inside could accept the outside, because bits are not exhaustible or exclusive, like atoms.

Cybernetic connectivity in bit-states might interlace everywhere and nowhere, remaking social structure, political power and cultural values to conform to the time sharing and connection speed of the networks they share. Following Marx, the telematicization of existence determines consciousness; and, 'since movement creates the event' in the electronic registers of cyberspace, one finds the telematic reality effect in cybernetic existence 'is *kinedramatic*' (Virilio 1995: 23). Computer user groups quickly become co-accelerants whose consciousness and action are framed by the ease and speed of their connectivity. Surfing the Net, grazing through websites, telepresencing in real time all over the planet, one becomes a mobile, distributed or threaded participant in many moving, dispersed and web-woven communities. Co-accelerants share conventional understandings in their acts and artifacts as their determinate social group coalesces out of parallel processing, time sharing and compression routines: bodies politic morph into bits politic. Such kinedramatic culture constitutes and enables an informational subjectivity whose conduct conforms to the demands of rapid connection, intense communication and frequent cooperation. Connectivity informs such subjectivities and societies, but 'to disconnect is to disinform oneself' (*ibid.*: 95).

Atom-states by contrast are mostly fixed in their time, and they fetishize territory, believing that no two or more states can occupy the same space simultaneously. One atomic bloc must displace another, and cause the loss of dimension in losing its dimensionality. Bit-states, however, are *n*-dimensional. They have no exhaustible, exclusive, exceptional dimensionalities to lose. Telemetricalities can, in fact, generate *n*-dimensional shapes and domains, assuring atomic sovereignty's paralysis. Atom-states share space held as the continuous exclusive occupation of many discrete places, but bit-states would share rates of interaction, acceleration and transmission accessed discontinuously and nonexclusively.

Bit-states may only be deterritorializing gridworks, netting together functionalities in machine languages, computer time and network topologies within and without the expanses of territorialized atom-states. Moving bits between nodes, bit-states would be powerpoints whose powers and capabilities might be determined by how expansive, swift and complex their switches are, as well as what they route to whom and why. The cybertectonics of bit-states could become datafied dwellings, linking thinking/acting/feeling/having/knowing in the emoticons and cognicons of speed-bodies to asynchronously transferred modes of being. Power would no longer move only as atoms between material volumes and surfaces, where atom-states are coercive figures whose powers and capabilities can be gauged by how extensive, measured and standardized their jurisdictive domains are as well as what they impose upon whom and how. Bit-states, like American dollars or Deutsch marks in the wild zones of the former Soviet Union, would pass through with full legitimacy and complete primacy in any and all domains of atom-state country, where atom-rulers still try to establish their now hollowed-out sovereignties in government language, historical time and legal topologies under the tracks left by bit-states (Mitchell 1995; Boyer 1996). Atom-states struggle to unformat modes of action synchronously in the single-sourced monologics of nationalized imagined community, while bit-states seep through asynchronously transferred modes of action in the cross-platform polynomials of transnationalized operating systems.

A world monetary system is already a regimen of speed-bodies whose dromologies link every nation's currencies in a perpetual regime of real-time exchange. Through their moneys, atom-states also now find their finances, taxes, markets and economies informationally and electronically bound together in embitted $24 \times 7 \times 52$ equivalence exchanges. Money is information, and information becomes money. Like the European Monetary System, which seeks to systematize the linked economies of the European Union by coding the euro out of many coins to carry its supranational sovereignty and economy through RL time and space, this Supraterranean Monetary System unifies all of the world's moneys in VL info/electro/dromo equivalents of dollarized credits and debits. The value of these data dollars, in turn, can be gauged by their acceleration, mobility and dispersion. Parked for a few hours in yen, then marks, then pounds, these data dollars valorize themselves by exploiting marginal variations in correlative values fluctuating by the minute as each business day begins on the hour in every global time zone. Global fast capitalism runs on fast money and quickened financial intelligence, which pulls currencies almost entirely away from the control of intranational fiscal, tax and monetary policies. Speed-bodies need fast money to spend in rapid-fire in quickened markets.

No longer exclusively national currencies, except as the legal tender to fill the basket on one's daily RL trips to the market, transnationalized moneys meld in the dromological, informational, electrified exchanges of telematic VL networks. Their atoms do not travel fast enough, so their bits break time and space barriers. Indeed, anyone with a secure cyberspatial network can issue electronic or on-line moneys to sustain exchange within its boundaries or anywhere in RL that accepts its VL value system as a convertible currency. Hence, many new Superterranean Monetary Systems, from e-cash, compu-credits or tele-marks to debit cards, infobucks or e-money, are circulating bits, mostly still denominated in data dollars, all over the world. Atom-states participate in the informationalization of money, but now anyone can issue any sort of currency as long as some linkage of convertibility links it into the terabyte reserves of the Superterranean Monetary System. Cyberspace is everywhere and nowhere, so all of the world's moneys are moving all of the time with essentially full acceptance, although not necessarily full value, into all sovereign territories as their digital (e)quality is channeled into volatile dromoscopic matrices of telematic value. These telemetrical imperias of monetized data, then, penetrate and obviate the mostly unenforceable claims of exclusivity asserted by any given national dominion of realist state authority.

The methodical movement of stuff is becoming 'the instantaneous and inexpensive transfer of electronic data that move at the speed of light . . . the change from atoms to bits is irrevocable and unstoppable' (Negroponte 1995: 3–4). Here the rites for celebrating a new power are captured in one megabyte of meaning: the bit world is irresistible, irreversible and utterly irreverent in its revolutionizing effects, because bits are fast, cheap and mobile. Like Marx's money, or 'fast capital,' Negroponte's information, or 'fast data,' vanish all that is solid into thin air. Not surprisingly, Bill Gates sees 'the road ahead' tunneling through wormholes of bits into 'a friction-free capitalism' out on the Net's 'global information market' (1995: 6–7). And, running flat out on the road ahead becomes, as Virilio claims, a work in applied 'dromology,' or the careful study and application of operational systems rooted in extraordinary speed, because being itself is organized around instantaneous, inexpensive transfers of time–space-compressed bits around the world at the speed of light.

The operational attributes of cyberspace in many ways merely parallel the free flow of transnational capital throughout the world marketplace of today's globalizing fast capitalist economy. Navigating through the WWW emulates Marx's equations of capital's self-valorization: $M + \Delta M = M'$. Information (I) + changes in information via hypertextually marked changes or connectively mediated bits (ΔI) = information plus (I'). The experience of agency in cyberspace often seems so empowering/enlightening because one is carried along with the high powers and bright lights of these informational forces. Simultaneously everywhere and nowhere, always in use and in storage, forever saved but also inexhaustible, neither exclusive nor inclusive, the WWW is the perfect superstructural representation of transnationalizing capital's infrastructural productive forces.

Given this insight, Negroponte's profile of the bit and the atom becomes quite helpful: 'a bit has no color, size, or weight, and it can travel at the speed of light. . . . It is a state of being: on or off, true or false, up or down, in or out, black or white. For practical purposes we consider a bit to be 1 or 0. The meaning of 1 or 0 is a separate matter' (1995: 14). Atoms, on the other hand, are bundles of material: 'world trade has traditionally consisted of exchanging atoms. . . . When you go through customs you declare your atoms, not your bits. . . . The change from atoms to bits is irrevocable and unstoppable' (*ibid.*: 4). The bit is an abstract state of mobile becoming, while atoms are about concrete conditions of immobilized being. Against the (eu)tope of the bit, the atom is a (nega)tope of government regulations, trade statistics or frozen fixities.

In Internet environments, these differences become decisive, because all media in these settings are digital. When everything is multimediated through digitization, two things happen: first, 'bits commingle effortlessly. They start to get up and can be used and reused together or separately'; and, second, 'a new king of bit is born – a bit that tells you about other bits' (*ibid.*: 18). Once bits can be combined in any fashion as well as coded with reflexive tags to refashion these combinations intelligently, then digitization can begin to operate on an unstable autopoietic basis. From within a world fixated upon moving, counting, owning or using atoms, the bit creates 'the potential for new content to originate from a whole new combination of sources' (*ibid.*: 19). One potential new content with many commingled points of origin is the utopia of 'Cyberia,' or new transnational imagined communities arising

> out there in 'cyberspace' – the territory of digital information. This
> apparently boundless universe of data breaks all the rules of physical

reality. People can interact regardless of time and location. They can fax 'paper' over phone lines, conduct twenty-party video-telephone conversations with participants in different countries, and even 'touch' one another from thousands of miles away through new technologies such as virtual reality. All this and more can happen in cyberspace.

(Rushkoff 1994: 2)

Commingling bits with bits regardless of time and location, then, is creating new kinds of bits that claim the life energies of people allegedly heedless of time and location.

In Rushkoff's infography, Cyberia is an imagined community, whose imaginative communitarianism expostulates the existential essences of digital cybernations where the bits of digital personae are both born, then borne, in computerized codes from boundless universes of data. Breaking all the rules of physical reality, where atom-states rule over piles of atomic matter, Cyberia becomes a state of being, according to Rushkoff, with basically 'psychedelic qualities' inasmuch as it leads its partisans to treat the accepted reality of moving atoms 'as an arbitrary one, and to envision the world unfettered by obsolete thought systems, institutions, and neuroses. Meanwhile, the cybernetic experience empowers children of all ages to explore a new, digital landscape. Using only a personal computer and a modem, anyone can now access the datasphere (a web of telecommunications and computer networks stretching around the world and into other space)' (*ibid.*: 5).

Of course, using 'only' a modem-equipped PC typically represents a capital investment of US$3,000 to US$5,000 plus monthly access fees of $10 to $100, so entry to this promised land is not cheap. Negroponte, however, dismisses those who worry about 'the social divide between the information-rich and the information-poor, the haves and the have-nots, the First and the Third Worlds,' because the 'real cultural divide is going to be generational' (1995: 6) as 'the young' integrate this technology into their lives first. This glib generational maneuver fails, however, to grasp all of the bigger questions about access, equality and distribution in the digerati's wired world. Like so many revolutionists before him, Negroponte addresses a small, privileged or wealthy elite: these cybercontras he addresses in universalist Jacobin terms as 'you, the reader.' Presuming that he speaks to/for all of humanity, he talks to that 35 per cent of American families that own computers, the 50 per cent of American teenagers who have PCs at home, or the 30 million people worldwide that surf the Net (*ibid.*: 5). To them, or 'you, the reader of *Being Digital,*' he can say that their computers will coevolve with them as part of their speed-bodies. Computers are now crawling 'into our laps and pockets,' so that 'early in the next millennium, your right and left cuff links or earrings may communicate with each other by low-orbiting satellites and have more computing power than your present PC ... Mass media will be redefined by systems for transmitting and receiving personalized information and entertainment. Schools will change to become more like museums and playgrounds for children to assemble ideas and socialize with other children all over the world. The digital planet will look and feel like the head of a pin. As we interconnect ourselves, many of the values of a nation-state will give way to those of both larger and smaller electronic communities. We will socialize in digital neighborhoods in which physical space will be irrelevant and time will play a different role' (*ibid.*: 6–7). Thus, the digerati tells would-be infographers 'to read yourself into this book [*Being Digital*],' just as you must with the digital revolution, so you will begin to 'feel and understand what "being digital" might mean to your life' (*ibid.*: 8).

What it means now is quite clear: the digital planet will remain an atom-world in which six billion people, minus the 40 million or so already on the Internet, will revolve around a much smaller bit-world snared in new worldwide webs of perpetual communicative interaction spun from within those select digital neighborhoods where physical space is irrelevant and time plays a different role. For those with info-cufflinks or cyber-earrings, personalized infotainment will make the digital planet one with the infographies of their larger and smaller electronic communities. Nation-states will become museum pieces as such digital neighborhoods function as playgrounds for web-wise interconnected selves to assemble ideas and socialize all over the world. But for that remaining 65 per cent of American families lacking any PC or that leftover 50 per cent of American teenagers without PCs at home, and the 5.96 billion other human beings still living off-line, they will begin to look and feel like pinheads. Being digital might mean one can become a netizen whose costume jewelry quickly signs you in and out at will of the cyberspatial telemetries projected from low Earth-orbit satellites. But for the nearly 70 per cent of humanity who do not have a plain old telephone service (POTS) on a personal/family/village level, such rich rhetoric about cyberchickens coming to nest in everyone's informatic pots is highly alarming. IBM, of course, is cooking up all of its many 'solutions for a small planet,' and maybe some Kenyan dance bands are composing on their PCs, while a few Italian octogenarians might take musicology PhDs at Indiana University over the Net from Tuscany. For the most part, however, actual physical access to a computer and/or the Internet is not a universally shared good. Any serious infography will show this telematic global village to be still quite small, sparsely populated and very up-scale.

Negroponte's profile of the bit promotes a new kind of helplessness, because his narratives of digitized modernization obscure a very problematic truth: a very material 'war of the worlds' is looming here (Slouka 1995). Even though we may soon live in a cybernetic world where making, moving and managing 'bits' will replace more embodied practices, which are rooted in crafting or coordinating 'atoms,' this bit-driven new cybernetic order will actually use bits to make, move and manage atoms. A bit-based economy is 'one in which fewer and fewer workers will be needed to provide the goods and services' that are required by global markets, which now take 'a drastic toll on the lives of millions of workers' (Rifkin 1995: xvi–xvii). Telepresent North American managers are able to staff virtual corporations, shop in virtual stores and interact in virtual neighborhoods because their digital personae have many more megabytes than those who work for them lower down or farther out in the software structures of Internet connectivity. Their bits make other people's atoms move to meet their digital administrative designs, which gives Foucault's vision of governmentality (1991) its most important, newest upgrade.

A conclusion

Conventional notions of Realpolitik may remain plausible only inasmuch as the subjects of some realist state regime's actions are corporeal presences captured as slow-bodies by the closed architectures of states, which are, in turn, discretely bounded containers of territoriality, branded with the identities of nationality, and ruled through the discursive directives of sovereignty. The RL space of atom-states is still mostly a three-dimensional geometric gridwork, their RL time is measured by clocks and calendars set to the hours of the sovereign's jurisdiction, and RL action follows the ruliness of a centrally appointed and continuously in-stated authority. Embodiment sustains and limits politics, so when embittedness invades embodiedness, the terms of order also shift. Digital beings are often telepresences, code formations or data effects. Their speed-bodies inhabit tele-metrical infostructures as clients of servers, time sharers in hypertextual domains, or software applicators on common hardware platforms. VL spaces are more *n*-dimensional, time is machine-driven in terms of daily process and generational duration, and action is often *ad hoc* patches placed between nonstandardized operating systems. Therefore, to the degree that political agency and structure are expressed in the bit worlds of digital beings, the old rules of politics for atom worlds and corporeal beings will be torn and tested (Luke 1993).

This *art modem* aesthetic of the digerati celebrates the 'digital nation' in deci-sively elitist terms. As Katz claims, the digital nation is quite special:

> Its citizens are young, educated, affluent. They inhabit wired institu-
> tions and industries – universities, computer and telecom companies,
> Wall Street and financial outfits, the media. . . . They are predominately
> male, although female citizens are joining in enormous – and increas-
> ingly equal – numbers. The members of the Digital Nation are not
> representative of the population as a whole: they are richer, better edu-
> cated, and disproportionately white. They have disposable income and
> available time. Their educations are often unconventional and continu-
> ous, and they have almost unhindered access to much of the world's
> information.
>
> (1997: 52)

Ignoring the on-line cybercultures of neo-Nazism and millennarian fundamen-talism, Katz also believes that the digital nation is intrinsically libertarian and democratic. At the same time, the digital nation is quite conservative in its economics, because it embraces entrepreneurialism, free markets, and personal responsibility (*ibid.*: 184).

At the end of the day, many digerati see the digital nation as inherently anti-statist and post-governmental (Kelly 1994). Their info-insurrection is 'founded on the ethos of individuality, not leadership. Information flows laterally, or from many to many – a structure that works against the creation of leaders' (Katz 1997: 184). All existing liberal capitalist democracies are unfriendly to netizens and their worldwide webs of electro-equality, on-line liberty and friction-free fraternity. Indeed, the atom-states are, at best, televisual gulags in which 'voters are now more concerned with imaginary threats than real ones,' because television gives them all 'a processed world, both eviscerated of context and artificially fortified toward no greater purpose than entrancing the audience' (Barlow 1996: 195). Thus, the info-insurrection attacks all nation-states, including the USA, because popular democracy in the television age is 'Government by Hallucinating Mob. . . . The U.S. government has broken, the victim of television and of connection crash in general,' as electronic televisual overload now paralyzes the inner life of republics (*ibid.*: 195).

The operational capabilities of any atom-state are embedded within a national system of industrial systems, which legitimates itself by producing, distributing and consuming material wealth. Ohmae argues that embitted informational flows, technologies and values essentially eviscerate these atomic nation-states as embitted investment, industry and individuals now gravitate to those global markets where they can obtain both products and prices for those products that are the best in the world. For Ohmae, 'the nation-state is increasingly a nostalgic fiction' as well as a 'remarkably inefficient engine of wealth distribution' (1995: 12). Digitization becomes an episode of global equalization inasmuch as it brings forth 'a world whose people, no matter how far-flung geographically or disparate culturally, are all linked to much the same sources of global information . . . the basic fact of linkage to global flows of information is a – perhaps, *the* – central, distinguishing fact of our moment in history. Whatever the civilization to which a particular group of people belongs, they now get to hear about the way other groups of people live, the kinds of products they buy, the changing focus of their tastes and preferences as consumers, and the styles of life they aspire to lead' (*ibid.*: 15).

Kanter characterizes this deterritorializing divide as a performative line falling between the 'world class,' which acts world-class in terms of its capabilities and assets and whose members could be considered 'cosmopolitans' with 'globalist' agendas, and the class of 'localists,' whose constituents are very local in terms of their abilities and resources, which also prompts them to behave as 'nativists.' Flowing through global markets, 'cosmopolitans are rich in three intangible assets, three C's that translate into preeminence and power in a global economy: *concepts* – the best and latest knowledge and ideas; *competence* – the ability to operate at the highest standards of any place anywhere; and *connections* – the best relationships, which provide access to the resources of other people and organizations around the world' (Kanter 1995: 23). Embedded at fixed sites, 'the local class are those whose skills are not particularly unique or desirable, whose connections are limited to a small circle in the neighborhood, and whose opportunities are confined to their own communities' (*ibid.*: 23). To sum up inequality succinctly in a fast capitalist digital economy, Kanter affirms the netizens' pleas for personal choice. One can distinguish between abstract choice and concrete loyalty to differentiate the ethos of these two antagonistic informational classes: 'cosmopolitans often value choices over loyalties – even in terms of which relationships deserve their loyalty. Local nativists value loyalties over choices, preferring to preserve distinctions and protect their own group. Cosmopolitans characteristically try to break through barriers and overcome limits; nativists characteristically try to preserve and even erect new barriers, most often through political means' (*ibid.*: 24).

Seconding Kanter's observations about the new class divisions arching across a rapidly informationalizing world system, Ohmae maintains that the nation-state mostly fails as a redistributive device as well as a guardian of sovereignty. Captured by the 'resource illusion' (insulating territory from world trade to conserve natural resources) or 'national interests' (labelling this land, factory, market as 'ours,' not 'theirs'), the nation-state finally reveals itself as a protection racket, mostly to protect the traffic in atoms held by the biggest, most active domestic racketeers. In the borderlessness of digitized information, 'the traditional national interest – which has become little more than a cloak of subsidy and protection – has no meaningful place. It has turned into a flag of convenience for those who, having left behind, want not so much a chance to move forward as to hold others back as well' (Ohmae 1995: 64). Nationalism, then, is essentially another manifestation of atom-state localism/nativism/ protectionism poised to prevent cosmopolitans from building fast informationalized capitalism on a global scale. This filtering/braking/retaining function of nation-states is significant on the digital planet only to the degree that national power distorts or enables the efficient circulation of information. As Ohmae asserts, 'economic borders have meaning, if at all, not as dividing lines between civilizations or nation states, but as contours of information flow' (*ibid.*: 25). Most importantly, as Gates argues, the digitization of everyday life within these global flows 'promises to make nations more alike and reduce the importance of national boundaries' (1995: 262).

The qualities of nationality, sovereignty and territoriality are attenuated in the Internet's bit worlds because the Net was created initially to operate after the erasure of nations, sovereigns and territories by superpower thermonuclear exchanges, which were denominated in the deadly codes of mutually assured destruction. The rhizomatic intelligence of IP/TCP packets was to patch together from any survival point all surviving points in a mutually assured (re)construction of new protocols for action. Strangely enough, World War III did not happen; but, in the aftermath of its RL non-eventuation, the network of networks is atomizing the atomic assumptions of embodied nationality, strongly centered sovereigns and fixed territoriality in contemporary world politics with new bit bases for identity, authority and community in VL.

As Haraway notes with regard to the biologics of the present, cybernetics is becoming an information system and economic system of a very specific sort. The rhetoric of digitization reprogram metaphor and matter in all of our technoscience. So we must recognize some new realities:

> We act and are inside this world, not some other. We are subject to, subjects in, and accountable for this world. The collapse of metaphor and materiality is a question not of ideology but of modes of practice among humans and nonhumans that configure the world – materially and semiotically – in terms of some objects and boundaries and not others. The world might be different but it is not.
>
> (1997: 97)

The road ahead, then, is a digital one where humans and nonhumans will reconfigure the world materially and semiotically around the embittedness of digitality (Stock 1993; Levy 1992).

Atom-states may capture the digital revolution within their ambit, and much of the Internet still bears the marks of its Cold War origins from the conflict of atomic wars/politics/states. As Strathern suggests, however, the ARPAnet origins of the Internet now represent 'a world made to Euro-American specifications' that expects, in turn, all interactions within it from this time onward to 'already be connected up in determined ways' (1992: 17). As the info-insurrectionalism of *Wired* netizens strongly suggests, the digital nation is testing the VL potentials of the bit-state against the RL of the atom-state. Time will tell how extensively bits will be free to be bits or how thoroughly atoms will simply be subjugated to new forms of domination and direction by bits whose origins, interests and goals merely rearticulate the old world disorders of inequality and unfreedom in the new world order of cybernations.

References

Anderson, B. (1991) *Imagined Communities: Reflections on the Origin and Spread of Nationalism*, second edition, London: Verso.

Barlow, J. P. (1996) 'The netizen: the powers that were,' *Wired*, 4(9)(September), 53–56, 195, 197, 199.

Brockman, J. (1996) *Digerati: Encounters with the Cyber Elite*, San Francisco: Hardwired.

Boyer, M. C. (1996) *Cybercities: Visual Perception in the Age of Electronic Communication*, New York: Princeton Architectural Press.

Edwards, P. N. (1996) *The Closed World: Computers and the Politics of Discourse in Cold War America*, Cambridge, Mass.: MIT Press.

Foucault, M. (1991) 'Governmentality,' in G. Burchell, C. Gordon and P. Miller (eds) *The Foucault Effect: Studies in Governmentality*, Chicago: University of Chicago Press.

Gates, B. with Myhrhold, N. and Rinearson, P. (1995) *The Road Ahead*, New York: Viking.

Haraway, D. (1997) *Modest Witness @ Second Millennium. Female Man© Meets OncoMouse™*, New York: Routledge.

Jameson, F. (1991) *Postmodernism, or the Cultural Logic of Late Capitalism*, Durham, NC: Duke University Press.

Jones, S. G. (ed.) (1995) *Cybersociety: Computer-Mediated Communication and Community*, London: Sage.

Kanter, R. M. (1995) *World Class: Thriving Locally in the Global Economy*, New York: Simon & Schuster.

Katz, J. (1997) 'The netizen: birth of a digital nation,' *Wired*, 5(4) (April), 49–52, 184, 186, 190–191.

Kelly, K. (1994) *Out of Control: The Rise of Neo-Biological Civilization*, Reading, Mass.: Addison-Wesley.

Lasch, C. (1995) *The Revolt of the Elites and the Betrayal of Democracy*, New York: W.W. Norton.

Levy, S. (1992) *Artificial Life*, New York: Pantheon.

Lucky, R. W. (1989) *Silicon Dreams: Information, Man and Machine*, New York: St Martin's Press.

Lukács, G. (1971) *History and Class Consciousness*, Cambridge, Mass.: MIT Press.

Luke, T. W. (1989) *Screens of Power: Ideology, Domination, and Resistance in Informational Society*, Urbana: University of Illinois Press.

Luke, T. W. (1993) 'Discourses of disintegration, texts of transformation: re-reading Realism,' *Alternatives*, 18: 229–258.

Luke, T. W. (1996) 'Liberal society and cyborg subjectivity: the politics of environments, bodies, and nature,' *Alternatives*, 21: 1–30

Lyotard, J. F. (1984) *The Postmodern Condition*, Minneapolis: University of Minnesota Press.

Mitchell, W. J. (1995) *City of Bits: Space, Place and the Infobahn*, Cambridge, Mass.: MIT Press.

Negroponte, N. (1995) *Being Digital*, New York: Knopf.

Ohmae, K. (1995) *The End of the Nation State: The Rise of Regional Economies*, New York: Free Press.

Ó Tuathail, G. (1996) *Critical Geopolitics*, Minneapolis: University of Minnesota Press.

Reich, R. (1991) *The Work of Nations: Preparing Ourselves for 21st Century Capitalism*, New York: Knopf.

Rheingold, H. (1993) *Virtual Reality*, New York: Summit.

Rifkin, J. (1995) *The End of Work: The Decline of the Global Labor Force and the Dawn of the Post-Market Era*, New York: G. P. Putnam & Sons.

Rushkoff, D. (1994) *Cyberia: Life in the Trenches of Hyperspace*, San Francisco: Harper.

Schwartau, W. (1994) *Information Warfare: Chaos on the Information Superhighway*, New York: Thunder Mouth Press.

Slouka, M. (1995) *War of the Worlds: Cyberspace and the High-Tech Assault on Reality*, New York: Basic Books.

Smith, N. (1984) *Uneven Development*, Oxford: Blackwell.

Soja, E. (1989) *Postmodern Geographies*, London: Verso.

Stock, G. (1993) *Metaman: The Merging of Humans and Machines into a Global Superorganism*, New York: Simon & Schuster.

Strange, S. (1996) *The Retreat of the State: The Diffusion of Power in the World Economy*, Cambridge: Cambridge University Press.

Strathern, M. (1992) *Reproducing the Future: Anthropology, Kinship, and the New Reproductive Technologies*, New York: Routledge.

Tapscott, D. (1996) *The Digital Economy: Promise and Peril in the Age of Networked Intelligence*, New York: McGraw-Hill.

Vattimo, G. (1992) *The Transparent Society*, Baltimore, Md: The Johns Hopkins University Press.

Virilio, P. (1990) *Popular Defense & Ecological Struggles*, New York: Semiotext(e).

Virilio, P. (1991) *The Lost Dimension*, New York: Semiotext(e).

Virilio, P. (1994) *The Vision Machine*, Bloomington: Indiana University Press.

Virilio, P. (1995) *The Art of the Motor*, Minneapolis: University of Minnesota Press.

Virilio, P. and Lotringer, S. (1983) *Pure War*, New York: Semiotext(e).

Wresch, W. (1996) *Disconnected: Haves and Have-Nots in the Information Age*, New Brunswick: Rutgers University Press.

Zuboff, S. (1988) *The Age of the Smart Machine*, New York: Basic Books.

14

GEOPOLITICS AND GLOBAL SECURITY

Culture, identity, and the 'pogo' syndrome

Simon Dalby

Geopolitics and global security

Geopolitics is a complex cultural matter, where identities are formulated, represented and repressed in contemporary political discourses. Geopolitics is also
about the crucially important power to define danger, and about the ability to
describe the world in ways that specify appropriate political behaviours in particular contexts to provide 'security' against those dangers. The power to construct
a popular understanding of the context is a crucial discursive task of geopolitics.
In the last decade, many of the dangers are understood as 'global' and the
response in policy-making circles has often been to discuss matters of newly
defined threats in terms of 'global security.' The phenomena of international
security, diplomacy and the 'high' politics of international relations between the
great powers have become very obviously conflated with concerns about popular culture and cultural identity in the discourses of geopolitics (Dijkink 1996).
Indeed, as the critical literature on these topics now makes clear, international
security and the practices of war are premised on some very powerful, taken-for-
granted cultural constructions, not the least of which are the cartographic
assumptions about nation-states (Walker 1993; Shapiro 1997).

Following the collapse of the Cold War scripts of 'ideological' geopolitics, quite
how the world should now be portrayed is a matter of political argument in the
capitals of what is still often called the Western world (Agnew 1998). In the
1990s, not only are 'we' in danger of possible civilizational clashes or failed states
on the small scale as identity is rearticulated in terms of tribalism (Kaplan 1996),
but the planet is endangered, we are now told, by various threats including global
warming, global population growth, ozone depletion and numerous other facets
of what is presented as a planetary environmental predicament (Myers 1993). The
identity that is now invoked as a threat in contemporary geopolitical discourses is
often the planetary biosphere itself. But, as the argument in this chapter suggests,
quite where the danger to this identity may lie is not often nearly so clear.

Recent critical and postmodern scholarship has made clear that the entities that are to be rendered sovereign and secure are best read as socially constructed identities rather than as the taken-for-granted objects to be used as the premises for thought, action or policy prescription (Biersteker and Weber 1995; Doty 1996; Manzo 1996). The culture of the experts, who are the designators of threats as well as the directors of societal responses to such designated threats, is also part of the analysis of a critical geopolitics. Much of the expertise for pronouncing on security is provided by Western political scientists. While economists, historians and the very occasional geographer have sometimes trespassed on this terrain in the past, now environmental science is also often invoked in the debates about new foreign policy priorities.

But, in contrast, matters of identity and culture are usually the subject for analysis by sociologists, anthropologists, interpreters of the humanities and practitioners of the amorphous arts of 'cultural studies.' Rather than engage in the finer points of meta-theoretical debate, in this chapter I take the categories of culture and identity seriously, and, following recent dissident international relations and critical geopolitics scholarship (Dalby 1996a, 1996b, 1996c), use them to ask practical questions about the categories that geopoliticians use to understand their own identities in ways that make possible the construction of a problematic called 'global security' in the first place.

Culture and identity

Culture is an especially difficult term to work with because of its multiple meanings. It variously refers to socially shared ideas, artistic accomplishments, both material and ideational practices in general and those pertaining to specific groups. Nations have cultures, although quite what these are and how they can be understood is a contested matter in the current literature in cultural studies (Hall 1993). Myths and histories are parts of any specific culture that works to define the identity of who 'we' or 'the people' are in a specific context. Now sometimes societies are posited as the entity with a culture that can be understood as insecure (McSweeny 1996). National cultures in turn are partly constituted by national 'geopolitical visions' (Dijkink 1996). But national culture, which is often that which practices of security provision are called upon to protect, is cut through by claims to 'high' culture and 'low' culture. The proliferation of terms continues, with mass culture and popular cultures being contrasted with high culture and with traditional, native, indigenous and vernacular cultures. Corporations apparently have 'cultures' these days, and not to be outdone there are a multiplicity of 'strategic cultures' that supposedly explain at least part of the unique approach taken by different strategic thinkers and military practitioners in different parts of the globe (Johnston 1995).

Thrown into sharp relief by the flourishing of post-colonial writing in the last few decades, the fields of knowledge that traditionally constructed the various 'cultures' around the world according to the 'objective' practices of anthropology and ethnology have been indicted on the grounds of their Eurocentrism and general failure to recognize the ethnocentric formulations of identity that structured anthropological discourse (Said 1993). As Smadar Lavie and Ted Swedenburg (1996) put it, the geopolitics of knowledge arranged things so that the world 'out there' was far removed from the 'in here' of 'the faculty club' – the supposed repository of universal high culture and accomplishment – which is now often understood as a bastion of white patriarchal heterosexism.

The nation-states that had faculty clubs were, they further argue, cut though with class prejudices that precluded clear evaluation of the exploitative relations between the imperial centres and the peripheries where the 'cultures' of Others were to be 'found.' Imperial cores and colonial peripheries continue to be important in the spatial metaphors of geopolitical imaginations (Agnew 1998). Critical geopolitical writing of the last decade has emphasized how important these spatial metaphors are in situating political actors and the discursive practices of security intellectuals, not least because of the irony that they work politically by removing the messy geographic complexity of places from cartographic imaginations (Ó Tuathail 1996).

The complexity of the term 'culture' initially caused critical analysts in the 1980s to have some difficulty in encompassing it either within Cold War security studies or as an avenue of critique of strategic thinking and the whole enterprise of international relations. The difficulties with culture were in part overcome by the turn to social theory and the exploration of poststructuralist and critical theories (George 1994). Feminist analyses also offered important ways into the concept by expanding the realms of inquiry that were considered part of the genre of international studies (Peterson 1992; Sylvester 1994). In the feminist literature, the voices of the victims of political practices are now being used to argue back against the practices of national security in particular places, and on the larger scale, against the international security practices securing a harshly divided global political economy undergoing processes of globalization (Enloe 1993). Geopolitics is all about the clash of these very different stories and the cultural representations of spaces that are parts of these practices.

Carol Cohn's (1987) work was especially clear that the 'scientific' culture of strategic studies during the Cold War constructed its own identity and practices in misogynist terms that, through a specific language, facilitated both a particular mode of 'knowing' the world and a method of excluding non-initiates. The important point here is not that a specific knowledge is 'wrong' but that it is subject to critique by a feminist cultural anthropologist who refused to use the categories of 'privileged' knowledge by which initiates defined themselves. As such, she was able to ask important questions about the political consequences of particular modes of knowledge and, albeit with considerable difficulty, sidestep the normal processes of discursive co-optation. This anthropological attitude, if not a detailed anthropological method, helped to ask new and revealing questions about the most important cultural identity that is involved in the problematic of this chapter – the identity that can produce a concept of global security in the first place.

In part, these critical reading strategies circumvented the difficulties of 'culture' by focusing instead on the questions of identity and specifically the discursive construction of difference and identity as crucial parts of the formulation of security (Lapid and Kratochwil 1996). Drawing from numerous intellectual sources, the construction of 'Others' as enemies allowed the formulation of the 'domestic' identity that was constructed as the antithesis of the external threat to be examined in ways that circumvented the conceptual straitjackets of both structural neo-realism and techno-strategic discourse (Campbell 1992).

These explorations have been facilitated by an inversion of the normal direction of application of the tools of academic anthropology. Instead of moving from the 'in here' of the faculty club to examine 'cultures' 'out there,' the new critical literature has reversed direction to examine the 'in here' of expert technical practices of security analysis from 'out there' in the new reflexive academic spaces of postmodern thinking. Culture and identity are now stuff 'we' have too. It is so because it is analysed in relation to that which has previously been specified as external and autonomous. Identity is defined in terms of difference; relations are now foregrounded, autonomous ontologies cast in doubt by a methodology that is sensitive to margins, boundaries and flows. These decentrings put sovereignty as a practice of both states, as well as individual knowing subjectivities, in doubt.

Cultures or the culture of globalization?

Samuel Huntingdon's (1993) much cited article on the likely future of global politics starts with the assumption that multiple cultures exist in this world that often collide in violent clashes. Conflict here is understood as the clash of 'civilizations' – the eternal lot of humanity according to the 'tragic' school of realist readings of human history. Primordial identities are posited as the premise for politics and the structural givens for the possibilities of thinking seriously about security. Ethnic identities in the Balkans are obviously important, but they are not the only place where conflict has been polarized by claims to specific ethnicities and communities. Ironically, ethnic separatists can find intellectual support from such prominent neo-realist theorists whose atomistic anarchy arguments dovetail nicely with claims to primordial differences. But at least some of these identity conflicts are aggravated by the simple matter of there being very little cultural difference between protagonists; the case of 'Serbs' and 'Croats' in the former Yugoslavia comes to mind, as does the ongoing 'troubles' in Northern Ireland, where similarity coupled with proximity apparently intensifies the claims to fundamental difference (Ignatieff 1993).

The converse argument is that there is an emerging global culture, accelerated by the process of globalization. If Russian workers in Siberia are watching the same Mexican soap opera on television as those in the Philippines or Rio de

Janeiro, so the argument goes, cultural barriers are coming down. If elites around the world watch CNN, the assumptions about what matters politically are converging. But these images are also brought into the living rooms of the middle classes and many others round the globe (Wark 1994). Disneyland, Coca Cola and McDonalds all function widely on planet Reebok. Modernization's powerful images are spread around the world by a powerful advertising industry that has appropriated the symbols of the globe to promote the products of modernity (McHaffie 1997).

Whether by television, or on the backs of elephants in India, as Gwynne Dyer's television documentary series 'The Human Race' so beautifully shows its audience, 'modern' commodities reach into remote villages and transform human subjectivities in many of even the most impoverished locales. In an era of global corporations, often led in the 'cultural' sphere by American movies, television and consumer items, what is for sale literally is a lifestyle and with it modes of subjectivity that are antithetical to traditional cultures. Identity is now what is purchased, and the images of modernity are the icons of the globalizing culture of the twentieth century. It follows that technology and culture are not separable in many important ways; automobiles and televisions are part of modernity, and identities generated by their use involve challenges to traditional ways of life.

Benjamin Barber's (1996) argument is especially interesting in this context of global and local cultures. He suggests that culture is about both the emergence of a global culture of consumption in a 'McWorld' and the specific 'tribal' reactions by specific identities. His use of the term 'Jihad' may not be helpful, but his argument about the reactions to global culture and the crucial point about growing disparities across the globe is important to consider. Global circulation of images and commodities, often of American origin, suggests a universal consumption culture in the making. But it also suggests subjectivities that identify to some degree with the characters in the movies and the commodities that they use in the television soap operas that are so popular in apparently unlikely places. Global culture can easily be construed as a crucial part of contemporary global hegemony (Taylor 1996).

But careful reflection on these ways of thinking about contemporary political arguments suggests that the geopolitical framework in play here is misleading. Specifically, the premise of 'one' or 'many' is not very helpful in understanding the political options available, much less the processes already in motion (Walker 1988). It is probably much more helpful to argue that there is both one culture and many cultures simultaneously as globalization involves both the reinvention of identities and their appropriation (Marden 1997). While global culture may be resisted in many ways, the weapons of resistance are often the tools of the culture that is being resisted. Iranian clerics used tape recorders to spread their messages across Iran; as Paul Routledge shows elsewhere in this volume, communications professors turned postmodern revolutionaries in Chiapas use laptop computers and the Internet to spread a new type of revolt carefully dressed up in earlier symbols. Its targets are only indirectly the state 'security' forces. More directly its targets are the Mexican government's scripts of national development and modernization, and specifically Mexican integration into the North American Free Trade Agreement as the road to happiness and individual fulfillment (Ó Tuathail 1997).

National liberation movements have long found the legitimating language of revolt and identity constituting practices in the universal aspirations of the United Nations and in the active encouragement of its cultural agencies. Wilsonian idealist principles of self-determination and the modern claims to sovereignty and liberation are at least part of the stuff of very many 'tribal' claims. As Tim Luke argues in his chapter in this volume, global and local are not so separate. International tourism actively promotes the commercial production of 'ethnic' artifacts for souvenir markets.

This argument is complemented by the other frequently heard, although often empirically questionable, suggestion that there is an emerging global civil society. Parallel conferences at the major United Nations meetings over the last few years have provided *forums* for debate and informal political activities that suggest that various forms of social movements and 'civil societies' are important parts of the unfolding political scene (Lipschutz 1996; Wapner 1996). In parallel with this is the apparently ever-growing power of the corporations, which are now often simply called 'global corporations' in recognition of their size and reach across numerous state boundaries (Barnett and Cavanagh 1994). States are no longer so easy to represent as the only political entities that matter. While they still have formidable powers to act in some ways, the consequences of their actions are so interconnected that simplistic claims to sovereignty are less than convincing.

From national security to global security

In an age of apparent globalizations, national security is no longer such an easy justification for state action. Coupled with the end of the Cold War and the demise of the Soviet Union, this has induced a crisis in the Western security community. The 'keepers of the threat' have been deprived of their principal 'threat'; the world can no longer be described in terms of a bipolar geopolitical division with a 'third world' to be struggled over and the Soviet Union geographically contained. Although the grand metaphor of the container is no longer so easily applied, its demise is not to be expected soon (Chilton 1996). The structure of the NATO discourse may now include partnerships with East European states, but it is often still focused on external threats and the differences between those within NATO and threats originating from 'out of area' sources. The Clinton administration's foreign policy has maintained some of the key spatial tropes of containment, only now partly reversed in their direction. Instead of containing the Soviet Union, American foreign policy now 'enlarges' the sphere of liberal democratic states.

How the geopolitical metaphors might be rethought to reorder the rituals of the community of security practitioners is a matter of pressing concern for many Western geopoliticians anxious to ensure that identity is once again coherently articulated. This is not least because of the failure of the profession to predict the geopolitical changes of the end of the Cold War (Gusterson 1993). In part this is also needed because, as Bradley Klein (1994) reminds us, the role of strategic studies and the practice of deterrence throughout the Cold War was much more than the spatial containment of the Soviet Union; it also involved maintaining an American-dominated geopolitical order in numerous places not directly related to the superpower rivalry. The construction of a liberal international order was intimately interconnected with the militarization of global politics (Latham 1997). The whole question of what exactly is being secured became an unavoidable matter for security analysts in the early 1990s (Dalby 1997). At the end of the Cold War, answers to this question were both very simple in the sense that through the period of the Cold War Western modernity was the 'referent object' of security (Buzan 1991), and very complicated, in that the stable assumptions about political order, threats and the purposes of security provision came unstuck dramatically in the years from 1989 to 1991.

In the absence of a threatening communist Other, numerous new threats to national security have been proposed. Rogue states and nuclear outlaws have offered some alternatives, but neither constitutes a threat of the magnitude of the Soviet one (Klare 1995). Concerns with low-intensity conflict as part of the Cold War has given way to more generalized concerns with violent 'internal' conflict and 'failed states' that may suddenly transcend limits, especially if internal conflict in some of the key larger states in the 'South' leads to spillover effects and regional instabilities (Brown 1996; Chase *et al.* 1996; Holsti 1996). The list of security issues goes on: transboundary criminal organizations, hostage incidents, non-state mediators and computer connections suggest, at least to those whose state-centric conceptual frameworks for dealing with order no longer fit the new situations, that 'global chaos' is, if not the new enemy, then a technically accurate description of a system in which small perturbations can have dramatic consequences elsewhere in the global system that 'need' to be managed (Crocker *et al.* 1996).

The term 'global security' traditionally referred to the end of warfare as a human danger. The epithet 'global' was often applied to security in the Cold War context of the possibilities of superpower warfare, because people everywhere on the globe would suffer as a result of the effects of fallout and climate disruptions. Negotiation and political compromise were understood as the only possibility for long-term survival. Hence one of the rationales for high-profile United Nations conferences. The response to the post-Cold War crisis of geopolitical representation and the anxiety caused by the realization that some of the 'new' threats cannot be specified as originating in a particular (hence containable) geographical location has frequently been to invoke the phrase 'global security' in new ways that operate to extend the mandate of security institutions to tackle all manner of newly 'threatening' changes.

Now the possibilities for 'threats' to global security come from numerous other sources. The possibilities of military intervention in 'failed states,' and the roles of the UN and other international organizations are a theme for discussion under the rubric of global security (Prins 1996). Writers in the Brookings Institute are concerned with the difficulties of *Global Engagement* and the role that American political and military institutions can play in the new circumstances (Nolan 1994). Population growth and forced migration, economic instabilities and resource flow disruptions, environmental changes, resurgent nationalisms, ethnic strife, drugs and international criminal organizations have recently been joined by diseases as matters of security threat. According to the editors of the journal *International Security*, the world faces new *Global Dangers* that will require whole new security strategies (Lynn-Jones and Miller 1995).

Environmental themes related to atmospheric change and stratospheric ozone depletion give the epithet some obvious, but contested, purchase on issues of contemporary importance, which can easily be portrayed as a 'global' problem apparently requiring a *Global Accord* to resolve many difficulties (Choucri 1993).

Viewed by people sympathetic to 'Southern' perspectives, global security also seems to suggest a situation requiring both the end of inter-state warfare and the possibility of political dialogue 'grand bargains' across states, civilizations, cultures and identities (Chubin 1996; Dalby 1998). The term 'global' has thus been further extended by the addition of a broad range of 'human rights,' which now sometimes appears as an explicit focus on 'human' security. This 'new' agenda for comprehensive human security includes these numerous additional facets of life under the rubric of security. The 1994 United Nations *Development Report* suggested no less than seven facets: economic, food, health, environmental, personal, community and political security as necessary conditions for 'human security.'

This approach has been expanded upon in the report of the Commission on Global Governance to outline numerous aspects of a much expanded governance agenda for *Our Global Neighbourhood*.

> Despite the growing safety of most of the world's states, people in many areas now feel more insecure than ever. The source of this is rarely the threat of attack from outside. Other equally important security challenges arise from threats to the earth's life support systems, extreme economic deprivation, the proliferation of conventional small arms, the terrorizing of civilian populations by domestic factions, and gross violations of human rights.
>
> (Commission on Global Governance 1995: 79)

This agenda suggests that traditional notions of security are inadequate and that social democratic aspirations will need to be universalized to ensure meaningful security to all the world's population. Human security implies dealing with more than the international political system; it also implies that both the causes of insecurity and the possible solutions to security problems are to be found only by including actors and institutions beyond the ambit of the state system.

Despite the optimistic phrasing of this new set of priorities, humane global governance, to use Richard Falk's (1995) phrasing, is not an easy matter of either inter-state bargaining or replacing states with democratized international civil societies. While states may be a security threat to other entities, and to their populations, they also 'cast a long shadow' over civil society in many post-colonial situations (Pasha 1996). States have not withered in the North either, but the point is that they have changed in some ways in the last few decades as the security situation has changed and the forces of economic globalization have narrowed their economic management possibilities. In the process, economic interventions as a policy option have been restricted and the welfare safety net has developed enough holes to render many poor and marginal peoples in the North very insecure in the face of rapid 'global' economic changes.

At the same time, the new security analyst is now supported by ever more powerful tools of geopolitical observation and global monitoring (Nye and Owens 1996). The 'information power' at the disposal of American policy makers in particular, which allows them to watch any part of the globe through satellite and electronic monitoring, and prescribe appropriate remedies and interventions from afar, promises to some the possibility of a panopticon of global surveillance that will facilitate management of matters at a global scale. The 'god's eye' view of detached observation now becomes a matter of global scale, and the technical expertise to manage is also supposedly available at a planetary scale. The positivist assumptions of objective knowledge leading to accurate prediction and management by a coterie of experts reaches its apotheosis in this new global surveillance. Once again security is tied to specific practices of envisioning and representing the world.

Whose globe? whose security?

This raises the persistent political question of who or what is being secured by these practices. In particular, the assumptions about 'global' are highly contested. It is more than coincidence that the epithet 'global' is currently applied to culture, security and the 'economy.' Andrew Ross (1991) analyses the managerial and economic metaphors in the concerns with 'global warming' in the early 1990s in these terms. He notes that it is the science of environment that links the problems of the 'South' to those of the 'North,' not any concern to think through the legacy of imperialism or neo-colonialism. In the processes of scientific discourse construction, the problem is attributed to humanity as a whole, a 'global' problem, rather than one with more specific causes. Human intervention in the planetary atmosphere is both the problem and that which is needed to solve the problem.

This modeling is governed by the new corporate logic of planetary management, with its centralized rationalization of the climate system's every conceivable component. The same cost–benefits logic is evident in new forms of global economic management, with its debt-for-nature swaps, and the growth of an international market in tradable pollutant-emission rights. While these developments are clear evidence of the political and economic impact of ecology at the highest levels of decision-making, there are reasons to be wary of a distributive system with such Olympian perspective.

(ibid.: 14)

But the Olympian detachment of the 'god's eye view' works to suggest a scientific and hence neutral way of dealing with objective phenomena. The privileging of the global and the assumption of a crisis scenario narrows the political possibilities by compelling adherence to the global crisis script (Roe 1993). 'In here' now specifies 'out there' in terms of the whole planet that can be known by the procedures created 'in here.' The neutrality of the language of science can only partly obscure the managerial ethos that it so easily connects to in terms of 'policy advice,' even in the rhetoric of 'think tanks' that are in many ways much more 'progressive' in outlook (Luke 1997).

Claims to the 'global' have a nasty habit of turning out to be 'local' interests. In the context of the 'global environment,' Vandana Shiva (1994) has warned that Northern elites have adopted the rhetoric of 'global' to suggest that problems largely of their making are matters of equal concern for all humanity. She argues that in the case of stratospheric ozone depletion the problem is caused by a few large corporations and a limited number of industrial plants, mostly located in the developed world. The attribution of 'global' suggests that the problem is a matter of wider responsibility, in the process obscuring the role of specific corporations. If the global environment is understood as a matter of threats to modernity by degradation of 'Southern' landscapes and the 'dangers' of rapid population growth, the analysis looks very different than if it is specified in terms of massive resource use and pollution generated by elites mainly living in the 'North' (Kaplan 1994; Dalby 1996b). If greenhouse gases are specified as a problem from 'Northern' consumption rather than 'Southern' forest destruction the policy suggestions that result are very different.

Likewise in the case of transnational corporations' attempts to patent seed varieties. The specific claims of corporations to global rights on particular strains obscures the widespread patterns of peasant innovation and experience with cultivating a wide diversity of hybrids. Again, claims to global interests obscure the very specific power structures in the attempts by corporations to monopolize food supplies that have traditionally been excluded from commodification. Claims to global solutions to food needs as a part of the process of development often overlook the crucial role of economic relationships of scarcity integral to the practices of development in the production of scarcity (Yapa 1996).

Peasant farmers demonstrating against global seed companies once again show that the political discourses of modernization are not politically neutral but part of the process of global power relations. Intellectual property rights are a matter of peasant survival even more than they are a mode of regulating corporate competition or inter-state trading practices. But the politics of which cultural frame of reference such issues are understood within is crucially important. Within the frames of 'global managerialism,' experts are called upon to pronounce on the nature of the 'problem' and a technical solution is duly formulated for administration by the managers (Sachs 1993). Rarely are the peasants or the workers involved consulted.

The construction of the identity of the poor as separate from the processes of development, indeed as the problem to be resolved, often both misses the causal relationships of social scarcity and removes the political processes of dispossession and displacement intrinsic to the process of modernization (Ferguson 1995). Within these frameworks, opposition to the spread of global agricultural corporations can easily be seen as a threat to the expansion of the modernity and technical progress supposedly 'needed' to solve many of the problems of poverty. But the assumptions that business as usual, Northern style, with a variety of technical fixes added to supposedly make development sustainable, are not assumptions that are likely to play out successfully in many Southern contexts (Chatterjee and Finger 1994).

Local and/or global security cultures?

While on the one hand globalism appears unstoppable, the processes by which the global economy appropriates resources in numerous locations often lead to the further marginalization of poor people in many locations (Pieterese 1996). In many cases, the state and its attempts to facilitate modernization are the cause of insecurities rather than their cure (Ayoob 1995). If state elites, in their rush to expand the economy, facilitate the enclosure of common lands and deforestation to generate revenues from exports, they cause numerous disruptions and the dispossessed people become 'ecological refugees' (Gadgil and Guha 1995). The identities that often find themselves in the way of these processes are those of specific ethnic groups and indigenous peoples who inhabit areas that supply resources for the commercial economy (Renner 1996). While the fate of the Ogoni people, and the high-profile execution of Ken Saro-Wiwa at the hands of the Nigerian military state, in an area where Shell Oil Corporation's activities have caused serious environmental degradation, occupies the attention of environmental and human rights activists as well as Commonwealth heads of state, this episode is far from unique (Sachs 1995).

Many cultural anthropologists, and a few geographers, are now involved in various campaigns to protect indigenous peoples and in attempts to portray these issues of securing cultural identity in terms of human rights (Johnston 1994). These struggles are important in their own right but they alone cannot provide a basis for conceptualizing 'global' security that is either comprehensive or practically useful in many other locations. These struggles at the margins of modernity raise the question of the role of the security expert directly. From the 'in here' of these specific places the threat is 'out there' in the formulations of modernity as security (Foerstel 1996). For many cultures around the world modernity in its numerous guises is the threat (Hipwell 1997). Bernard Nietschmann (1994) has gone so far as to argue that modern states are effectively at war with indigenous minorities and aboriginal peoples in many parts of what can be called the 'Fourth World.'

This is the point that the grand dreams of global panopticons often cannot easily incorporate or understand. The ethical dilemmas faced by anthropologists and cultural geographers are poignant in that they might best act as security experts in explaining to indigenous peoples how to prepare themselves to adapt to the onslaughts of modernity as 'development' 'opens up' rural areas to its economic practices and cultural dislocations. But such a role is rarely taught in the curriculum of security studies or in survey courses of contemporary geopolitics. Neither are courses in organizing ecological resistance movements (Taylor 1995). Terrains of resistance are important in understanding the power plays of contemporary geopolitics, and in some crucial places actions to prepare indigenous peoples to resist modernization might be the most appropriate activity to facilitate local security and protect important parts of the ecosphere (Routledge 1993).

But cultural anthropologists are not the only experts that we need to understand contemporary changes. Neo-populist celebrations of 'local knowledge' are a valuable corrective to the ethnocentric preoccupations of development experts and the effective depoliticization that often follows from the stories that development discourses tell. But they are not the only or, in terms of overall impact, perhaps the most important dimension of the accelerating political economy of environmental destruction. As Piers Blaikie (1996: 85) so eloquently puts it, 'the circumstances in which multinational corporations gain access to timber concessions so large that their eventual impact dwarfs all the world's neo-populist projects put together are not published in learned journals much, perhaps because they are not intellectually interesting.'

Modern culture is premised on this powerful political economy that is now effectively operating at a global scale. Its technical practices of control are premised on detachment and manipulation of an environment understood as external to its activities. Modernity is environment tamed and manipulated – that which is not modern is beyond control. The culture of modernity and the world order that it has constructed is built on a productive system of global dimensions. The 'dark side' of modern economic capabilities is a legacy of destruction in many parts of the world. Resources are needed for industrial production, and they come from somewhere (Redclift 1996). These geographical dimensions of environmental destruction are often forgotten. Japanese forests may be in relatively good shape, but the forests of Southeast Asia are devastated as a consequence. Malaysian corporations are consequently searching out forestry concessions in Latin America, with the obvious dangers of more devastated landscapes.

These concerns are compounded by Western assumptions about the existence of 'failed' states, which often rely on assumptions that the political entities designated as newly independent actually have the attributes of statehood and overlooks the long history of colonial practices, which practically undermine such claims while offering the judicial and discursive space in which such political assumptions become the points of departure for analysis and policy prescription. The structural and historical problems of decolonization are overlooked in privileging Western norms of international law and the Western way as the only possibility for international peace and security. Other political possibilities are not canvassed by most accounts of the crises in Africa. But, as Grovogui (1996) points out, the possibility of doing so is discounted by the larger assumptions that modernity has the only possible answers to political difficulties that are to be accounted for by its absence.

Some 'failed' states have already contracted out their security to transnational mercenary corporations like 'Executive Outcomes' (Rubin 1997). Effectively, other states have also handed over at least some medical provision to nongovernmental organizations like Médecines sans Frontières and welfare provision to emergency aid providers. Reading journalistic accounts of African mining communities working for international diamond companies and guarded by corporate mercenaries dovetails all too closely with geopolitical speculations of contemporary science fiction dystopias where private armies guard corporate resource mining outposts in the badlands of remote areas, which are necessary to provide the 'tame zones' of affluence with their commodities. 'Security' maybe. But this vision is global only as far as the corporations and their ('security') experts are concerned.

Geopolitical identity and the 'POGO syndrome'

Most cultural identities include a crucial ethnocentric formulation in their celebrations of the value of being a 'we.' In their formulations of security problems as caused by Others, what is silenced is that possibility that, in the Walt Kelly cartoon character Pogo's famous formulation, 'the enemy is us.' I irreverently call this problem the 'POGO syndrome' because it suggests that the security problematic can be understood in geopolitical terms as the persistence of (P)olitical (O)rganizations to (G)enerate (O)thers! A sensitivity to these processes is crucial to the further understanding of the contradictions in the policies of 'global security' premised on an uncritical adoption of the identities of modernity. The appellation 'global' does not provide an exception to these matters, however fondly some practitioners of the term might hope that it assigns them the universal moral 'high' ground. Modern geopolitical practices of division, administration and rule are intrinsically violent; the popular representations of Others are part of the larger processes whereby geopolitical imagi-nations are used in the process of foreign policy formulation.

A sensitivity to these formulations might suggest that a self-reflexive geopolitics could produce security identities that are simultaneously more far-reaching and less prone to violence. Reconceptualized estrangements may lead to reformulated diplomatic practices to mediate differences. If difference is understood less as radical otherness and more in terms of complex interconnectedness, the possibilities of a Levinasian ethics of responsibility for Others is given more room to shape cultural responses to 'danger' (Dillon 1996). In parallel but rather different terms, sympathetic engagements with activists involved in working in 'terrains of resistance' in various parts of the world suggest a very different identity for intellectuals concerned about security. The politics of expert knowledge is crucial here. Can intellectuals of geopolitics, in recognizing the connectedness of contemporary insecurities, understand knowledge across the divide between 'North' and 'South' in terms of respect, recognition and sensitivity, as David Slater (1997) hopes?

These reflections raise the possibilities of thinking about what Thom Kuehls (1996) calls 'ecopolitical space' beyond the constraints of either conventional conservationist assumptions about Nature or conventional sovereignty-bound reflections on the possibilities for geopolitical action. The search for political space is a complex matter that requires not only practical efforts, often at the local level, but also a change of categorical understandings by political scientists about what acting politically means and the possibility of thinking politically without assuming the necessity of sovereignty (Magnusson 1996). If politics is not what it used to be understood to be then neither is security (Dillon 1996). As the essays in this volume suggest, the terms in which we understand the related matters of geopolitics are also changing.

If political communities can now be understood as other than states then the provision of security is also a matter of more than states. The implications are clear in that institutions of governance, in the general sense, need to be the subject of investigation. One important implication is that 'rights' agendas like those in *Our Global Neighbourhood* will need to be more sensitive to cultural differences (Baxi 1996). They will also have to take seriously the contradictions of the ethnocentric assumptions of modern states, premised on the wealth generated by an unsustainable political economy, as the model on which conceptions of comprehensive human security are formulated.

All of which also suggests that Ole Waever's (1995) argument for the 'desecuritization' of many issues is an appropriate response to many themes in contemporary global security literature. Understanding many facets of political life as routine rather than as matters of threats and danger requires participation and dialogue rather than expert diagnosis and (violent) intervention. This understanding in the current global economy inevitably raises questions of what the modern practices of security are rendering safe. Modern lifestyles of consumption for some of the world's population at the expense of the majority is the unsettling answer (Alexander 1996). So too is the implication that intellectual property rights, expert knowledge and the finer points of the World Trade Organization's practices may be the most important focal point for a consideration of many of the practical causes of 'global' insecurity.

None of this is to deny that there are many dangers in the modern world, or that violence is an unfortunate fact of political life that cannot be ignored. But it makes a huge difference how such matters are represented, who is defined as a threat to what geopolitical entity and the how the possibilities for political action are defined (Dalby 1996a). Getting beyond the current formulations of global security will require reconsiderations of the cultures of both consumption and expertise, and a reconsideration of the modern geopolitical identities whose consumption can be secured by this technical expertise. It requires a recognition that the simplifications of geopolitical representations are usually an important part of the practices that make violence possible.

By asking questions about the production of geopolitical representations that portray the poor and the marginal as threats to the affluent, critical geopolitics makes the crucial point that conceptualizing matters in terms of 'global' priorities, as current discourses so frequently do, elides the specific contexts of insecurities and obscures the causal dimensions of contemporary violence. Which is why this chapter concludes with the POGO syndrome: It emphasizes the importance of understanding identity as problematic and potentially self-destructive. It also points to the intellectual necessity of distancing oneself from one's fondestheld fears to look again at one's identity in the light of its being rendered strange.

References

Agnew, J. (1998) *Geopolitics: Re-Visioning World Politics*, London: Routledge.

Alexander, T. (1996) *Unravelling Global Apartheid: An Overview of World Politics*, Cambridge: Polity.

Ayoob, M. (1995) *The Third World Security Predicament*, Boulder, Colo.: Lynne Rienner.

Barber, B. (1996) *Jihad vs. McWorld: How Globalism and Tribalism are Reshaping the World*, New York: Ballantine.

Barnett, R. J. and J. Cavanagh (1994) *Global Dreams: Imperial Corporations and the New World Order*, New York: Simon & Schuster.

Baxi, U. (1996) '"Global neighbourhood" and the "universal otherhood": Notes on the Report of the Commission on Global Governance,' *Alternatives*, 21(4): 525–549.

Biersteker, T. J. and C. Weber (eds) (1995) *State Sovereignty as Social Construct*, Cambridge: Cambridge University Press.

Blaikie, P. (1996) 'Post-modernism and global environmental change,' *Global Environmental Change*, 6(2): 81–85.

Brown, M.E. (ed.) (1996) *The International Dimensions of Internal Conlict*, Cambridge, Mass.: Massachusetts Institute of Technology Press.

Buzan, B. (1991) *People, States and Fear: An Agenda for International Security Studies in the Post Cold-War Era*, Boulder, Colo.: Lynne Rienner.

Campbell, D. (1992) *Writing Security: United States Foreign Policy and the Politics of Identity*, Minneapolis: University of Minnesota Press.

Chase, R. S., E. B. Hill and P. Kennedy (1996) 'Pivotal States and U.S. Strategy,' *Foreign Affairs*, 75(1): 33–51.

Chatterjee, P. and M. Finger (1994) *The Earth Brokers: Power, Politics and World Development*, New York: Routledge.

Chilton, P. (1996) *Security Metaphors: Cold War Discourse from Containment to Common House*, New York: Peter Lang.

Choucri, N. (ed.) (1993) *Global Accord: Environmental Challenges and International Responses*, Cambridge, Mass.: MIT Press.

Chubin, S. (1996) 'The South and the New World Order,' in B. Roberts (ed.) *Order and Disorder after the Cold War*, Cambridge, Mass.: MIT Press, 429–449.

Cohn, C. (1987) 'Sex and death in the rational world of defence intellectuals,' *Signs: Journal of Women in Culture and Society*, 12: 687–718.

Commission on Global Governance (1995) *Our Global Neighbourhood*, Oxford: Oxford University Press.

Crocker, C. A., F. O. Hampson and P. Aall (eds) (1996) *Managing Global Chaos: Sources and Responses to International Conflict*, Washington: United States Institute of Peace.

Dalby, S. (1996a) 'Continent adrift?: dissident security discourse and the Australian geopolitical imagination,' *Australian Journal of International Affairs*, 50(2): 59–75.

Dalby, S. (1996b) 'The environment as geopolitical threat: reading Robert Kaplan's "Coming Anarchy",' *Ecumene*, 3(4): 472–496.

Dalby, S. (1996c) 'Crossing disciplinary boundaries: political geography and international relations after the Cold War,' in Eleonore Kofman and Gillian Youngs (eds) *Globalization: Theory and Practice*, London: Pinter, 29–42.

Dalby, S. (1997) 'Contesting an essential concept: reading the dilemmas in contemporary security discourse,' in Keith Krause and Michael Williams (eds) *Critical Security Studies*, Minneapolis: University of Minnesota Press, and London: Pinter, 3–31.

Dalby, S. (1998) 'The threat from the South: geopolitics, equity and environmental security,' in Daniel Deudney and Richard Matthew (eds) *Contested Grounds: Security and Conflict in the New Environmental Politics*, Albany: State University of New York Press.

Dijkink, G. (1996) *National Identity and Geopolitical Visions*, London: Routledge.

Dillon, M. (1996) *Politics of Security: Towards a Political Philosophy of Continental Thought*, London: Routledge.

Doty, R.L. (1996) *Imperial Encounters*, Minneapolis: University of Minnesota Press.

Enloe, C. (1993) *The Morning After: Sexual Politics at the End of the Cold War*, Berkeley: University of California Press.

Falk, R. (1995) *On Humane Governance: Toward a New Global Politics*, Pennsylvania: Pennsylvania University Press.

Ferguson, J. (1995) *The Anti-Politics Machine: Development, Depoliticization and Bureaucratic Power in Lesotho*, Minneapolis: University of Minnesota Press.

Foerstel, L. (ed.) (1996) *Creating Surplus Populations: The Effect of Military and Corporate Policies on Indigenous Peoples*, Washington: Maisonneuve Press.

Gadgil, M. and R. Guha (1995) *Ecology and Equity: The Use and Abuse of Nature in Contemporary India*, London: Routledge.

George, J. (1994) *Discourses of Global Politics: A Critical (Re)Introduction to International Relations*, Boulder, Colo.: Lynne Rienner.

Grovogui, S. N. (1996) *Sovereigns, Quasi Sovereigns and Africans: Race and Self Determination in International Law*, Minneapolis: University of Minnesota Press.

Gusterson, H. (1993) 'Realism and the international order after the Cold War,' *Social Research*, 60: 279–300.

Hall, S. (1993) 'Culture, community, nation,' *Cultural Studies*, 7(3): 349–363.

Hipwell, W. T. (1997) 'They've got no stake in where they're at': Ecology, the Fourth World and local identity in the Bella Coola region, Ottawa: Carleton University, MA thesis.

Holsti, K. (1996) *The State, War, and the State of War*, Cambridge: Cambridge University Press.

Huntingdon, S. (1993) 'The Clash of Civilizations,' *Foreign Affairs*, 72(3): 22–49.

Ignatieff, M. (1993) *Blood and Belonging: Journeys into the New Nationalism*, Toronto: Penguin.

Johnston, A. I. (1995) 'Thinking about Strategic Culture,' *International Security*, 19(4): 32–64.

Johnston, B. R. (ed.) (1994) *Who Pays the Price? The Sociocultural Context of Environmental Crisis*, Washington: Island Press.

Kaplan, R. D. (1994) 'The coming anarchy,' *The Atlantic Monthly*, 273(2): 44–76.

Kaplan, R. D. (1996) *The Ends of the Earth: A Journey at the Dawn of the 21st Century*, New York: Random House.

Klare, M. (1995) *Rogue States and Nuclear Outlaws: America's Search for a New Foreign Policy*, New York: Hill & Wang.

Klein, B. S. (1994) *Strategic Studies and World Order: The Global Politics of Deterrence*, Cambridge: Cambridge University Press.

Kuehls, T. (1996) *Beyond Sovereign Territory: The Space of Ecopolitics*, Minneapolis: University of Minnesota Press.

Latham, R. (1997) *The Liberal Moment: Modernity, Security and the Making of Postwar International Order*, New York: Columbia University Press.

Lapid, Y. and F. Kratochwil (eds) (1996) *The Return of Culture and Identity in IR Theory*, Boulder, Colo.: Lynne Rienner.

Lavie, S. and T. Swedenburg (1996) 'Between and among the boundaries of culture: bridging text and lived experience in the third timespace,' *Cultural Studies*, 10(1): 154–79.

Lipschutz, R. D. (1996) *Global Civil Society and Global Environmental Governance*, Albany: State University of New York Press.

Luke, T. (1997) *Ecocritique*, Minneapolis: University of Minnesota Press.

Lynn-Jones, S. M. and S. E. Miller (eds) (1995) *Global Dangers: Changing Dimensions of International Security*, Cambridge, Mass.: MIT Press.

McHaffie, P. (1997) 'Decoding the globe: globalism, advertising and corporate practice,' *Environment and Planning D: Society and Space*, 15(1): 73–86.

McSweeney, B. (1996) 'Identity and security: Buzan and the Copenhagen school,' *Review of International Studies*, 22(1): 81–93.

Magnusson, W. (1996) *The Search for Political Space*, Toronto: University of Toronto Press.

Manzo, K. (1996) *Creating Boundaries: The Politics of Race and Nation*, Boulder, Colo.: Lynne Rienner.

Marden, P. (1997) 'Geographies of dissent: globalization, identity and the nation,' *Political Geography*, 16(1): 37–64.

Myers, N. (1993) *Ultimate Security: The Environmental Basis of Political Stability*, New York: Norton.

Nietschmann, B. (1994) 'The Fourth World: nations versus states,' in G. Demko and W. Wood (eds) *Reordering the World: Geopolitical Perspectives on the 21st Century*, Boulder, Colo.: Westview, 225–242.

Nolan, J. E. (ed.) (1994) *Global Engagement: Cooperation and Security in the 21st Century*, Washington: Brookings Institute.

Nye, J. S. and W. A. Owens (1996) 'America's information edge,' *Foreign Affairs*, 75(2): 20–36.

Ó Tuathail, G. (1996) *Critical Geopolitics: The Politics of Writing Global Space*, Minneapolis: University of Minnesota Press.

Ó Tuathail, G. (1997) 'Emerging markets and other simulations: Mexico and the Chiapas revolt in the geo-financial opticon,' *Ecumene*, 4: 300–317.

Pasha, M. K. (1996) 'Security as hegemony,' *Alternatives*, 21(3): 283–302.

Peterson, V. S. (ed.) (1992) *Gendered States: Feminist (Re)visions of International Relations Theory*, Boulder, Colo.: Lynne Rienner.

Pieterese, J. N. (1996) 'The development of development theory: towards critical globalism,' *Review of International Political Economy*, 3(4): 541–564.

Prins, G. (1996) 'Global security and military intervention,' *Security Dialogue*, 27(1): 7–16.

Redclift, M. (1996) *Wasted: Counting the Costs of Global Consumption*, London: Earthscan.

Renner, M. (1996) *Fighting for Survival: Environmental Decline, Social Conflict, and the New Age of Insecurity*, New York: Norton.

Roe, E. (1993) 'Global warming as analytical tip,' *Critical Review*, 6(2&3): 411–426.

Routledge, P. (1993) *Terrains of Resistance: Nonviolent Social Movements and the Contestation of Place in India*, Westport, Conn.: Praeger.

Ross, A. (1991) 'Is global culture warming up?,' *Social Text*, 28: 3–30.

Rubin, E. (1997) 'An army of one's own,' *Harpers Magazine*, February, 44–55.

Sachs, A. (1995) *Eco-Justice: Linking Human Rights and the Environment*, Washington, Worldwatch Institute: Worldwatch Paper 127.

Sachs, W. (ed.) (1993) *Global Ecology: A New Arena of Political Conflict*, London: Zed Books.

Said, E. (1993) *Culture and Imperialism*, New York: Knopf.

Shapiro, M. (1997) *Violent Cartographies: Mapping Cultures of War*, Minneapolis: University of Minnesota Press.

Shiva, V. (1994) 'Conflicts of global ecology: environmental activism in a period of global reach', *Alternatives*, 19(2): 195–207.

Slater, D. (1997) 'Spatialities of power and postmodern ethics – rethinking geopolitical encounters,' *Environment and Planning D: Society and Space*, 15(1): 55–72.

Sylvester, C. (1994) *Feminist Theory and International Relations in a Post-Modern Era*, Cambridge: Cambridge University Press.

Taylor, B. R. (ed.) (1995) *Ecological Resistance Movements: The Global Emergence of Radical and Popular Environmentalism*, Albany: State University of New York Press.

Taylor, P. (1996) *The Way the Modern World Works: World Hegemony to World Impasse*, London: John Wiley.

Waever, O. (1995) 'Securitization and Desecuritization,' in R. Lipschutz (ed.) *On Security*, New York: Columbia University Press, 46–86.

312

Walker, R. B. J. (1988) *One World/Many Worlds: Struggles for a Just World Peace*, Boulder, Colo.: Lynne Rienner.

Walker, R. B. J. (1993) *Inside/Outside: International Relations as Political Theory*, Cambridge: Cambridge University Press.

Wapner, P. (1996) *Environmental Activism and World Civic Politics*, Albany: State University of New York Press.

Wark, M. (1994) *Virtual Geography: Living with Global Media Events*, Bloomington: Indiana University Press.

Yapa, L. (1996) 'Improved seeds and constructed scarcity,' in R. Peet and M. Watts (eds) *Liberation Ecologies: Environment, Development, Social Movements*, London: Routledge, 69–85.

INDEX

Printed and bound by CPI Group (UK) Ltd, Croydon, CR0 4YY

01/11/2024

01782635-0011